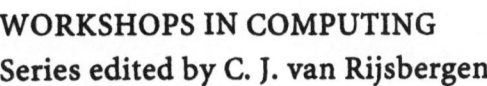

WORKSHOPS IN COMPUTING
Series edited by C. J. van Rijsbergen

W0055359

Also in this series

Logic Program Synthesis and Transformation
Proceedings of LOPSTR 93, International
Workshop on Logic Program Synthesis and
Transformation, Louvain-la-Neuve, Belgium,
7–9 July 1993
Yves Deville (Ed.)

Database Programming Languages (DBPL-4)
Proceedings of the Fourth International
Workshop on Database Programming Languages
– Object Models and Languages, Manhattan, New
York City, USA, 30 August–1 September 1993
Catriel Beeri, Atsushi Ohori and
Dennis E. Shasha (Eds)

**Music Education: An Artificial Intelligence
Approach,** Proceedings of a Workshop held as
part of AI-ED 93, World Conference on Artificial
Intelligence in Education, Edinburgh, Scotland,
25 August 1993
Matt Smith, Alan Smaill and
Geraint A. Wiggins (Eds)

Rules in Database Systems
Proceedings of the 1st International Workshop
on Rules in Database Systems, Edinburgh,
Scotland, 30 August–1 September 1993
Norman W. Paton and
M. Howard Williams (Eds)

Semantics of Specification Languages (SoSL)
Proceedings of the International Workshop on
Semantics of Specification Languages, Utrecht,
The Netherlands, 25–27 October 1993
D.J. Andrews, J.F. Groote and
C.A. Middelburg (Eds)

Security for Object-Oriented Systems
Proceedings of the OOPSLA-93 Conference
Workshop on Security for Object-Oriented
Systems, Washington DC, USA,
26 September 1993
B. Thuraisingham, R. Sandhu and
T.C. Ting (Eds)

Functional Programming, Glasgow 1993
Proceedings of the 1993 Glasgow Workshop on
Functional Programming, Ayr, Scotland,
5–7 July 1993
John T. O'Donnell and Kevin Hammond (Eds)

Z User Workshop, Cambridge 1994
Proceedings of the Eighth Z User Meeting,
Cambridge, 29–30 June 1994
J.P. Bowen and J.A. Hall (Eds)

6th Refinement Workshop
Proceedings of the 6th Refinement Workshop,
organised by BCS-FACS, London,
5–7 January 1994
David Till (Ed.)

**Incompleteness and Uncertainty in Information
Systems**
Proceedings of the SOFTEKS Workshop on
Incompleteness and Uncertainty in Information
Systems, Concordia University, Montreal,
Canada, 8–9 October 1993
V.S. Alagar, S. Bergler and F.Q. Dong (Eds)

**Rough Sets, Fuzzy Sets and
Knowledge Discovery**
Proceedings of the International Workshop on
Rough Sets and Knowledge Discovery
(RSKD'93), Banff, Alberta, Canada,
12–15 October 1993
Wojciech P. Ziarko (Ed.)

Algebra of Communicating Processes
Proceeedings of ACP94, the First Workshop on
the Algebra of Communicating Processes,
Utrecht, The Netherlands,
16–17 May 1994
A. Ponse, C. Verhoef and
S.F.M. van Vlijmen (Eds)

Interfaces to Database Systems (IDS94)
Proceedings of the Second International
Workshop on Interfaces to Database Systems,
Lancaster University, 13–15 July 1994
Pete Sawyer (Ed.)

Persistent Object Systems
Proceedings of the Sixth International Workshop
on Persistent Object Systems,
Tarascon, Provence, France, 5–9 September 1994
Malcolm Atkinson, David Maier and
Véronique Benzaken (Eds)

Functional Programming, Glasgow 1994
Proceedings of the 1994 Glasgow Workshop on
Functional Programming, Ayr, Scotland,
12–14 September 1994
Kevin Hammond, David N. Turner and
Patrick M. Sansom (Eds)

East/West Database Workshop
Proceedings of the Second International
East/West Database Workshop,
Klagenfurt, Austria, 25–28 September 1994
J. Eder and L.A. Kalinichenko (Eds)

continued on back page...

G. Birtwistle and A. Davis (Eds)

Asynchronous Digital Circuit Design

Springer-Verlag London Ltd.

Graham Birtwistle, BSc, PhD, DSc
Department of Computer Science,
University of Calgary, 2500 University Drive,
Calgary, Alberta, T2N 1N4, Canada

Alan Davis, BSEE, PhD
Department of Computer Science,
University of Utah, Salt Lake City,
UT 84112, USA

ISBN 978-3-540-19901-4

British Library Cataloguing in Publication Data
A catalogue record for this book is available from the British Library

Library of Congress Cataloging-in-Publication Data
Asynchronous digital circuit design / G. Birtwistle and A. Davis. eds.
    p. cm. – (Workshops in computing)
    Based on a workshop held in Banff, Alberta, August 28–September 3, 1993.
    "Published in collaboration with the British Computer Society."
    Includes bibliographical references and index.
    ISBN 978-3-540-19901-4   ISBN 978-1-4471-3575-3 (eBook)
    DOI 10.1007/978-1-4471-3575-3
        1.   Digital integrated circuits–Design and construction–Data processing.
2.   Asynchronous circuits–Design and contruction–Data processing.   3.   Computer-
aided design.   I.   Birtwistle, G.M. (Graham Mark), 1946–   .
II.   Davies A. (Alan Lynn), 1946–   .   III.   British Computer Society.   IV.   Series.
TK7874.65.A88 1995   95-1009
621.39'5–dc20   CIP

© Springer-Verlag London 1995
Originally published by Springer London Berlin Heidelberg New York in 1995

Typesetting: Camera ready by contributors

34/3830-543210 Printed on acid-free paper

Preface

The contents of this book are an expanded and thorough treatment of a set of presentations made at a workshop held in Banff, Alberta, 28 August – 3 September 1993 by leading practitioners in asynchronous hardware design. The papers cover a wide range of current practice from practical design, through silicon compilation, to the applications of formal specifications.

Jo Ebergen, John Segers and Igor Benko (Waterloo) describe work on the formal specification of asynchronous circuits in a CSP-like notation which they then go on to analyze for delay, safety, progress, and performance issues.

Al Davis (Utah) describes a set of automatic asynchronous design synthesis tools that were used to implement a large, compact and fast industrial design (a message passing post-office). These tools have been integrated into an existing commercial VLSI CAD framework. The tools, their usage, algorithms, and interfaces are presented.

Kees van Berkel (Philips) and Martin Rem (Eindhoven) view VLSI design as a programming activity using the CSP-based programming language Tangram. They show how Tangram descriptions can be compiled into handshake circuits and thence into layout.

Steve Furber (Manchester) describes and motivates the design of a working asynchronous implementation of the ARM microprocessor developed at Manchester University. The design is based upon Sutherland's micropipelines. The architecture is described in detail and an evaluation of the first silicon is presented.

The collection is rounded out by a paper by Steven Nowick (Columbia) and Al Davis which gives a state-of-the-art survey of asynchronous hardware design. This paper was not presented at the workshop.

Alain Martin (CalTech) also lectured at the workshop but was precluded from contributing to this volume by pressures of work.

The workshop was held at the Banff Conference Center and ran, as ever, as though by clockwork. It is a pleasure to record our thanks to their well-organised, and ever helpful and friendly staff.

Graham Birtwistle   Al Davis
Calgary   Utah

Contents

Contents

Asynchronous Circuit Design: Motivation, Background, & Methods

Al Davis

Department of Computer Science, University of Utah
Salt Lake City, UT 84112, USA

Steven M. Nowick

Department of Computer Science, Columbia University
New York, NY 10027, USA

Abstract

The purpose of this book is to present a current view of the state of the art for the field of asynchronous circuit design and analysis which was the topic of a workshop in Banff in the fall of 1993. Asynchronous circuits have been studied in one form or another since the early 1950's [64] when the focus was primarily on mechanical relay circuits and when the differences between the asynchronous and clocked circuit design styles were somewhat indistinct. A number of theoretical issues were studied in detail by Muller and Bartky as early as 1956 [92]. Since then, the field of asynchronous circuits has gone through a number of high-interest cycles. In the last 5 years there has been an unprecedented level of interest in both academic and industrial settings [56]. This historical trend continues today with the majority of the current research effort focused more on theory than on practice. Nonetheless, the advance of practical asynchronous circuit design techniques also has an unusual level of interest. The work presented at the Banff workshop was concerned more with practice than theory and provided a reasonable coverage of the current approaches to asynchronous circuit design. Similarly this chapter will primarily focus on practical design issues. Prior to introducing the four chapters which follow, we present an introduction to the basic concepts and motivations behind asynchronous circuit design. This will hopefully enable those not already familiar with asynchronous circuit design to better understand the subsequent chapters.

1 Motivation and Basic Concepts

Circuit design styles can be classified into two major categories, namely synchronous and asynchronous. It is worthwhile to note that neither is independent of the other and that there are many designs that have been produced using a hybrid design style which mixes aspects of both categories. There is also considerable debate within the asynchronous circuit community as to what constitutes a *pure* asynchronous circuit. It may be difficult to understand the motivation for asynchronous circuit techniques when the bulk of commercial practice and considerable experience, artifact, and momentum exists for the synchronous circuit design style. For some, the motivation to pursue the study of asynchronous circuits is based on the simple fact that they are different. Others find that asynchronous circuits have a particular modular elegance that is amenable to theoretical analysis. However, for those interested in the practical aspects of asynchronous circuit design, the motivation often comes from some concern with the basic nature of synchronous circuits.

Of common concern are the cost issues associated with the global, periodic, and common clock that is the temporal basis for all purely synchronous circuits. The *fixed* clock period of synchronous circuits is chosen as a result of worst-case timing analysis. It is not adaptive and therefore does not take advantage of average- or even best-case computational situations. Asynchronous circuit proponents view this as an opportunity to achieve increased performance since asynchronous methods are inherently adaptive. Arithmetic circuits are a good example in this regard. Arithmetic circuit performance is typically dominated by the propagation delay of carry or borrow signals. The worst-case propagation situation rarely occurs, yet synchronous arithmetic circuits must be clocked in a manner that accommodates this rare worst-case condition. Some asynchronous circuit designers have made the mistake of generalizing this observation into a view that the inherent adaptivity of asynchronous circuits implies that they are capable of achieving higher performance in general, but this is not necessarily the case.

All asynchronous circuits have additional operational constraints when compared to their synchronous counterparts. All forms of asynchronous circuits are concerned with providing hazard or *glitch* free outputs under some timing model. In order to achieve this behavior, the asynchronous circuit will often contain more gates than a functionally equivalent synchronous circuit. Therefore in terms of the number of basic components, asynchronous circuits are often somewhat larger than synchronous circuits. More gates implies more wires, and this often results in slower rather than faster circuit latencies. Furthermore in order to achieve their inherently adaptive nature, asynchronous circuits must generate their own control signals such as a request and an acknowledge signal. The acknowledge signal indicates completion of a previously requested action. In synchronous circuits much of this type of control signaling is implicit in the common clock signal. The generation of these additional control signals further exacerbates the complexity of asynchronous circuits, and may lead to a further performance degradation.

The adaptive potential remains where the worst-case situation is rare and when the difference between the worst-case and average-case latencies is significant. However, synchronous circuit designers are also well aware of this situation and take considerable care to create a clock model and circuit structure that can take advantage of these differences. The most notable example of this tactic is in the finely-grained pipeline structures of modern floating point units. Yet, for very large circuits, such as microprocessors, balancing all the timing constraints of a large computational space so that there is little difference between the worst and average case timing models is a difficult task. The work by Mark Dean on the STRiP processor [39] provides an interesting example. Dean showed that even a well-balanced and well-designed processor such as the MIPS-X CPU could be sped up if the instruction set were split into three classes, and the clock period adjusted appropriately to match the

3

temporal needs of each class.

Dean also demonstrated that an even greater performance enhancement could be achieved due to the tighter margins which are possible with adaptive clocking. Synchronous systems usually rely on an externally-generated clock signal which is distributed as the common timing reference to all of the system components. The speed at which integrated circuits operate varies with the circuit fabrication process, and fluctuations in operating temperature and supply voltage. In order to achieve a reasonable shield against these variables, the clock period is extended by a certain *margin*. In current practice, these margins are often 100% or more for high-speed systems. Adaptive clocking cannot be generated externally, and therefore must be provided internally to each device. The fact that the clock generator is affected by the same process, temperature, and supply variations as the rest of the chip permits the safety margin to be reduced significantly.

Clock distribution is becoming an increasingly costly component of large modern designs. Today's microprocessors contain over two million transistors and their clock rates are around 200 MHz. The clock period is determined by adding the worst case propagation delay, the margin, and the maximum clock skew. Clock skew is simply the maximum difference in the clock arrival as seen by all clocked points in the circuit. The latency of the clock pulse to the reception points is not a concern. With today's large VLSI circuits exceeding 15 mm per side, several nanoseconds of skew is easily possible. However with a 5 nanosecond clock period, several nanoseconds of skew is a disaster. Clock distribution and deskewing methods are abundant but they share the common characteristic of being extremely expensive. A common method is to distribute the clock via a balanced H-tree configuration [6] with amplifying buffers placed at the fanout points. The problem with this approach is that as more buffers are added to a clock path, larger skew results. The designers of the DEC Alpha CPU [119] took the opposite approach. The Alpha contains 1.68 million transistors and is fabricated in a .75 micron, 3.3 volt CMOS process. Even with three layers of metal, the chip is 16.8 mm by 13.9 mm. In order to keep clock skew to within 300 picoseconds, the Alpha's designers localized the clock buffering to minimize process induced variations and therefore the skew induced by the buffers. Details of the method can be found in [45] but the result is a clock driver circuit that occupies about 10% of the chip area, and consumes over 40% of the 30 watts of power dissipated by the chip. 19 mm^2 of area and over 12 watts of power is a very high price to pay for keeping the skew under control. Clearly this technique will be difficult to extend to a domain where circuit speeds and transistor counts double.

A similar skew problem exists for circuit boards as well as chips. The literature contains an abundance of methods for deskewing clocks [2, 26] on a board but most of them are costly in either area or complexity, and some will probably not be robust enough for use in commercial circuits. An interesting example is the Monarch [110] processor chip which used active signal selection on each input pad. In this instance, a five slot delay line was used to skew signals to match the clock skew. The appropriate tap in the delay line was selected based on analyzing the clock vs. the incoming signal. While the technique did work, its cost and complexity are probably more instructive in a pathological sense. The bottom line is that clock management is a difficult problem and solving it in today's high-speed, highly complex designs is costly. Asynchronous circuit proponents advocate a simple solution, namely throw away the whole concept of a global clock. This is not a free solution since global absolute timing must be replaced with the relative and sequential mechanisms which lie at the heart of asynchronous circuit signaling protocols. Chuck Seitz wrote an excellent introduction to this general topic in his chapter on *System Timing* in the classic VLSI book by Mead and Conway [85]. The next section of this treatise presents some of the more commonly used protocols and terminology.

Another common motivation for pursuing the asynchronous circuit option is the quest for low-power circuit operation. The consumer market's hunger for powerful yet portable digital systems which run on lightweight battery packs is growing at a rapid rate. Hence there is a strong commercial interest in low-power design methods which extend the operational life of a particular battery technology. CMOS circuits have a particular appeal since they consume negligible power when they are idle. This would not be true however if the clock of a synchronous circuit were to continue running. Therefore, low power synchronous circuits usually involve some method of shutting down the clock to most of the system. The exception is the subcomponent that must monitor the environment for the next call to action. However these techniques often result in increased clock skew problems. Asynchronous circuits have the advantage that they go into idle mode for free since, by nature, when there is nothing to do there are no transitions on any wire in the circuit. Another advantage is that even for an active system, only the subsystems that are required for the computation at hand will dissipate any power. Researchers such as Kees van Berkel [133] and Steve Furber [51] are pursuing asynchronous circuit designs where they attempt to exploit this feature.

The final motivation is based on the inherent ease of composing asynchronous subsystems into larger asynchronous systems. While there is still room for doubt about whether asynchronous circuits can achieve their potential advantages in terms of higher performance or lower power operation than synchronous circuits, there is little doubt that asynchronous circuits do have a definite advantage with respect to composability. Asynchronous circuits are functional modules in that they contain both their timing and data requirements *explicitly* in their interfaces. In a sense they "keep time for themselves", hence the term self-timed circuits. Synchronous circuit modules contain only data requirements in their interfaces and share *the clock*. However, important temporal issues such as when data must be valid to avoid set-up and hold time violations between modules are *implicit* at best. Composing asynchronous modules is almost trivial. If the interfaces match and observe the same signaling protocol then they can simply be connected. More detailed knowledge of module internals is required before synchronous subsystems can be connected.

The problem of combining synchronous systems is exacerbated when each module has a separate clock, each running at a different frequency. The effects of this problem are numerous and usually involve some variant of metastability failure [25]. It is commonly accepted, although not definitively proven to the authors' knowledge, that it is impossible to build a perfect synchronizer. Many of the subsystems in today's computers run on clocks which are not synchronized with the CPU. A good example is the I/O subsystem. In this case, techniques must be used which trade increased latency for more reliable synchronization. The reliability is adjusted to meet the MTBF (Mean Time Before Failure) requirements of the system, and the resulting decreased performance is simply viewed as the price that must be paid for the required reliability.

The ease of composing asynchronous subsystems is a tremendous advantage. It allows components from previous designs to be reused, it permits modification of slower components to permit incremental performance improvements without impacting the overall design, and facilitates behavioral analysis by formal methods. However, asynchronous circuits are presently not the mainstay of commercial practice. The definite advantage of composability is not a strong enough factor to counter the significant synchronous circuit momentum, and the promises of improved performance and decreased power consumption remain to be generally realized. There is also a clear gap in the quality of the design infrastructure, e.g. CAD tools, libraries, etc. In addition, the level of synchronous design experience dwarfs the small experience base in asynchronous circuit design. The individual chapters in this book are indications that this gap is narrowing. The asynchronous circuit discipline is becoming more viable even though much work remains to be done before we can hope to compete in the commercial sector with synchronous design styles.

2   Signaling Protocols

Most asynchronous circuit signaling schemes are based on some sort of protocol involving *requests*, which are used to initiate action, and *acknowledgments*, which are used to signal completion of an action. These control signals provide all of the necessary timing information to properly sequence the computational events in the system. The resulting computational model is very much like the dataflow model [36, 1] where the arrival of the necessary operand data triggers the operation. Similarly there is a concept of a sender of information and a corresponding receiver. From the circuit perspective, and ignoring data transmission issues for now, these request and acknowledge control signals typically pass between two modules of an asynchronous system. For example let there be two modules, a sender **A** and a receiver **B**. A request is sent from A to B to indicate that A is requesting some action by B. When B is done with either the action or has stored the request, it acknowledges the request by asserting the acknowledge signal which is sent from B to A. Most asynchronous signaling protocols require a strict alternation of request and acknowledge events.

There are several choices of how these alternating events are encoded onto specific control wires. Two choices have been so pervasive that they will be described here to illustrate the concept. One common choice is the *4-cycle* protocol shown in Figure 1. *RZ* (return to zero), *4-phase*, and *level-signaling*

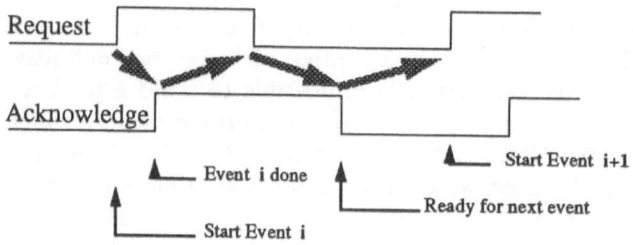

Figure 1: 4-cycle Asynchronous Signaling Protocol

are other names that have been used to identify this protocol. In Figure 1, the waveforms appear periodic for convenience but they do not need to be so in practice. The bold gray arrows indicate the required sequence of events. There is no implicit assumption about the delay between successive events. Note that in this protocol there are typically 4 transitions (2 on the request and 2 on the acknowledge) required to complete a particular event transaction. Proponents of this scheme argue that typically 4-cycle circuits are smaller than they are for 2-cycle signaling, and that the time required for the falling transitions on the request on acknowledge lines do not usually cause a performance degradation since they happen in parallel with other circuit operations, or are useful for controlling the transmission of the answer back to the requester.

The other common choice is *2-cycle* signaling shown in Figure 2, also called *transition* or *NRZ* (non-return to zero) signaling. In this case the waveforms are the same as for 4-cycle signaling with the exception that every transition on the request wire, both falling and rising, indicates a new request. The same is true for transitions on the acknowledge wire. 2-cycle proponents argue that that 2-cycle signaling is better from both a power and a performance standpoint, since every transition represents a meaningful event. The disadvantage is that most 2-cycle interface implementations require more logic than their 4-cycle equivalents. 2-cycle signaling is particularly useful for high-speed micropipelines as pointed out by Ivan Sutherland in his Turing Award paper [124].

So far, the discussion has only addressed control signals. There are also choices for how to encode data. A common choice is the use of a *bundled*

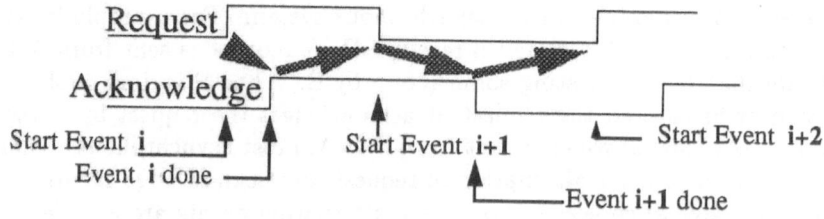

Figure 2: 2-cycle Asynchronous Signaling Protocol

protocol which can be either 2- or 4-cycle. In this case if an n-bit data value is to be passed from the sender to the receiver then n+2 wires will be required (n bits of data, 1 request bit, and 1 acknowledge bit). While this choice is conservative in terms of wires, it does contain an implied timing assumption. Namely the assumption is that the propagation times of the control and data lines are either equal or that the control propagates slower than the data signals. A sending module will assert the data wires and when they are valid will assert the request. It is important that the same relationship of data being valid prior to request assertion be observed at the receiving side. If this were not the case the receiver could initiate the requested action with incorrect data values. This requirement is often simply called the *bundling constraint*. Most asynchronous circuits have been designed with bundled data protocols because the logic and wires required to implement bundled data circuits is significantly less than with non-bundled approaches. However, in order for bundled data asynchronous circuits to work properly, the bundling constraint must be met. Antagonists of this approach note that these timing assumptions are similar to those made for synchronous circuit design.

The common alternative to the bundled data approach is *dual rail* encoding. In this case each bit of data is encoded with its own request onto 2 wires. A typical 4-cycle dual rail encoding has four states:

1. 00 - Idle, data is not valid

2. 10 - Valid 0

3. 01 - Valid 1

4. 11 - Illegal

After a valid data value is asserted the wires must return to the idle state prior to assertion of the next valid datum. In this case for an n-bit data value the link between sender and receiver must contain 2n+1 wires: 2n for the n-bits of data and the associated requests and 1 for the acknowledge. In this dual rail protocol, sending a bit requires the transition from the Idle state to either the valid 0 or valid 1 state and then after receiving the acknowledge it must transition back to the idle state. The illegal state is not used. If recognized by the receiver, it should cause an error.

A 2-cycle variant would still require 2 wires per bit but could signal a valid 0 by a single transition of the left bit, while a valid 1 would be signaled by a transition on the right bit. Concurrent transitions on both the left and right bits are illegal. Sending a 0 or a 1 must be followed by a transition on the acknowledge wire before another bit can be transmitted. Alternative encoding schemes have been proposed as well [139, 41]. Dual rail signaling is insensitive to the delays on any wire and therefore is more robust when assumptions like the bundling constraint cannot be guaranteed. The receiver will need to check for validity of all n-bits before using the data or asserting the acknowledge.

3 Completion Signals

One of the added complexities of asynchronous circuits is the need to generate completion signals that often correspond to the acknowledge signal in the various signaling protocols. There are many methods, none of which is universally satisfactory. A common approach is to design an asynchronous module in a manner that is similar to a synchronous circuit. Namely the arrival of the request starts the modules internal clock generator and after a certain number of internal clocks the circuit is done and an acknowledge can be generated. The idea was originally suggested by Chuck Seitz and was used during the construction of the DDM1 dataflow computer [36]. This technique works well when the size of the module is large, but when the module is small, the additional logic required for the internal clock generator represents an overhead that is too costly.

Another choice for completion signal generation is the use of a *model delay*. In this case, conventional synchronous timing analysis of the datapath is used to determine how long the circuit will take to compute a valid result after the request has been received. A delay element such as an inverter chain is then used to turn the request into the appropriately delayed acknowledge signal. Note that this method works equally well for both 2- and 4-cycle signaling protocols.

Special functions often have unique opportunities. For example, arithmetic circuits can be built to generate completion signals based on carry propagation patterns [62]. Other functions can independently compute both F and $\overline{F}$ and use the exclusive-OR of their outputs to generate the acknowledge signal. Note that this technique will only work directly in a 4-cycle signaling protocol. If used for a 2-cycle protocol, then additional logic such as a T flip-flop will be required.

Another common case is when multiple subsystems are activated in parallel by a master controller. When all of their completion signals are asserted then the master controller should be acknowledged. A tree of C-elements is commonly used to provide the appropriate control in this situation.

A novel technique was proposed by Mark Dean [40] where completion detection was performed by observing the power consumption of the circuit. When activated the circuit consumes power and when it is done the power consumption falls below a particular threshold.

The study of completion signal generation methods in asynchronous circuits could be the topic of an entire book. For now, it is only necessary to realize that some method must be chosen and that the need for completion signals and related signaling protocols is a necessary overhead of asynchronous circuit design.

4   Delay Models and Hazards

4.1   Delay Models, Circuits and Environments

There is a wide spectrum of asynchronous designs. One way to distinguish among them is to understand the different underlying models of delay and operation. Every physical circuit has inherent delay. Yet, while synchronous circuits process inputs between fixed clock ticks, asynchronous circuits can often be regarded as computing continuously through time intervals. Therefore, a delay model is critical in defining the dynamic behavior of an asynchronous circuit.

There are two fundamental models of delay: the *pure delay* model and the *inertial delay* model [128, 115]. A pure delay can delay the propagation of a waveform, but does not otherwise alter it. An inertial delay may alter the shape of a waveform by eliminating short glitches. More formally, an inertial delay has a threshold period, δ. Pulses of duration less than δ are filtered out.

Delays are further characterized by timing models. In a *fixed delay* model, a delay has a fixed value. In a *bounded delay* model, a delay may have any value in a given time interval. In an *unbounded delay* model, a delay can take on any finite value. A delay model is used to describe the behavior of a circuit. A delay is usually associated with each wire. In a *simple-gate* (or *gate-level*) model, a delay is associated with each gate or component in the circuit. In a *complex-gate* model, a single delay is associated with a network of basic components. That is, the network is assumed to behave as a single operator, with no internal delays. A *circuit model* is then defined using delay models for the individual wires and components. Typically, the functionality of a gate is modeled by an instantaneous operator with an attached delay.

Given a circuit model, it is important to define the interaction of the circuit with its *environment*. A circuit and its environment form a closed system, called a *complete circuit* (see Muller in [88]). If an environment can respond to a circuit without any timing constraints, the two interact in *input/output mode* [19]. Otherwise, environmental timing constraints are assumed. The most common example is *fundamental mode* [82, 128] where the environment must wait for a circuit to stabilize before responding to circuit outputs. Such a requirement can be seen as the hold time of a simple latch or flipflop [83].

4.2   Classes of Asynchronous Circuits

Given these models for a circuit and its environment, asynchronous circuits can be classified into a hierarchy.

A *delay-insensitive (DI)* circuit is one which operates correctly regardless of delays on gates and wires. That is, an unbounded gate and wire model is assumed. The concept of a delay-insensitive circuit grows out of work by Clark and Molnar in the 1960's on *Macromodules* [31]. DI systems have been formalized by Udding [127] and Dill [43]. The class of DI circuits built out of simple gates and operators is quite limited [78, 20]; however, practical circuits can be built from more complex components [46, 63].

A *quasi-delay-insensitive (quasi-DI or QDI)* circuit is delay-insensitive except that "isochronic forks" are assumed [22]. An isochronic fork is a forked wire where all branches have the same delay.[1] In contrast, in a delay-insensitive circuit, delays on the fork branches may be different. The motivation of QDI circuits is that they are the weakest compromise to pure delay-insensitivity needed to build practical circuits using simple gates and operators. Martin [79] and van Berkel [133] have used QDI circuits extensively and have described their advantages and disadvantages [78, 135].

A *speed-independent (SI)* circuit is one which operates correctly regardless of gate delays; wires are assumed to have zero or negligible delay. SI circuits were introduced by David Muller in the 1950's (see [88]). Muller's formulation only considered deterministic input and output behavior. This class has recently been extended to include circuits with a limited form of non-determinism [9, 67].

A *self-timed* circuit, described by Seitz [85], contains a group of self-timed "elements". Each element is contained in an "equipotential region", where wires have negligible or well-bounded delay. An element itself may be an SI circuit, or a circuit whose correct operation relies on localized timing assumptions. However, no timing assumptions are made on the communication between elements.

All of the above circuits operate in input/output mode: there are no timing assumptions on how the environment responds to the circuit. The most general category is an *asynchronous circuit* [128]. These circuits contain no global clock. However, they may make use of timing assumptions both within the circuit and in the interaction between circuit and environment. Latches and flip-flops, with setup and hold times, belong to this class.

4.3   Hazards

In a synchronous system, glitches on wires are not usually a problem. Computation occurs between clock ticks, and transitions on wires must stabilize before the next clock tick. However, since an asynchronous system has no global clock, computation is not bound to discrete time intervals and any glitch may cause the system to fail. The potential for a glitch in an asynchronous design is called a *hazard* [128]. Hazards were first studied in the context of asynchronous state machines, and much of the original work focused on combinational logic. Sequential hazards are also possible in asynchronous state machines; these are called *critical races* and *essential hazards*, and are discussed subsequently.

There are several approaches to eliminating combinational hazards. First, inertial delays may be used to filter out undesired "spikes". Much of the early work in asynchronous synthesis relied on the use of inertial delays (see Unger [128]). Second, synthesis techniques can be used to avoid hazards; this approach is discussed below. Third, if bounded delays are assumed, hazards may be "fixed" by adding delays to slow down certain paths in a circuit. A final approach is to tolerate hazards where they will do no harm; this approach was also used in some early work.

---

[1] Other formulations allow bounded skew between the different branches of a fork.

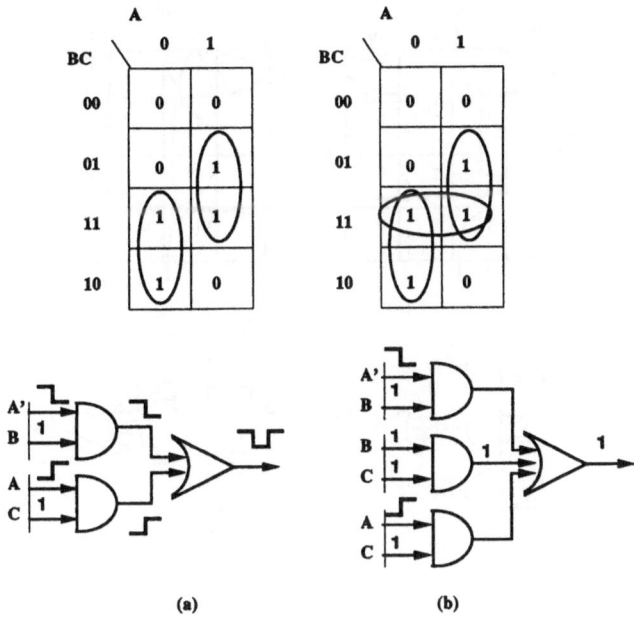

Figure 3: Combinational hazard example: SIC transition

Hazards are temporal phenomena: they are manifest during the dynamic operation of a circuit. As an example, consider the Karnaugh map in Figure 3(a), defining a Boolean function with 3 inputs: A, B, C. A minimum-cost sum-of-products realization is given by the expression: $f = A'B + AC$. The corresponding AND-OR circuit is shown in the figure. Consider the behavior of the circuit during the *single-input change (SIC)* from $ABC = 011$ to $ABC = 111$. Initially, AND-gate $A'B$ is 1, AND-gate AC is 0 and the OR-gate output is 1. When A changes to 1, AND-gate $A'B$ goes to 0 and AC goes to 1. However, if AND-gate AC is slower than $A'B$, the result is a glitch on the OR-gate output: $1 \rightarrow 0 \rightarrow 1$. That is, the circuit has a hazard for this transition.

The Karnaugh map in Figure 3(b) shows an alternative realization of the function. A third product, BC, has been added to the map. For the same transition as above, AND-gate BC holds its value at 1, and the OR-gate output remains at 1 without glitches. That is, the new circuit is *hazard-free* for the transition. The added product, BC, is redundant in terms of function f, but is necessary to eliminate the hazard. This product is used to cover the K-map transition, $ABC : 011 \rightarrow 111$.

The original theory of combinational hazards for SIC transitions was developed by Huffman, Unger and McCluskey (see [128]). The preceding example indicates how to eliminate an SIC *static-1* hazard, that is, for an input change where the function makes a $1 \rightarrow 1$ transition. For the remaining transition types, no hazards will occur in any AND-OR realization: $0 \rightarrow 0$, $0 \rightarrow 1$, and

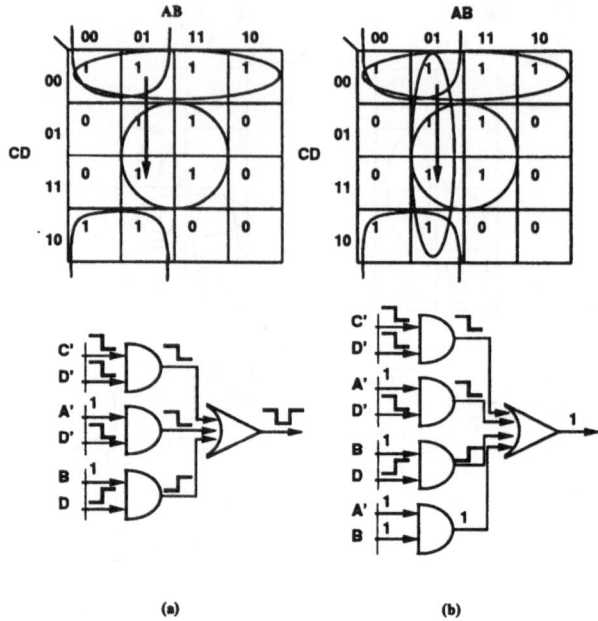

Figure 4: Combinational hazard example: static MIC transition

$1 \rightarrow 0$ [128]. [2]

The case of a *multiple-input change* (*MIC*) is more complex: both static and dynamic hazards must be eliminated. Static-1 hazards are eliminated by a similar approach to the SIC case [48]. For example, Figure 4(a) shows a Karnaugh map for a function with 4 inputs: A, B, C, D. A minimum-cost sum-of-products realization requires 3 products: $f = C'D' + A'D' + BD$. An MIC transition from $ABCD = 0100$ to $ABCD = 0111$ is indicated by an arrow. AND-gates $C'D'$ and $A'D'$ each make a $1 \rightarrow 0$ transition, and AND-gate BD makes a $0 \rightarrow 1$ transition. If BD is slow, the result is a glitch on the OR-gate output: $1 \rightarrow 0 \rightarrow 1$. This hazard is eliminated by adding a fourth product term, $A'B$, which holds its value at 1 throughout the transition. As shown in Figure 4(b), the product covers the entire transition, $ABCD : 0100 \rightarrow 0111$.

The above approach eliminates MIC static hazards but does not work for MIC *dynamic hazards*. Figure 5(a) contains the same Karnaugh map as in Figure 4(b), but with a new multiple-input change: from $ABCD = 0111$ to $ABCD = 1110$. In this case, the function makes a dynamic transition from 1 to 0. The implementation is hazardous. AND-gates BD and $A'B$ each make a $1 \rightarrow 0$ transition. At the same time, AND-gate $A'D'$ has inputs changing from 10 to 01, and therefore may glitch: $0 \rightarrow 1 \rightarrow 0$. If $A'D'$ is slow, this glitch will propagate to the OR-gate *after* the other AND-gates have gone to 0, and the OR-gate output will glitch: $1 \rightarrow 0 \rightarrow 1 \rightarrow 0$.

---

[2]More precisely, these realizations will be hazard-free as long as no AND gate contains a pair of complementary literals.

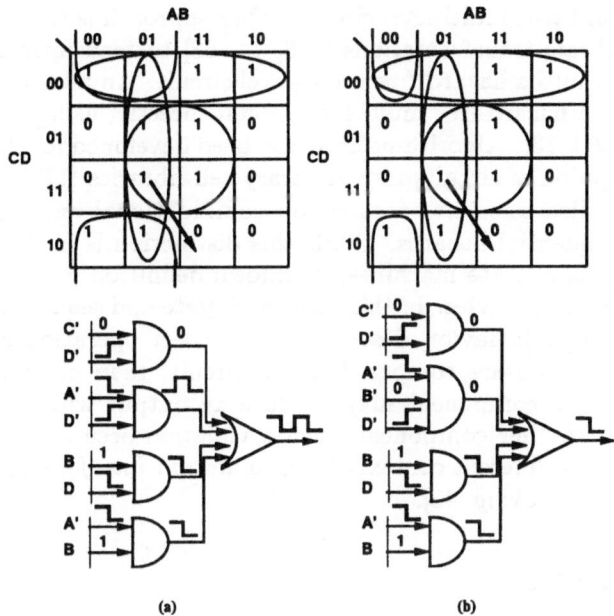

Figure 5: Combinational hazard example: dynamic MIC transition

To prevent a dynamic MIC hazard, no AND-gate may temporarily turn on during the transition. In this example, $A'D'$ becomes enabled, then disabled, as inputs A and D change value. This phenomenon can be seen in the Karnaugh map: product $A'D'$ intersects the transition from $ABCD : 0111 \rightarrow 1110$, but intersects neither the start point ($ABCD = 0111$) nor the end point ($ABCD = 1110$) of the transition [15]. Beister[11] proposed a solution: product $A'D'$ is *reduced* to a smaller product $A'B'D'$ which no longer intersects the transition. Note that this product is non-prime. The final cover is shown in Figure 5(b). AND-gate $A'B'D$ remains at 0 throughout the transition, and the transition is hazard-free.

The preceding examples indicate how to eliminate hazards for any MIC transition. Hazard elimination can be viewed as a covering problem on a Karnaugh map. For the $1 \rightarrow 1$ case, the entire transition must be covered by some product. For the $1 \rightarrow 0$ and $0 \rightarrow 1$ cases, every product which intersects the transition must also contain its start or end point. For the remaining case, $0 \rightarrow 0$, no hazard will occur in any AND-OR realization [128]. Such conditions suffice to eliminate any single MIC hazard. Unfortunately, when attempting to eliminate hazards for several MIC transitions simultaneously, these covering conditions may be unsatisfiable. That is, for an arbitrary set of MIC transitions, a hazard-free cover may not exist [128, 49, 101].

A hazard-free two-level minimization algorithm was developed by Nowick and Dill [101]. The algorithm finds a minimum-cost cover which is hazard-free for a set of MIC transitions, if a solution exists. Other methods have focused on hazard-free multi-level circuits. One approach is to use hazard-non-increasing algebraic transformations [128, 14, 68] to transform a hazard-free two-level circuit into a hazard-free multi-level circuit. An alternative approach is to synthesize a hazard-free multi-level circuit directly, using binary decision diagrams *(BDDs)* [73]. Algorithms have also been developed for the hazard-free technology mapping of circuits into arbitrary cell libraries [118].

So far, the discussion has focused on combinational hazards, treated as distinct from sequential hazards. While this distinction is useful for the analysis and synthesis of state machines, a uniform definition of hazards is better suited for other design styles. In this view, each gate and sequential component has a specified legal behavior, describing the correct operation of the component. As components are composed into a circuit, their composite behavior is determined. If a component may produce an output which cannot legally be accepted by another component, then a violation occurs. This notion has been formalized in different contexts as *computation interference* [46], *stability violation* [79] and *choking* [43].

5 Arbitration

In order to avoid non-deterministic behavior, asynchronous circuits must be hazard-free under some circuit delay model. As discussed above, certain forms of MIC behavior can be tolerated; but the most general signal concurrency must be controlled by arbitration in order to avoid circuit hazards. For example, if some circuit is to react one way if it sees a transition on signal A and react differently for a transition on signal B, then some guarantee must be provided that this circuit will see mutually exclusive transitions on inputs A and B. Nondeterministic behavior will occur if this guarantee cannot be provided. Such mutually exclusive signal conditioning is usually provided by *arbitration.*

Latches and flip-flops cannot be used for arbitration due to the inherent possibility of them entering their metastable regions [25, 24]. Arbiter circuits are typically constructed to adhere to a particular signaling protocol and therefore vary somewhat. However all arbiters rely on a *mutual exclusion*, or *ME*, element to separate possible concurrent signal transitions. The ME element is essentially a latch with a metastability detector on its outputs. If sufficient signal separation exists between the two inputs then the first one wins. However, if both inputs occur within a device-specific time, then the latch will go metastable but the metastability detector will prevent the outputs of the ME circuit from changing until the metastability condition is resolved.

Chuck Seitz proposed an ME circuit that is particularly useful for MOS based designs, shown in Figure 6. The cross-coupled inverters form the usual SR latch. The outputs of the latch are connected to a pair of transistors which form the metastability detector. When the latch is in its metastable region, V1 and V2 will differ by less than the threshold voltage of the N-type transistors. In this case, both T1 and T2 will be off, since the gate-to-source voltage will be less than the threshold. If T1 and T2 are off then the outputs of the ME circuit will remain high. When V1 and V2 differ by more than the threshold voltage,

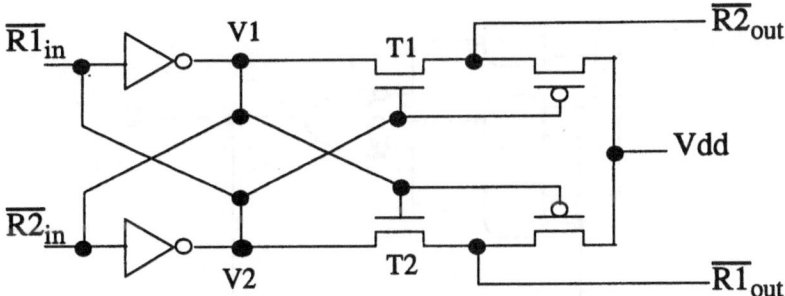

Figure 6: The Mutual Exclusion Element

then the latch will stabilize into either of its stable states. At this point, either T1 or T2 will turn on and the respective output will fall to its asserted level.

Once the mutual exclusion property is provided by an ME element, constructing the rest of an arbiter circuit to conform to a particular signaling protocol is relatively straightforward. An example of a 4-cycle arbiter originally proposed by Chuck Seitz and modified by David Dill and Ed Clarke is shown in Figure 7. Four-cycle arbitration is relatively simple since the race which must be detected is for both signal trajectories going in the same direction (typically a low to high transition). The use of the C elements in the arbiter prevents another pending request from passing through the arbiter until after the active request cycle has cleared. A C element (originally described in the pioneering work of David Muller) is a commonly used asynchronous circuit component. The component produces an output of 1 only after all of its inputs are 1, and produces an output 0 only when all of its inputs are 0. For all other inputs, the component holds its output value. A C element serves as a 4-cycle protocol-preserving rendezvous.

A simple combinational logic implementation of a 2-input C element is shown in Figure 8. C-elements are somewhat problematic since they inherently contain an output which is fed back internally to the C element. This case represents the worst form of isochronic fork, since one of the forks is contained

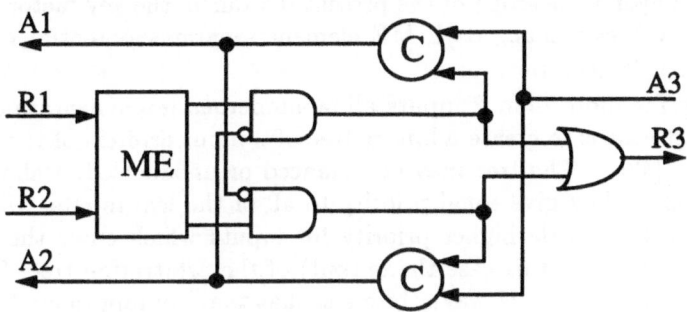

Figure 7: A 4-cycle Arbiter

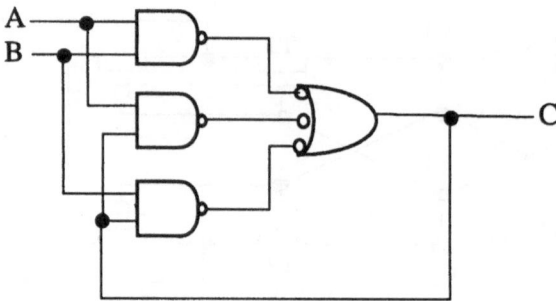

Figure 8: A 2-input C element

within the C element circuit module while the other is exported to outside modules. The feedback signal is also an indication that the C element is itself a form of latch and as such acts as a synchronization point which may result in reduced performance. It is rare that large asynchronous circuits can be built using no C-elements, but the existence of many C-elements in the circuit is often an indication that the performance of the circuit will be reduced.

Two cycle arbitration is somewhat more complex, since the inputs of the arbiter may race in all possible combinations of signal trajectories. Ebergen [19], for example, has reported on a particular 2-cycle arbiter known as the *RGD* (Request, Grant, Done) arbiter.

Another interesting arbitration problem was posed by Davis and Stevens during the development of the Post Office chip [38]. One potential performance difficulty with asynchronous signaling protocols is that waiting for the next event is the normal mode of operation. Hence if two requesters want to share some resource, the loser must wait until the winner is finished before access to that resource can be granted. However, if the loser wants to do something else if it does not win arbitration, then the previously discussed arbitration methods will be insufficient. The need is for a NACK'ing arbiter which provides the requester with an acknowledge if the resource is available, and a negative acknowledge or *NACK* if the resource is busy. Several versions of this NACK'ing arbiter have been designed by various researchers; while these designs are beyond the scope of the present discourse, the key factor is the use of multiple ME elements. Each ME element separates potential races for 2 distinct signal trajectories.

Arbiters for more than 2 inputs allow numerous implementation options. The simplest case is to create a binary tree of 2-input arbiters of the appropriate size (see [43]) . The tree may be balanced or unbalanced. Balanced trees are *fair* in that they give equal priority to all of the leaf inputs. Unbalanced trees inherently provide higher priority for inputs which enter the structure closest to the root (in this case the output) of the arbitration tree. The problem with tree-structured N-way arbiters is that they contain many C-elements and therefore suffer from decreased performance. Another approach is to use redundant ME elements to provide mutually exclusive assertion of 1 of the N input signals. This approach was also used by Ken Stevens in the design of the Post Office chip [38] and is currently under investigation by Charles Molnar at Sun Laboratories for the *COP* component of the counterflow pipeline processor [121].

While the methods are diverse, there is little doubt that the design of efficient arbitration structures is a key aspect of any complex asynchronous system design.

6 An Overview of Prior Work

6.1 Pioneering Efforts

In the mid 1950's asynchronous circuits were first studied by analyzing the nature of input restrictions on sequential circuits. These efforts were part of the general interest in switching theory. Huffman postulated [59, 60] that there must be a *minimum* time between input changes in order for a sequential circuit to be able to recognize them as being distinct. There must then be two critical periods, δ_1 and δ_2, where $\delta_1 < \delta_2$. Signals which occur within a time that is less than or equal to δ_1 cannot be distinguished as being separate events. Signals which are separated by a time of δ_2 or greater are distinguishable as a sequence of separate events. Signal events separated by a time between δ_1 and δ_2 cause nondeterministic sequential circuit behavior. This leads to a class of circuits that became known as *Huffman* circuits. This work was extended in the 1950's and 60's by the fundamental contributions of Unger, McCluskey and others.

Muller [92, 93] proposed a different class of circuits which are more closely related to modern asynchronous circuits. In particular he proposed the use of a *ready* signal. Input signals to Muller circuits were only permitted when the ready signal was asserted. In some sense, the concept is similar to that of a simple 4-cycle circuit. The unasserted acknowledge serves as a ready indication. When the circuit is not ready to accept additional input then it can merely hold its acknowledge to indicate that no further requests can be tolerated.

The efforts of Muller and Huffman spurred considerable theoretical debate in the switching circuit literature. The next notable event from a modern perspective was the seminal work by Stephen Unger that resulted in the publication of his classic text [128]. In this book Unger provided a detailed method for synthesizing single input change asynchronous sequential switching circuits. He also provided a more partial view of what would be required for multiple input change circuits. This textbook had a significant influence on much of the practical work that followed in the next decade. For example, the subsequent work of both of these authors was heavily influenced by Unger's work. Additionally, several early mainframe computers were constructed as entirely asynchronous systems, notably the MU-5 and Atlas computers.

Another noteworthy effort, the *Macromodule Project* [31], conducted at Washington University in St. Louis and sponsored by DARPA, provided an early demonstration of the composition benefits of asynchronous circuit modules. This project created a digital Lego kit of modules. These modules could (and were) rapidly used to configure special purpose computing engines as well as general purpose computers. The project took a significant step forward and provided a sound foundation for the numerous macromodular synthesis approaches being investigated today [18, 132, 46].

Yet another noteworthy pioneer was Chuck Seitz, whose MIT dissertation [117] introduced a Petri Net like formalism which proved to be extremely useful in the design and analysis of asynchronous circuits. In his subsequent academic career, Prof. Seitz taught numerous courses at the University of Utah and then later at CalTech where he infected a large number of students with what proved to be an incurable interest in asynchronous circuits. His influence directly resulted in the asynchronous implementation of the first operational dataflow computer [36], and the first commercial system containing asynchronous hardware. This system was the Evans & Sutherland LDS-1, which also had the distinction of being the first commercial special purpose computer graphics computer. Prof. Seitz's role as an educator is also significant in that his courses on asynchronous circuits, starting as early as 1970, inspired many of the field's current researchers.

The influence of these pioneering efforts is still seen in most of the asynchronous circuit work that is in progress today.

6.2   Synthesis of Controllers.

6.2.1   Asynchronous state machines.

The most traditional approach to building asynchronous controllers is as asynchronous finite state machines. This view of computation is state-based: a machine is in some state, it receives inputs, generates outputs, and moves to a new state. Such specifications are naturally described by a *flow table* or *state table* [128]. These tables define the behavior of outputs and next state as a function of the inputs and current state. Current and next states are described symbolically (see Figure 9).

The earliest asynchronous state machine implementations were *Huffman machines* (see [128]). These machines consist of combinational logic, primary inputs, primary outputs and fed-back state variables. No latches or flip-flops are used: state is stored on feedback loops, which may have added delay elements. A block diagram of a Huffman machine is shown in Figure 10.

Synthesis methods for asynchronous state machines usually follow the same general outline as synchronous methods [83]. A flow table is reduced through *state minimization*. Symbolic states are assigned binary codes using *state assignment*. Finally, the resulting Boolean functions are implemented in combinational logic using *logic minimization*.

There are many ways to operate an asynchronous state machine. Unger [129] proposed a hierarchy, based on the kinds of input changes that a machine can accept. In a *single-input change (SIC)* machine, only one input may change at a time. Once the input has changed, no further inputs may change until the machine has stabilized. This operating mode is highly restrictive, but simplifies

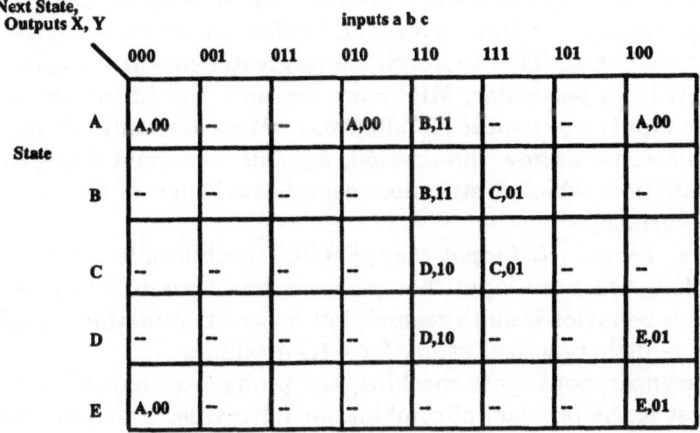

Next State, Outputs X, Y	inputs a b c							
	000	001	011	010	110	111	101	100
State A	A,00	–	–	A,00	B,11	–	–	A,00
B	–	–	–	–	B,11	C,01	–	–
C	–	–	–	–	D,10	C,01	–	–
D	–	–	–	–	D,10	–	–	E,01
E	A,00	–	–	–	–	–	–	E,01

Figure 9: An asynchronous flow table

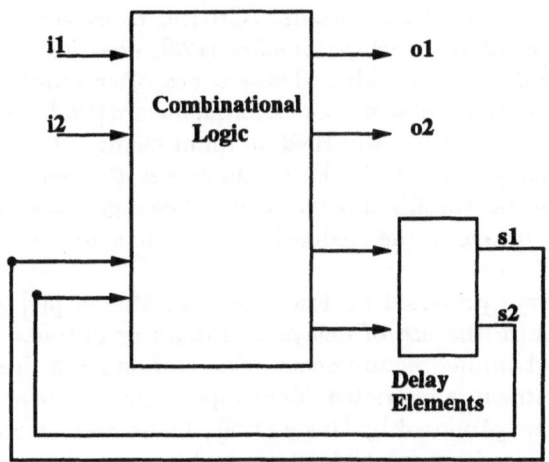

Figure 10: Block diagram of a Huffman machine

the elimination of hazards. A summary of SIC asynchronous state machines can be found in Unger [128].

A *multiple-input change (MIC)* machine allows several inputs to change concurrently. Once the inputs change, no further inputs may change until the machine has stabilized. This approach allows greater concurrency, but it is still quite restricted. In particular, MIC machines have the added constraint that the multiple-inputs are *almost simultaneous.* More formally, all inputs must change within some narrow time period, δ. This constraint helps to simplify hazard elimination, which is still more complicated than in the SIC case, but it limits concurrency.

Finally, an *unrestricted-input change (UIC)* machine allows arbitrary input changes, as long as no one input changes more than once in some given time interval, δ. This behavior is quite general, but hazard elimination is problematic. Hence there is little practical utility for UIC machines.

In any asynchronous state machine, the problem of hazards must be addressed. First is the problem of combinational hazards. The difficulty of combinational hazard elimination depends on whether the machine operates in SIC or MIC mode. As mentioned earlier, SIC hazards are easier to eliminate. Hazards are eliminated by hazard-free synthesis or by using inertial delays to filter out glitches. Alternatively, many traditional synthesis methods ignore hazards on outputs, and only eliminate hazards in the next-state logic.[3]

Second, since asynchronous state machines have state, sequential hazards must be addressed. When a state machine goes from one state to another, several state bits may change. If the machine may stabilize incorrectly in a transient state, a *critical race* occurs. Critical races are eliminated using specialized state encodings, such as *one-hot* [128], *one-shot* [128], Liu [75] or Tracey [126] *critical race-free* codes. These codes often require extra bits. A second type of sequential problem is an *essential hazard* [128]. Essential hazards arise if a machine has not fully absorbed an input change at the time the next-state begins to change. In effect, the machine sees the new state before the combinational logic has stabilized from the input change. Essential hazards are avoided by adding delays to the feedback path or, in some cases, using special logic factoring [5].

MIC designs were proposed by Friedman and Menon [50] and Mago [76]. These designs require the use of delays on inputs or outputs, special "delay boxes", and careful timing requirements. The usefulness of these designs in a concurrent environment is restricted, since inputs must be near-simultaneous.

UIC designs were proposed by Unger [129]. These designs are not currently practical: they require large inertial delays, and have not been proven to avoid metastability problems.

Because of the complexity of building correct Huffman machines, an alternative approach was introduced, called *self-synchronized machines.* These machines are similar to Huffman machines, but have a local self-synchronization unit which acts like a clock on the machine's latches or flip-flops. Unlike a synchronous design, the clock is *aperiodic*, being generated as needed for the given

---

[3]Such machines are called *S-proper* or *properly-realizable* [128].

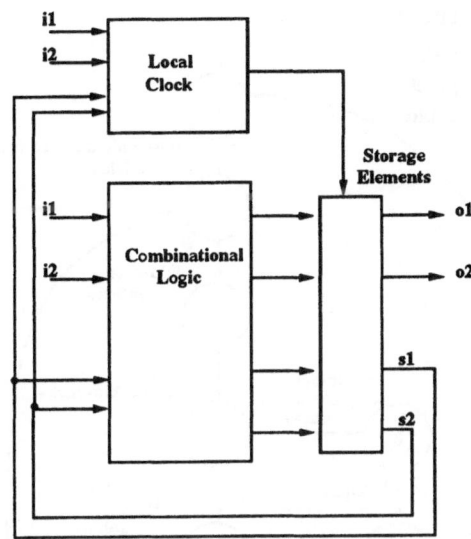

Figure 11: Block diagram of self-synchronized machine

computations. A block diagram of a self-synchronized machine is shown in Figure 11. Both SIC [57, 125] and MIC [3, 30, 111, 130] self-synchronized machines have been proposed. In a related approach, the local clock is replaced by an explicit external completion signal [69]. Other researchers have developed hybrid *mixed-operation mode* machines [143, 27]. Self-synchronized machines tend to have a simpler construction but a greater overhead than Huffman machines.

Asynchronous state machines offer a number of attractive features. First, input-to-output latency is low: if no delays are added to inputs or outputs, the delay is combinational. Second, since the machines are state-based, sequential and combinational optimization algorithms can be used, similar to those which have been effective in the synchronous domain. However, asynchronous state machine design is subtle: it is difficult to design hazard-free implementations which (i) allow reasonable concurrency and also (ii) have high-performance.

Much of the recent work on asynchronous state machines is centered on *burst-mode machines*. These specifications were introduced to allow more concurrency than traditional SIC machines, and therefore to be more effective in building concurrent systems. A further motivation for burst-mode implementations was to insure freedom from hazards while maintaining high-performance. Burst-mode specifications are based on the work of Davis on the *DDM Machine* [37]. In this dataflow machine, Davis used state machines which would wait for a collection of input changes ("input burst"), and then respond with a collection of output changes ("output burst"). The key difference between this data-driven style and MIC-mode is that, unlike MIC machines, inputs within a burst could be *uncorrelated*: arriving in any order and at any time. As a result, these machines could operate more flexibly in a concurrent environment.

22

INPUTS: OUTPUTS:

req-send   tack
treq   peack
rd-iq   adbld
adbld-out
ack-pkt

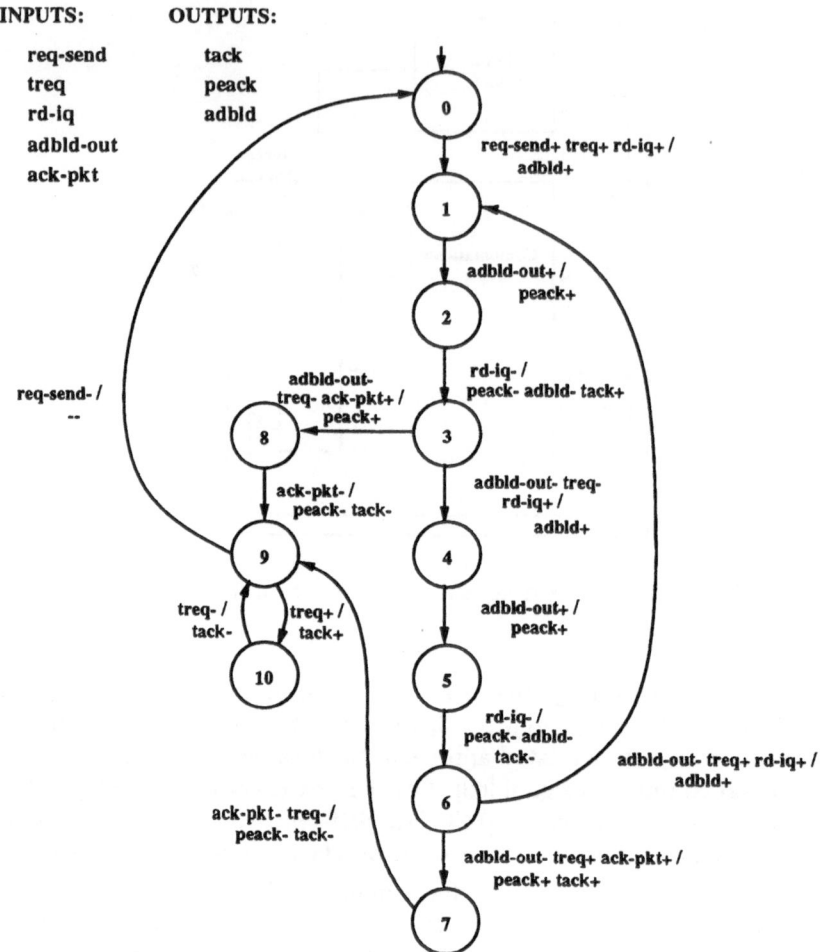

Figure 12: Burst-mode specification for HP controller *pe-send-ifc*

More recently, Davis, Coates and Stevens implemented this approach in the MEAT synthesis system at HP Labs [35]. The synthesis method was applied to the design of controllers for the Post Office routing chip for the Mayfly project. However, although it produced high-performance implementations, it relied on a verifier to insure hazard-free designs.

An example of a burst-mode specification is shown in Figure 12. Each transition is labeled with an input burst followed by an output burst. Input and output bursts are separated by a slash, /. A rising transition is indicated by a "+" and a falling transition is indicated by a "−". The specification describes a controller, *pe-send-ifc*, which has been implemented in the Post Office chip.

Nowick and Dill [100, 98] made two main contributions. First, they constrained and formalized the specifications used in MEAT into the final form called *burst-mode* [100]. Second, they introduced a new self-synchronized design style called a *locally-clocked state machine* [100, 98]. The synthesis method guarantees a hazard-free gate-level implementation for any burst-mode specification. In addition, unlike several previous self-synchronized design methods, this method produces low-latency designs, where the latency of the machine is primarily combinational delay. The design method has been automated and applied to a number of significant designs: a high-performance second-level cache controller [99], a DRAM controller and a SCSI controller [103].

Yun and Dill [145] later proposed an alternative implementation style for burst-mode machines, called a *3D machine*. These machines are named after the 3-dimensional flow table used in their synthesis. Unlike locally-clocked machines, these are Huffman machines, with no local clock or latches. The synthesis method has been fully automated and applied to several large designs.

Burst-mode specifications allow greater concurrency than MIC designs, but they still have two main limitations. First, they require strictly alternating bursts of inputs and outputs: concurrency occurs only within a burst. Second, as in many asynchronous design styles, there is no notion of "sampling" of level signals which may or may not change. Yun, Dill and Nowick [147, 146] introduced *extended burst-mode* specifications to eliminate these two restrictions. These generalized specifications allow a limited form of intermingled input and output changes, providing greater concurrency. These designs also allow the sampling of level signals. Yun *et al.* extended the 3D machine synthesis method to handle extended burst-mode operation.

Davis, Marshall, Coates and Siegel [77] have built a CAD framework to incorporate all of the burst-mode synthesis methods. The framework includes tools for simulation and layout as well. Their tools have been applied to several significant designs, including a low-power infrared communications chip for portable communication. The chip has been fabricated; the measured current consumption of the core receiver (without pads) is less than 1 mA at 5 volts when the receiver is actually receiving data, and less than 1 μA when it is waiting for data. Details of this work are contained in the chapter by Davis in this volume.

6.2.2 *Petri-net and graph-based methods.*

Petri nets and other graphical notations are a widely-used alternative to specify and synthesize asynchronous circuits. In this model, an asynchronous system is viewed not as state-based, but rather as a partially-ordered sequence of events. A Petri net [107] is a directed bipartite graph which can describe both concurrency and choice. The net consists of two kinds of vertices: *places* and *transitions*. *Tokens* are assigned to the various places in the net. An assignment of tokens is called a *marking*, which captures the state of the concurrent system. Starting from an initial marking, tokens flow through the net, transforming the system from one marking to another. As tokens flow, they *fire* transitions in their path according to certain *firing rules*. Since the firing of a transition in a Petri net corresponds to the execution of an event, each such simulation or

24

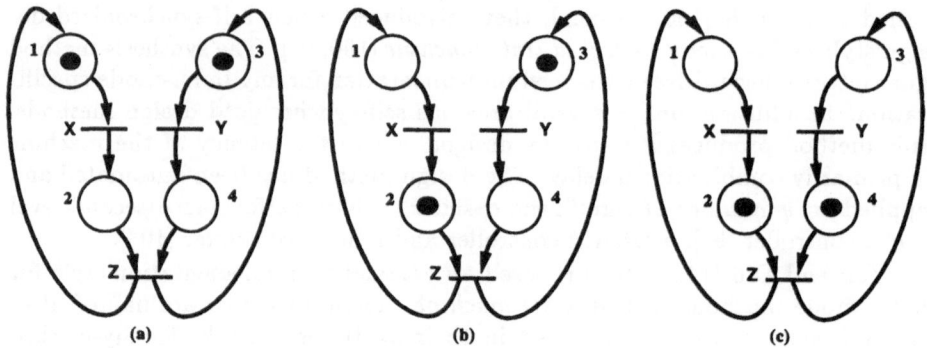

Figure 13: Petri-net example

token game describes a different possible interleaved execution of the system.

A Petri net is shown in Figure 13(a). Places are drawn as circles, and transitions as bars. The initial marking is indicated by black dots in two of the places. If a place is connected by an arc to a transition, the former is called an *input place* of the transition. Likewise, if a transition is connected by an arc to a place, the latter is called an *output place* of the transition. In this example, transition X has input place 1 and output place 2; transition Y has input place 3 and output place 4.

Two transitions are enabled in the figure: X and Y. Each transition is enabled because there is a token in each input place. An enabled transition may fire at any time, removing a token from each input place and moving one to each output place. The result of firing transition X is shown in Figure 13(b). The firing of a transition corresponds to the occurrence of an event. In this example, events X and Y can occur concurrently: both transitions are enabled and may fire in any order. Figure 13(c) indicates the result of firing transition Y after X. After both events have fired, transition Z is enabled and may fire.

Patil [104] proposed the synthesis of Petri nets into *asynchronous logic arrays*. In this approach, the structure of the Petri net is mapped directly into hardware. Most modern synthesis methods use a Petri net as a behavioral specification only, not as a structural specification. Using *reachability analysis*, the Petri net is typically transformed into a *state graph*, which describes the explicit sequencing behavior of the net. An asynchronous circuit is then derived from the state graph.

Several approaches use a constrained class of Petri net called a *marked graph* [32]. Marked graphs are used to model concurrency, but not choice. That is, a marked graph cannot model that one of several possible inputs (or outputs) may change in some state. Examples include Seitz's *M-Nets* [116] and Rosenblum and Yakovlev's *Signal Graphs* [114]. Vanbekbergen *et al.* [136] introduced the notion of a *lock class* to synthesize designs from marked graphs.

More general classes of Petri nets include Molnar *et al.*'s *I-Nets* [90] and Chu's *Signal Transition Graphs* or *STGs* [28, 29]. These nets allow both concurrency and a limited form of choice. Chu developed a synthesis method which transforms an STG into a speed-independent circuit, and applied the method to a number of examples, such as an A-to-D controller and a resource locking module. This work was extended by Meng [86], who produced an automated synthesis tool for speed-independent designs from STGs. Meng also explored design tradeoffs to allow greater concurrency in the resulting circuits.

Recent work on Petri-net and graph-based asynchronous synthesis is proceeding in three major directions: (i) extending specifications; (ii) optimizing synthesis algorithms; and (iii) improving hazard elimination. Several extensions have been proposed to describe more general behavior than is possible with the original STG's. These include the use of "epsilon" and "dummy" transitions [28], "don't-care" and "toggle" transitions [91], *OR-causality* [142] and semaphore transitions [33]. Sutherland and Sproull have introduced a notation for composite Petri nets called "snippets". Others allow timing constraints for specification and synthesis, using a related *Event-Rule* formalism [94].

Several optimized synthesis algorithms have been developed, most using modifications of synchronous techniques. Lavagno *et al.* [71], Vanbekbergen *et al.* [137] and Puri and Gu [108] have proposed algorithms for state minimization and state assignment from STG specifications. Lin and Lin [74] have developed algorithms which avoid expensive intermediate representations during synthesis, instead performing synthesis directly on an STG representation.

Recent STG methods are also addressing the problem of gate-level hazards. Earlier STG synthesis methods usually assumed a complex-gate model, where a combinational circuit is regarded as a monolithic block, rather than a collection of separate gates with individual delays. Several recent methods assume a simple-gate model, which can model hazards due to delays in a collection of individual gates and wires. Moon *et al.* [91] and Yu and Subrahmanyam [144] propose heuristic techniques for gate-level hazard elimination for speed-independent design. Lavagno *et al.* [70] use logic synthesis algorithms, hazard analysis and added delays to avoid hazards, assuming bounded gate delays.

Finally, other researchers are using state graphs for specifications, rather than Petri nets [138, 9, 67]. State graphs allow the direct specification of interleaved behavior, avoiding some of the structural complexity of Petri nets. The target designs are speed-independent gate-level implementations. Originally, this work focused on determinate specifications, having no input or output choice, based on Muller's semi-modular lattice formulations (see [88]). More recent research allows generalized behavior with choice.

6.2.3 Transformation methods.

While STG-based methods view computation as partially-ordered sequences of events, a different approach is to view an asynchronous system as a collection of communicating processes. A system is specified as a program in a high-level language of concurrency. Typically, the program is based on a variant of Hoare's *CSP* [58], such as *occam* or *trace theory* [109]. The program is then transformed, by a series of steps, into a low-level program which maps directly to a circuit. Such transformation methods use algebraic or compiler techniques to carry out the translation.

Ebergen [46] introduced a synthesis method for delay-insensitive circuits using specifications called *commands*. A command is a concise program notation to describe concurrent computation based on trace theory. Several operations are used to construct a complex command from simpler commands, such as *concatenation*, *repetition* and *weave*.

Figure 14 illustrates commands for several basic DI components. A *wire* is a component with one input, $a?$, and one output, $b!$. The symbol "?" indicates an input to the wire, and "!" indicates an output of the wire. In a delay-insensitive system, a wire may have arbitrary finite delay. As a result, if two successive changes occur on input $a?$, the output behavior is unpredictable: $b!$ may glitch. To insure correct operation, input and output events must strictly alternate: once input $a?$ changes value, no further change on $a?$ is permitted until output event $b!$ occurs. A command for *wire* is given in Figure 14(a). The notation $a?; b!$ indicates that input event $a?$ must be followed by output event $b!$; ";" is the *concatenation* operator. No distinction is made between a *rising* or *falling* event on a wire; $a?$ simply means a change in value on the wire. An asterisk (∗) indicates *repetition*: $a?$ and $b!$ may alternate any number of times. Finally "pref" is the *prefix-closure* operator, indicating that any prefix of a permitted behavior is also permitted. The final command describes the permitted interaction of a wire and its environment when it is properly used.

Figure 14(b) illustrates a more complex component called a *toggle*. A toggle has one input, $a?$, and two outputs, $b!$ and $c!$. Each input event, $a?$, results in exactly one output event. Output events alternate or toggle: the first input event $a?$ results in output event $b!$ (as indicated by the black dot); the next input event results in output event $c!$; and so on. The resulting command is shown in the figure.

Another important component is a *C-element*, shown in Figure 14(c) (also known as a Muller C-element, DI C-element, rendezvous, or join element). The component has two inputs, $a?$ and $b?$, and one output, $c!$. The component waits for events on *both* inputs. When both inputs arrive, the component produces a single event on output $c!$. Each input may change only once between output events, but the input events $a?$ and $b?$ may occur in any order. Such parallel behavior is described in a command by the *weave operator*: $a? \parallel b?$. The final command for a C-element, allowing repeated behavior, is shown in the figure.

A final component, called a *merge*, is shown in Figure 14(d). The component is basically an *exclusive-or* gate, but its operation is restricted so that no glitching occurs. The component has two inputs, $a?$ and $b?$, and one output, $c!$. The component waits for *exactly one* input event: either $a?$ or $b?$. Once an input event occurs, the component responds with output event, $c!$. The component can be thought of as "joining" two input streams to a single output stream, where only one input stream is active at a time. Such an exclusive choice between inputs is described in a command by the *union operator*: $a? \mid b?$. The final command for a join element, allowing for repeated behavior, is shown in the figure.

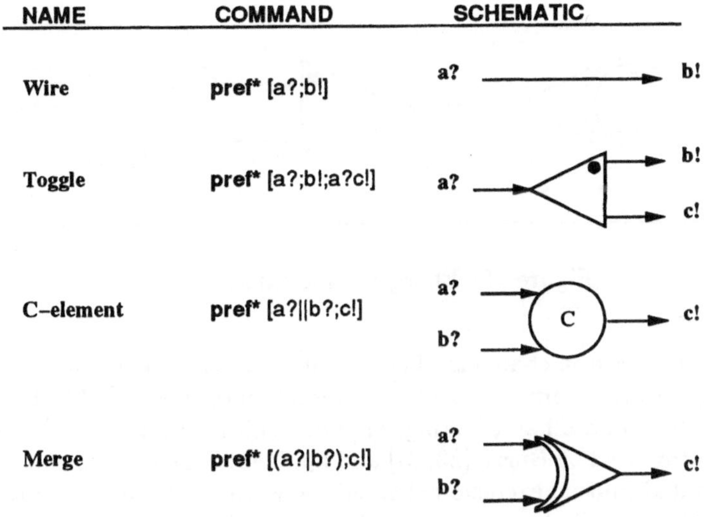

NAME	COMMAND	SCHEMATIC
Wire	pref* [a?;b!]	a? ⟶ b!
Toggle	pref* [a?;b!;a?c!]	a? ⟶ (toggle) b! / c!
C–element	pref* [a?||b?;c!]	a? / b? ⟶ C ⟶ c!
Merge	pref* [(a?|b?);c!]	a? / b? ⟶ (merge) ⟶ c!

Figure 14: Commands for some simple components

A command can be used to specify a complex circuit or system. The command is then *decomposed* in a series of steps into an equivalent network of components, using a "calculus of decomposition". As an example, a *modulo-3 counter* can be specified by the following command [46]:

$$MOD3 = \mathbf{pref}*[a?; q!; a?; q!; a?; p!]$$

This command describes a counter with one input, $a?$, and two outputs, $p!$ and $q!$. The counter receives events on input $a?$. Each input event must be acknowledged by one output event before the next input event can occur. The first and second input events are acknowledged on $q!$, while the third input event is acknowledged on $p!$. This behavior repeats, hence the command describes a modulo-3 counter. Using techniques for delay-insensitive decomposition, this command can be decomposed into a network of 2 toggles and 1 merge which implements equivalent behavior, as shown in Figure 15. Ebergen has applied his decomposition method to a number of designs, including modulo-n counters, stacks, committee schedulers [12] and token ring arbiters (see his chapter in this volume).

A related algebraic approach was proposed by Udding and Josephs [127, 63]. Their method is based on a *delay-insensitive algebra* which formally characterizes a delay-insensitive system. Using axioms and lemmas, a specification is transformed into a provably correct delay-insensitive circuit. The method has been used to design a delay-insensitive stack, a routing chip, and an up-down counter.

While the above methods use algebraic calculi to derive asynchronous circuits, other transformation methods rely on compiler-oriented techniques. Martin [79] specifies an asynchronous system in CSP as a set of concurrent processes

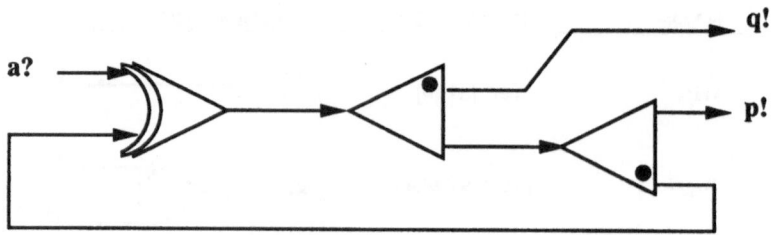

Figure 15: Ebergen's modulo-3 counter

which communicate on channels. The specification is then translated into a collection of gates and components which communicate on wires. Most of Martin's designs use four-phase handshaking for communication. The synthesis method has been automated by Burns [23, 21] and applied to substantial examples, such as a distributed mutual exclusion element, lazy stack, and multiply-accumulate unit [96]. The compiler includes several optimization steps, such as reshuffling, symmetrization, and transistor sizing [22].

Martin's work has been extended by Akella and Gopalakrishnan in a system called *SHILPA* [4]. This method allows global shared variables, and uses flow analysis techniques to optimize resource allocation.

A different compiler-based approach was developed by van Berkel, Rem and others [132, 133] at Philips Labs and Eindhoven, using the *Tangram* language. Tangram, based on CSP, is a specification language for concurrent systems. A system is specified by a Tangram program, which is then compiled by syntax-directed translation into an intermediate representation called a *handshake circuit*. A handshake circuit consists of a network of *handshake processes*, or *components*, which communicate asynchronously on channels using handshaking protocols. The circuit is then improved using peephole optimization and, finally, components are mapped to VLSI implementations.

As an example, the following is a Tangram program for a 1-place buffer, *BUF1*:

$$(a?W \& b!W) \cdot \mid [x : \textbf{var } W \mid \#[a?x; b!x]] \mid$$

The buffer accepts input data on a and produces output data on b. The expression in parentheses is a declaration of the external ports of the module. The buffer has an *input port*, a, and an *output port*, b, handling data of some type, W. The remainder of the program, structured as a block, is called a *command*. A local variable x is defined for internal storage of data. The statement $\#[a?x; b!x]$ indicates that data is received on port a and stored in internal variable x; this data is then sent out on port b. The ";" operator indicates sequencing, and "#" indicates infinite repetition.

This Tangram program is translated into the handshake circuit of Figure 16. Each circle represents a handshake process or component. Each arc represents a channel, which connects an *active port* (indicated by a black dot) to a *passive port* (indicated by a white dot). Communication on a channel is by

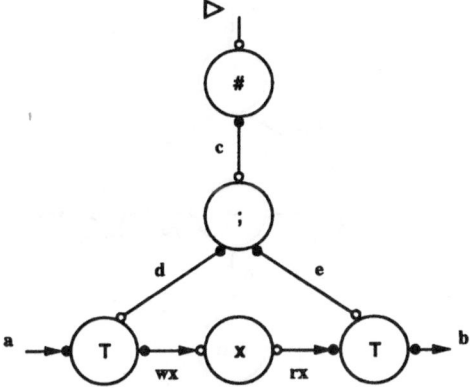

Figure 16: Handshake circuit for BUF1 example

handshaking: an active port initiates a request and a passive port returns an acknowledgment.

In this example, port ▷ is the top-level port for the circuit, called *go*. The environment activates the buffer by an initial request on this passive port. This port is connected to a *repeater* process, which implements the repetition operator, "#". This process repeatedly initiates handshaking on channel c. Channel c is connected to a *sequencer* process, which implements the ";" operator. The sequencer first performs handshaking on channel d. When handshaking is complete, it then performs handshaking on channel e.

Channels d and e in turn are each connected to *transferrers*, labelled T. When the sequencer process initiates a request on channel d, the corresponding transferrer actively fetches data on input channel a and then transfers it to storage element x. Once the transfer is complete, the sequencer initiates a request on channel e, causing the second transferrer to fetch the data from x and transfer it to output channel b.

A more complex example is 2-place buffer, *BUF2*, which can be described in terms of two 1-place buffers:

$$(a?W \& c!W) \cdot \mid [b : \textbf{chan } W \mid (BUF_1(a, b) \parallel BUF_2(b, c))] \mid$$

The program defines the buffer by the parallel composition, $\parallel$, of the two 1-place buffers, which are connected by an internal channel, b. The corresponding handshake circuit is shown in Figure 17. A *parallel* component implements the composition operator, $\parallel$. An initial request on its passive *go* port, results in parallel communication on channels l and r. These channels are both connected to a 1-place buffer B (indicated by double circles). The two buffers communicate through a *synchronizer* process (indicated by a black dot). If active requests arrive on both of its channels, lb and rb, the synchronizer first performs handshaking on channel b, then returns parallel acknowledgments on channels lb and rb. The attached *run* process is used to hide channel b; it simply acknowledges every request it receives.

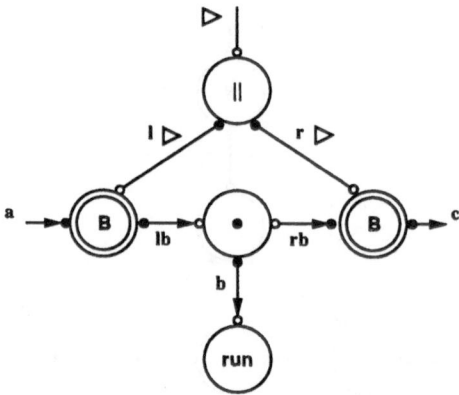

Figure 17: Handshake circuit for BUF2 example

The Tangram compiler has been successfully used for several significant DSP designs for portable electronics, including a systolic RSA Converter, counters, decoders, image generators, and an error corrector for a digital compact cassette player [134]. A major goal this work is rapid turnaround time and low-power implementation (see the chapter by van Berkel and Rem in this volume).

Brunvand and Sproull [16, 18] proposed an alternative compiler using occam specifications. Unlike the approaches of Martin and van Berkel, communication between processes is through two-phase handshaking, or transition-signaling. In their method, an occam specification is first compiled into an unoptimized circuit using syntax-directed translation. Peephole optimization techniques are then applied to improve the resulting circuits. The circuits are then mapped to a library of transition-signaling components.

6.3   Design of Asynchronous Datapaths.

As in synchronous design, different techniques and structures are used when designing datapaths and controllers.

Asynchronous datapaths are often built using a pipelined structure. In synchronous pipelines, data advances in lock-step through the pipe at a fixed clock rate. Since stages may have different delays, the clock must be set to the slowest stage. Furthermore, since a stage's performance may vary with actual temperature, voltage, process and computation, increased margin is typically added to the clock. Finally, clock speed is further reduced to alleviate clock skew. As a result, under nominal conditions, a synchronous pipeline may operate far slower than its potential performance.

In an asynchronous pipeline, there is no clock. Therefore, in principle, each stage may pass data to its neighbor whenever the stage is done and the neighbor is free. Such *elastic* pipelines promise improved performance: different stages may operate at different speeds, and stages may complete early depending on the actual data. Of course, new overhead may be introduced, since each stage must now tell its neighbor when it is ready.

Sutherland introduced an elegant and influential approach to building asynchronous pipelines, which he called *micropipelines* [124]. A micropipeline has alternating computation stages separated by storage elements and control circuitry. This approach uses transition-signaling for control along with bundled data. Sutherland describes several designs for the storage elements, called "event-controlled registers", which respond symmetrically to rising and falling transitions on inputs. Such pipelines have been used by several researchers in the design of asynchronous microprocessors. Sutherland, Sproull, Molnar, and others at Sun Labs have recently designed a "counterflow microprocessor" based on micropipelines [121]. Micropipelines also form the basis for the Manchester ARM microprocessors, developed by Furber and the AMULET group [51, 52, 105]. A more detailed account of this work is contained in Furber's chapter in this volume.

Figure 18 illustrates the operation of a micropipeline with 4 stages. For simplicity, only the control is indicated. In practice, a bundled datapath is also used, along with event-controlled registers to store the data as it propagates down the pipe. A control stage of the pipeline consists of a C-element (described above). A C-element with two inputs and one output behaves as follows. If both inputs are 1, the output is 1; if both inputs are 0, the output is 0. Otherwise, if inputs have different values, the output holds its current value. The C-elements in the micropipeline behave similarly, except that each has one inverted input.

Initially, all wires in the micropipeline are at 0, as shown in Figure 18(a). When new data arrives, a request $R(in)$ is asserted ($R(in)$ goes to 1). The first C-element, C_1, becomes enabled, and its output makes a transition (to 1). This event has two consequences: the request is acknowledged on the left interface (transition on $A(in)$), and the request is forwarded to the right interface (transition on $R(1)$). The same behavior is repeated at the second stage: the request is acknowledged on the left interface (transition on $A(1)$), and the request is forwarded to the right interface (transition on $R(2)$). This process continues through stages 3 and 4, and a final request appears on the rightmost interface, $R(out)$. Effectively, the initial request propagates to the right through the pipe, and acknowledges are generated to the left. The resulting micropipeline configuration is shown in Figure 18(b). Note that since transition-signaling is used, only one request and one acknowledge are generated between each pair of stages. In contrast, a 4-phase (RZ) protocol would have required a second request/acknowledge sequence to reset the wires to their original values.

Since the initial request was acknowledged at the leftmost interface, $A(in)$, new data may now arrive and a second request, $R(in)$, can occur. Since $R(in)$ is currently 1, a request is asserted by changing $R(in)$ to 0. This request propagates through the micropipeline as before. The left interface is acknowledged ($A(in)$ goes to 0), and the request is forwarded to the right interface ($R(1)$ goes to 0). This process repeats at the 2nd and 3rd stages. However, once the second stage is acknowledged ($A(2)$ goes to 0) and the request is made to the 4th stage ($R(3)$ goes to 0), the propagation halts. Although $R(3)$ made a transition (request from stage 3), stage 4 still contains the earlier data that was

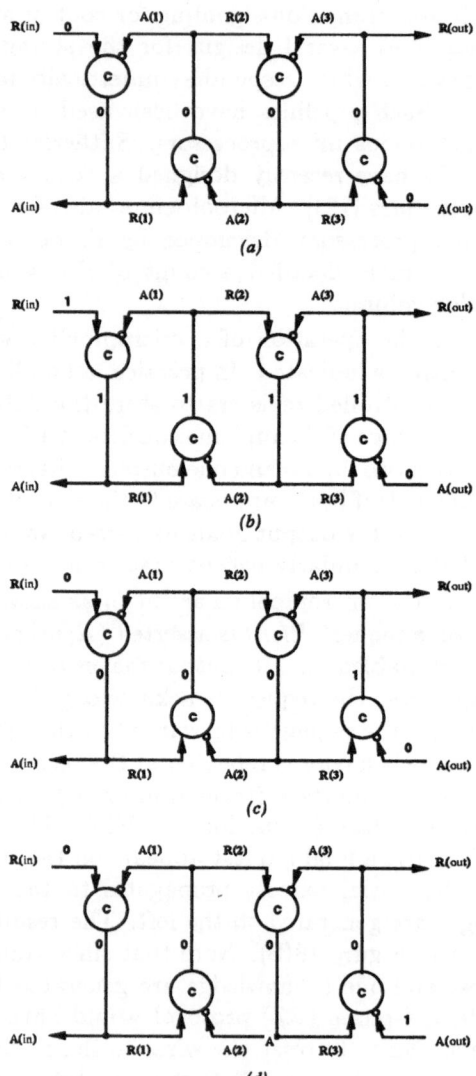

Figure 18: Micropipeline example

entered into the pipe. This data was not removed, since no $A(out)$ transition has occurred. Since $A(out)$ is still 0, no new transition can occur on $R(out)$. Instead, the new data is held in the 3rd stage. The resulting micropipeline configuration is shown in Figure 18(c). The micropipeline now contains data in the 3rd and 4th stages.

Finally, the original data in the 4th stage can be removed from the right interface, which then issues an acknowledge ($A(out)$ goes to 1). At this point, the 4th stage issues a new $R(out)$ transition (since $R(3)$ is 0 and $A(out)$ is 1), as data in stage 3 is moved to stage 4. The 3rd stage is acknowledged as well ($A(3)$ goes to 0). The resulting configuration is shown in Figure 18(d). In practice, more complicated scenarios are possible, since data may be added and removed from the pipeline concurrently.

Although micropipelines use transition-signaling, other signaling conventions have been used in asynchronous pipelines as well. Williams [140], Martin [80] and van Berkel [133] have used 4-phase handshaking (the "return-to-zero" protocol) between stages. An alternative two-phase signaling scheme, called *LEDR* (level-encoded dual-rail), was introduced to combine advantages of both transition-signaling and four-phase [41].

Asynchronous pipelines have been designed for numerous applications: multiplication [124, 96], division [141], and DSP [133, 87]. The Williams and Horowitz self-timed divider [141] is especially impressive: the fabricated chip was twice as fast as comparable synchronous designs.

Research on asynchronous pipelines and datapaths is now proceeding in many directions. Several generalizations to asynchronous pipeline structures have been proposed: *rings* [140], *multi-rings* [120] and *2-dimensional micropipelines* [54]. Techniques to reduce the communication overhead between stages have been developed [140, 40]. Liebchen and Gopalakrishnan have proposed a *reordering pipeline* [72] which allows the freezing and dynamic reordering of data within the pipe using "LockC" elements. Finally, low-power micropipeline structures have been introduced using adaptive scaling of supply voltage [97]. A number of other efforts are focused on low-power asynchronous datapaths [80, 77, 51, 134].

6.4 Asynchronous Processor Design.

Perhaps the greatest challenge in large-scale asynchronous design is to combine the techniques for asynchronous controller and datapath synthesis, and build asynchronous processors.

Asynchrony has in fact been present in processors from the early days, but it is only recently that such techniques have been applied systematically to the design of asynchronous microprocessors. The first asynchronous microprocessor was developed at Caltech by Martin *et al.* [80] in the late 1980's. The 16-bit design is almost fully quasi-delay-insensitive except for the memory interface. A 2μ CMOS version consumed 145mW at 5V and 6.7mW at 2V. A 1.6μCMOS version consumed 200mW at 5V and 7.6mW at 2V. The architecture was later re-implemented in GaAs.

Furber and the AMULET group at Manchester University have fabricated an asynchronous implementation of the ARM microprocessor [51, 106, 52, 105]. The design is based on micropipelined datapaths, and is part of a large-scale investigation of low-power techniques. The project addresses issues such as caching, exceptions and architectural optimization which are critical to the development of production-quality asynchronous machines. This work is described in greater detail in Furber's chapter in this text.

Sutherland, Sproull, Molnar, and other at Sun Labs have been developing an asynchronous *Counterflow Pipeline Processor* [121]. The architecture is based on a novel looped micropipeline, which synchronizes instruction and data flowing in opposite directions. The processor makes careful use of arbiters to regulate the synchronization.

Brunvand developed the *NSR* RISC microprocessor [17] at the University of Utah, using transition-signaling for control, bundled data, and a micropipelined datapath. Other micropipelined-based RISC designs have been proposed by David *et al.* [34] and Ginosar and Michell [53].

A delay-insensitive microprocessor, *TITAC*, has been developed by Nanya *et al.* at Tokyo Institute of Technology [95]. The designers introduce several optimizations to improve performance.

A different approach was proposed by Unger at Columbia University [131]. His "computers without clocks" use traditional asynchronous state machines for control logic, and a building block approach to design rather than compilation schemes. This approach allows a spectrum of timing assumptions to insure correct designs.

Finally, Dean's *STRiP* (self-timed RISC) processor at Stanford University combines synchronous and asynchronous features [39]. The design uses synchronous functional units in a globally-clocked pipeline. However, the clock rate may change dynamically based on the current contents of the pipeline, using a technique called *dynamic clocking*. The clock also is suspended during off-chip operations, such as input/output or access to a second-level cache. Using careful simulation, the design was shown to be almost twice as fast as a comparable synchronous design due solely to its asynchronous features.

6.5   Verification, Timing Analysis and Testing.

The above survey indicates an impressive surge of activity in the design of asynchronous controllers, datapaths and processors. However, design techniques alone cannot make asynchronous circuits commercially viable. In synchronous design, many ancillary techniques are needed to insure the correctness of designs, including verification, timing analysis and testing. These techniques are especially critical for asynchronous design because of their inherent subtlety. This section sketches briefly some of the recent work on validation of asynchronous designs.

6.5.1 Formal Verification and Timing Analysis

Because of the large variety of asynchronous design approaches, it is difficult to find a unified approach to the analysis and verification of all asynchronous circuits. For speed-independent and delay-insensitive systems, though, Hoare's *CSP* [58] and Milner's *CCS* [89] have been especially effective as formal underpinnings.

Rem, Snepscheut and Udding's *trace theory* [109], based on CSP, has been used for both specification and formal verification. In trace theory, the behavior of a concurrent system is described by the set of possible *traces*, or sequences of events, which may be observed. Each trace describes one possible interleaved behavior of the system. The traces are combined into a set, which defines the observable behavior of the system.

Dill [43, 44] and Ebergen [47] have built effective verification tools for SI and DI circuits based on trace theory. In Dill's theory, an implementation and specification are each modeled by trace sets. These sets are compared using a formal relation called *conformance*, which defines precisely when an implementation meets its specification. Dill has uncovered bugs in published circuits using the verifier. More efficient algorithms for approximate verification (allowing occasional false negatives) have been developed as well [8].

Dill's verifier effectively checks for safety violations (where a design has incorrect behavior), but does not check for liveness violations (where a design has deadlock or live-lock). Dill also introduced a theory of *complete trace structures* [43], based on Buchi automata, which can model general liveness properties. Although these general verification algorithms may be too expensive to apply in practice, a verifier has been developed for a constrained class of specifications [122]. Other methods use a restricted notion of liveness that can be easily checked [47, 55].

An alternative verification method based on CCS has been proposed by Birtwistle, Stevens, *et al.* [123, 122]. CCS has been successfully used for the specification of several asynchronous designs, including a token ring arbiter and SCSI controller. Specifications can then be checked for deadlock, safety and liveness properties using a modal logic. A substantial specification has been developed for the AMULET processor [13], with detailed models for the different instruction classes.

The above verification techniques handle SI and DI circuits, and therefore are not concerned with timing. However, timing is critical for the analysis and verification of many asynchronous systems. Recently, algorithms for verification of asynchronous state machines with bounded delays were introduced by Devadas *et al.* [42]. Williams [140] and Burns [22] have developed analysis methods for system-level performance. Other work has focused on timing analysis to determine minimum and maximum separation of events in a concurrent circuit or system [84, 61]. Such analysis can aid in both the optimization and verification of asynchronous designs.

6.5.2 Testing and Synthesis-for-Testability.

While formal verification is used to validate designs, testing is needed to validate the correctness of fabricated implementations. Testing and synthesis-for-testability play a major role in the industrial production of synchronous chips. The testing of asynchronous circuits is complicated by their special design constraints. For example, asynchronous circuits may use redundant logic to eliminate hazards, but redundant logic makes testing more difficult.

Initial results on the testing of speed-independent circuits include work by Beerel and Meng [7] and Martin and Hazewindus [81]. These papers indicate that certain classes of speed-independent circuits are "self-testing" with respect to stuck-at faults, where certain faults will cause the circuit to halt. Beerel and Meng generalized their approach to handle stuck-at faults in timed control circuits [10].

A general synthesis-for-testability method has been proposed by Keutzer, Lavagno and Sangiovanni-Vincentelli [65] which considers both stuck-at and path-delay faults. This work was extended by Nowick, Jha and Cheng [102]. Other work has focused on handshake circuits [113, 112] and micropipelines [66].

7 A Preview Of Subsequent Chapters

This book is an expanded treatise on the material covered at a workshop held in Banff, Alberta Canada in August of 1993. The speakers were selected both for their experience, present activity, and for their diversity. Sadly, Alain Martin was unable to contribute to this volume.

The work by Jo Ebergen shows that asynchronous circuits can be synthesized using an approach that is very similar to parallel programming. His method has an algebraic basis that provides a number of elegant mathematical benefits, such as semantic-preserving manipulation and formal verification. This chapter treats two design problems in depth, and provides a variety of implementations. These designs are used to illustrate a number of tradeoffs, such as area complexity, power consumption and performance.

The chapter by Steve Furber describes the redesign and fabrication of an existing synchronous CPU, the ARM microprocessor, as an asynchronous circuit. The goal of this effort was not to push a particular methodology but rather to use whatever method was appropriate for the task at hand. The chapter provides an excellent practical treatise of the design of an asynchronous processing element for low-power operation.

Kees van Berkel and Martin Rem review the potential of asynchronous circuits for low-power CMOS ICs, using a variety of simple, yet practical examples. They apply Tangram, a CSP-based VLSI-programming language, for their design and they use so-called handshake circuits as a general-purpose asynchronous architecture. Due to the transparent translation scheme, tradeoffs between power and costs (and performance) can be addressed at the Tangram level.

The work of Al Davis was heavily influenced by the work of Chuck Seitz and Stephen Unger. His work focused on starting with a finite state machine style specification that is commonly used by commercial synchronous designers. Recent efforts have been focused on creating an automatic synthesis system for burst-mode state machines. His efforts were sponsored by Hewlett-Packard corporation and saw significant collaboration with members of David Dill's group at Stanford. The focus of these efforts was on the synthesis of high-performance asynchronous circuits. This is in contrast with van Berkel's focus on low power properties.

Each chapter represents the state of the art in asynchronous circuit synthesis or analysis. Although there are many noteworthy efforts not covered directly in this volume, the material in these chapters provides enough depth and breadth to allow the serious reader to understand the research literature. It also provides a starting place for those new to the field to begin a serious research program of their own. As authors, we hope that readers will find as many questions in this book as they find answers, and furthermore that they will be inspired to answer those questions more thoroughly. We are all committed to expanding the knowledge of our chosen asynchronous circuit discipline, and if we succeed in inspiring new talent then we view this as a good start.

References

[1] W. B. Ackerman and J. B. Dennis. VAL - A Value-Oriented Algorithmic Language Preliminary Reference Manual. Technical Report LCS/TR-218, Massachusetts Institute Technology, Computer Science Department, 1979.

[2] M. Afhahi and C. Svensson. Performance of Synchronous and Asynchronous Schemes for VLSI Systems. *IEEE Transactions on Computers*, 41(7):858–872, July 1992.

[3] F. Aghdasi. Synthesis of asynchronous sequential machines for VLSI applications. In *Proceedings of the 1991 International Conference on Concurrent Engineering and Electronic Design Automation (CEEDA)*, pages 55–59, March 1991.

[4] V. Akella and G. Gopalakrishnan. SHILPA: a high-level synthesis system for self-timed circuits. In *Proceedings of the IEEE/ACM International Conference on Computer-Aided Design*, pages 587–91. IEEE Computer Society Press, November 1992.

[5] D.B. Armstrong, A.D. Friedman, and P.R. Menon. Realization of asynchronous sequential circuits without inserted delay elements. *IEEE Transactions on Computers*, C-17(2):129–134, February 1968.

[6] H. B. Bakoglu. *Circuits, Interconnections, and Packaging for VLSI*. Addison-Wesley, 1990.

[7] P. Beerel and T. Meng. Semi-Modularity and Self-Diagnostic Asynchronous Control Circuits. In Carlo H. Sequin, editor, *Proceedings of the 1991 University of California/Santa Cruz Conference*, pages 103–117. The MIT Press, 1991.

[8] P.A. Beerel, J. Burch, and T. Meng. Efficient verification of determinate speed-independent circuits. In *Proceedings of the IEEE/ACM International Conference on Computer-Aided Design*, pages 261–267. IEEE Computer Society Press, November 1993.

[9] P.A. Beerel and T. Meng. Automatic gate-level synthesis of speed-independent circuits. In *Proceedings of the IEEE/ACM International Conference on Computer-Aided Design*, pages 581–586. IEEE Computer Society Press, November 1992.

[10] P.A. Beerel and T. H.-Y. Meng. Testability of asynchronous timed control circuits with delay assumptions. In *Proceedings of the 28th ACM/IEEE Design Automation Conference*, pages 446–451. ACM, June 1991.

[11] J. Beister. A unified approach to combinational hazards. *IEEE Transactions on Computers*, C-23(6):566–575, June 1974.

[12] I. Benko and J.C. Ebergen. Delay-insensitive solutions to the committee problem. In *Proceedings of the International Symposium on Advanced Research in Asynchronous Circuits and Systems (Async94)*, pages 228–237. IEEE Computer Society Press, November 1994.

[13] G. Birtwistle and Y. Liu. Specification of the Manchester Amulet 1: Top level Specification. Computer Science Department Technical Report, University of Calgary, December 1994.

[14] J.G. Bredeson. Synthesis of multiple-input change hazard-free combinational switching circuits without feedback. *International Journal of Electronics (GB)*, 39(6):615–624, December 1975.

[15] J.G. Bredeson and P.T. Hulina. Elimination of static and dynamic hazards for multiple input changes in combinational switching circuits. *Information and Control*, 20:114–224, 1972.

[16] E. Brunvand. Translating concurrent communicating programs into asynchronous circuits. Technical Report CMU-CS-91-198, Carnegie Mellon University, 1991. Ph.D. Thesis.

[17] E. Brunvand. The NSR processor. In *Proceedings of the Twenty-Sixth Annual Hawaii International Conference on System Sciences*, volume I, pages 428–435. IEEE Computer Society Press, January 1993.

[18] E. Brunvand and R. F. Sproull. Translating concurrent programs into delay-insensitive circuits. In *Proceedings of the IEEE International Conference on Computer-Aided Design*, pages 262–265. IEEE Computer Society Press, November 1989.

[19] J.A. Brzozowski and J.C. Ebergen. Recent developments in the design of asynchronous circuits. Technical Report CS-89-18, University of Waterloo, Computer Science Department, 1989.

[20] J.A. Brzozowski and J.C. Ebergen. On the delay-sensitivity of gate networks. *IEEE Transactions on Computers*, 41(11):1349–1360, November 1992.

[21] S. M. Burns. Automated compilation of concurrent programs into self-timed circuits. Technical Report Caltech-CS-TR-88-2, California Institute of Technology, 1987. M.S. Thesis.

[22] S.M. Burns. Performance analysis and optimization of asynchronous circuits. Technical Report Caltech-CS-TR-91-01, California Institute of Technology, 1991. Ph.D. Thesis.

[23] S.M. Burns and A.J. Martin. Syntax-directed translation of concurrent programs into self-timed circuits. In J. Allen and T.F. Leighton, editors, *Advanced Research in VLSI: Proceedings of the Fifth MIT Conference*, pages 35–50. MIT Press, Cambridge, MA, 1988.

[24] T. J. Chaney and C. E. Molnar. Anomalous Behaviour of Synchronizer and Arbiter Circuits. *IEEE Transactions on Computers*, C–22(4):421–422, 1973.

[25] T.J. Chaney, S.M. Ornstein, and W.M. Littlefield. Beware the synchronizer. In *IEEE 6th International Computer Conference*, pages 317–319, 1972.

[26] V. L. Chi. Salphasic Distribution of Clock Signals for Synchronous Systems. *IEEE Transactions on Computers*, 43(5):597–602, May 1994.

[27] J.-S. Chiang and D. Radhakrishnan. Hazard-free design of mixed operating mode asynchronous sequential circuits. *International Journal of Electronics*, 68(1):23–37, January 1990.

[28] T.-A. Chu. Synthesis of self-timed vlsi circuits from graph-theoretic specifications. Technical Report MIT-LCS-TR-393, Massachusetts Institute of Technology, 1987. Ph.D. Thesis.

[29] T.-A. Chu. Automatic synthesis and verification of hazard-free control circuits from asynchronous finite state machine specifications. In *Proceedings of the IEEE International Conference on Computer Design*, pages 407–413. IEEE Computer Society Press, 1992.

[30] H.Y.H. Chuang and S. Das. Synthesis of multiple-input change asynchronous machines using controlled excitation and flip-flops. *IEEE Transactions on Computers*, C-22(12):1103–1109, December 1973.

[31] W.A. Clark. Macromodular computer systems. In *Proceedings of the Spring Joint Computer Conference, AFIPS*, April 1967.

[32] F. Commoner, A. Holt, S. Even, and A. Pnueli. Marked directed graphs. *Journal of Computer and System Sciences*, 5(5):511–523, October 1971.

[33] J. Cortadella, L. Lavagno, P. Vanbekbergen, and A. Yakovlev. Designing asynchronous circuits from behavioural specifications with internal conflicts. In *Proceedings of the International Symposium on Advanced Research in Asynchronous Circuits and Systems (Async94)*, pages 106–115. IEEE Computer Society Press, November 1994.

[34] I. David, R. Ginosar, and M. Yoeli. Self-timed implementation of a reduced instruction set computer. Technical Report 732, Technion and Israel Institute of Technology, October 1989.

[35] A. Davis, B. Coates, and K. Stevens. Automatic synthesis of fast compact self-timed control circuits. In *1993 IFIP Working Conference on Asynchronous Design Methodologies (Manchester, England)*, 1993.

[36] A.L. Davis. The architecture and system method of DDM-1: A recursively-structured data driven machine. In *Proc. Fifth Annual Symposium on Computer Architecture*, 1978.

[37] A.L. Davis. A data-driven machine architecture suitable for VLSI implementation. In C.L. Seitz, editor, *Proceedings of the Caltech Conference on Very Large Scale Integration*, pages 479–494, January 1979. ·

[38] A.L. Davis, B. Coates, and K. Stevens. The post office experience: Designing a large asynchronous chip. In *Proceedings of the Twenty-Sixth Annual Hawaii International Conference on System Sciences*, volume I, pages 409–418. IEEE Computer Society Press, January 1993.

[39] M.E. Dean. STRiP: A self-timed RISC processor architecture. Technical report, Stanford University, 1992. Ph.D. Thesis.

[40] M.E. Dean, D.L. Dill, and M. Horowitz. Self-timed logic using current-sensing completion detection (CSCD). In *Proceedings of the IEEE International Conference on Computer Design*. IEEE Computer Society Press, October 1991.

[41] M.E. Dean, T.E. Williams, and D.L. Dill. Efficient self-timing with level-encoded 2-phase dual-rail (LEDR). In Carlo Sequin, editor, *Advanced Research in VLSI : Proceedings of the 1991 University of California Santa Cruz Conference*, pages 55–70. The MIT Press, 1991. *ISBN 0-262-19308-6.*

[42] S. Devadas, K. Keutzer, S. Malik, and A. Wang. Verification of asynchronous interface circuits with bounded wire delays. In *Proceedings of the IEEE/ACM International Conference on Computer-Aided Design*, pages 188–195. IEEE Computer Society Press, November 1992.

[43] D.L. Dill. *Trace Theory for Automatic Hierarchical Verification of Speed-Independent Circuits*. MIT Press, Cambridge, MA, 1989.

[44] D.L. Dill, S.M. Nowick, and R.F. Sproull. Specification and automatic verification of self-timed queues. *Formal Methods in System Design*, 1(1):29–60, July 1992.

[45] D. W. Dobberpuhl and et al. A 200-MHz 64-bit Dual-issue CMOS Microprocessor. *Digital Technical Journal*, 4(4):35–50, 1993.

[46] J.C. Ebergen. A formal approach to designing delay-insensitive circuits. *Distributed Computing*, 5(3):107–119, 1991.

[47] J.C. Ebergen. A verifier for network decompositions of command-based specifications. In *Proceedings of the Twenty-Sixth Annual Hawaii International Conference on System Sciences*, volume I, pages 310–318. IEEE Computer Society Press, January 1993.

[48] E.B. Eichelberger. Hazard detection in combinational and sequential switching circuits. *IBM Journal of Research and Development*, 9(2):90–99, 1965.

[49] J. Frackowiak. Methoden der analyse und synthese von hasardarmen schaltnetzen mit minimalen kosten I. *Elektronische Informationsverarbeitung und Kybernetik*, 10(2/3):149–187, 1974.

[50] A.D. Friedman and P.R. Menon. Synthesis of asynchronous sequential circuits with multiple-input changes. *IEEE Transactions on Computers*, C-17(6):559–566, June 1968.

[51] S. B. Furber, P. Day, J. D. Garside, N. C. Paver, and J. V. Woods. A micropipelined ARM. In *Proceedings of VLSI 93*, pages 5.4.1 –5.4.10, September 1993.

[52] S.B. Furber, P. Day, J.D. Garside, N.C. Paver, S. Temple, and J.V. Woods. The design and evaluation of an asynchronous microprocessor. In *Proceedings of the IEEE International Conference on Computer Design*, pages 217–220. IEEE Computer Society Press, October 1994.

[53] R. Ginosar and N. Michell. On the potential of asynchronous pipelined processors. Technical Report UUCS-90-015, VLSI Systems Research Group, University of Utah, 1990.

[54] G. Gopalakrishnan. Micropipeline wavefront arbiters using lockable C-elements. *IEEE Design and Test*, 11(4):55–64, Winter 1994.

[55] G. Gopalakrishnan, E. Brunvand, N. Michell, and S.M. Nowick. A correctness criterion for asynchronous circuit validation and optimization. *IEEE Transactions on Computer-Aided Design of Integrated Circuits and Systems*, 13(11):1309–1318, November 1994.

[56] S. Hauck. Asynchronous design methodologies: An overview. *Proceedings of the IEEE*, 83(1):69–93, January 1995.

[57] A.B. Hayes. Stored state asynchronous sequential circuits. *IEEE Transactions on Computers*, C-30(8):596–600, August 1981.

[58] C.A.R. Hoare. Communicating sequential processes. *Communications of the ACM*, 21(8):666–677, August 1978.

[59] D. A. Huffman. The synthesis of sequential switching circuits. *Journal of the Franklin Institute*, 257(3):161–190, March 1954.

[60] D. A. Huffman. The synthesis of sequential switching circuits. *Journal of the Franklin Institute*, 257(4):275–303, April 1954.

[61] H. Hulgaard, S.M. Burns, T. Amon, and G. Borriello. Practical applications of an efficient time separation of events algorithm. In *Proceedings of the IEEE/ACM International Conference on Computer-Aided Design*, pages 146–151. IEEE Computer Society Press, November 1993.

[62] K. Hwang. *Computer Arithmetic: Principles, Architecture, and Design.* John Wiley and Sons, 1979.

[63] M.B. Josephs and J.T. Udding. An overview of D-I algebra. In *Proceedings of the Twenty-Sixth Annual Hawaii International Conference on System Sciences*, volume I, pages 329–338. IEEE Computer Society Press, January 1993.

[64] W. Keister, A. E. Ritchie, and S. ·H. Washburn. *The Design of Switching Circuits.* Van Nostrand, Princeton, New Jersey, 1951.

[65] K. Keutzer, L. Lavagno, and A. Sangiovanni-Vincentelli. Synthesis for testability techniques for asynchronous circuits. In *Proceedings of the IEEE International Conference on Computer-Aided Design*, pages 326–329. IEEE Computer Society Press, November 1991.

[66] A. Khoche and E. Brunvand. Testing micropipelines. In *Proceedings of the International Symposium on Advanced Research in Asynchronous Circuits and Systems (Async94)*, pages 239–246. IEEE Computer Society Press, November 1994.

[67] A. Kondratyev, M. Kishinevsky, B. Lin, P. Vanbekbergen, and A. Yakovlev. Basic gate implementation of speed-independent circuits. In *Proceedings of the 31st ACM/IEEE Design Automation Conference*, pages 56–62. ACM, June 1994.

[68] D.S. Kung. Hazard-non-increasing gate-level optimization algorithms. In *Proceedings of the IEEE/ACM International Conference on Computer-Aided Design*, pages 631–634. IEEE Computer Society Press, November 1992.

[69] M. Ladd and W. P. Birmingham. Synthesis of multiple-input change asynchronous finite state machines. In *Proceedings of the 28th ACM/IEEE Design Automation Conference*, pages 309–314. ACM, June 1991.

[70] L. Lavagno, K. Keutzer, and A. Sangiovanni-Vincentelli. Algorithms for synthesis of hazard-free asynchronous circuits. In *Proceedings of the 28th ACM/IEEE Design Automation Conference*, pages 302–308. ACM, June 1991.

[71] L. Lavagno, C.W. Moon, R.K. Brayton, and A. Sangiovanni-Vincentelli. Solving the state assignment problem for signal transition graphs. In *Proceedings of the 29th IEEE/ACM Design Automation Conference*, pages 568–572. IEEE Computer Society Press, June 1992.

[72] A. Liebchen and G. Gopalakrishnan. Dynamic reordering of high latency transactions using a modified micropipeline. In *Proceedings of the IEEE International Conference on Computer Design*, pages 336–340. IEEE Computer Society Press, 1992.

[73] B. Lin and S. Devadas. Synthesis of hazard-free multi-level logic under multiple-input changes from binary decision diagrams. In *Proceedings of the IEEE/ACM International Conference on Computer-Aided Design*, pages 542–549. IEEE Computer Society Press, November 1994.

[74] K.-J. Lin and C.-S. Lin. Automatic synthesis of asynchronous circuits. In *Proceedings of the 28th ACM/IEEE Design Automation Conference*, pages 296–301. ACM, June 1991.

[75] C.N. Liu. A state variable assignment method for asynchronous sequential switching circuits. *Journal of the ACM*, 10:209–216, April 1963.

[76] G. Mago. Realization methods for asynchronous sequential circuits. *IEEE Transactions on Computers*, C-20(3):290–297, March 1971.

[77] A. Marshall, B. Coates, and P. Siegel. The design of an asynchronous communications chip. *IEEE Design and Test*, 11(2):8–21, Summer 1994.

[78] A.J. Martin. The limitation to delay-insensitivity in asynchronous circuits. In W.J. Dally, editor, *Advanced Research in VLSI: Proceedings of the Sixth MIT Conference*, pages 263–278. MIT Press, Cambridge, MA, 1990.

[79] A.J. Martin. Programming in VLSI: From communicating processes to delay-insensitive circuits. In C.A.R. Hoare, editor, *Developments in Concurrency and Communication*, UT Year of Programming Institute on Concurrent Programming, pages 1–64. Addison-Wesley, Reading, MA, 1990.

[80] A.J. Martin, S.M. Burns, T.K. Lee, D. Borkovic, and P.J. Hazewindus. The design of an asynchronous microprocessor. In *1989 Caltech Conference on Very Large Scale Integration*, 1989.

[81] A.J. Martin and P.J. Hazewindus. Testing delay-insensitive circuits. In Carlo H. Séquin, editor, *Advanced Research in VLSI: Proceedings of the 1991 UC Santa Cruz Conference*, pages 118–132. MIT Press, 1991.

[82] E.J. McCluskey. *Introduction to the Theory of Switching Circuits.* McGraw-Hill, New York, NY, 1965.

[83] E.J. McCluskey. *Logic Design Principles: with emphasis on testable semi-custom circuits.* Prentice-Hall, Englewood Cliffs, NJ, 1986.

[84] K. McMillan and D.L. Dill. Algorithms for interface timing verification. In *Proceedings of the IEEE International Conference on Computer Design*, pages 48–51. IEEE Computer Society Press, October 1992.

[85] C. Mead and L. Conway. *Introduction to VLSI Systems*, chapter 7. Addison-Wesley, Reading, MA, 1980. C.L. Seitz, System Timing.

[86] T. H.-Y. Meng, R.W. Brodersen, and D.G. Messerschmitt. Automatic synthesis of asynchronous circuits from high-level specifications. *IEEE Transactions on Computer-Aided Design of Integrated Circuits and Systems*, 8(11):1185–1205, November 1989.

[87] T.H. Meng. *Synchronization Design for Digital Systems.* Kluwer Academic Publishers, Boston, MA, 1991.

[88] R.E. Miller. *Switching Theory. Volume II: Sequential Circuits and Machines.* John Wiley and Sons, New York, NY, 1965.

[89] R. Milner. *Communication and Concurrency.* Prentice Hall, London, 1989.

[90] C.E. Molnar, T.-P. Fang, and F.U. Rosenberger. Synthesis of delay-insensitive modules. In Henry Fuchs, editor, *Proceedings of the 1985 Chapel Hill Conference on Very Large Scale Integration*, pages 67–86. CSP, Inc., 1985.

[91] C.W. Moon, P.R. Stephan, and R.K. Brayton. Synthesis of hazard-free asynchronous circuits from graphical specifications. In *Proceedings of the IEEE International Conference on Computer-Aided Design*, pages 322–325. IEEE Computer Society Press, November 1991.

[92] D. E. Muller and W. S. Bartky. A theory of asynchronous circuits I. Digital Computer Laboratory 75, University of Illinois, November 1956.

[93] D. E. Muller and W. S. Bartky. A theory of asynchronous circuits II. Digital Computer Laboratory 78, University of Illinois, March 1957.

[94] C. Myers and T. Meng. Synthesis of timed asynchronous circuits. In *Proceedings of the IEEE International Conference on Computer Design*, pages 279–284. IEEE Computer Society Press, October 1992.

[95] T. Nanya, Y. Ueno, H. Kagotani, M. Kuwako, and A. Takamura. TITAC: design of a quasi-delay-insensitive microprocessor. *IEEE Design and Test*, 11(2):50–63, Summer 1994.

[96] C.D. Nielsen and A. Martin. The design of a delay-insensitive multiply-accumulate unit. In *Proceedings of the Twenty-Sixth Annual Hawaii International Conference on System Sciences*, volume I, pages 379–388. IEEE Computer Society Press, January 1993.

[97] L.S. Nielsen, C. Niessen, J. Sparso, and K. van Berkel. Low-Power Operation Using Self-Timed Circuits and Adaptive Scaling of the Supply Voltage. *IEEE Transactions on VLSI*, 2(4):7, 1994.

[98] S.M. Nowick. Automatic synthesis of burst-mode asynchronous controllers. Technical report, Stanford University, 1993. Ph.D. Thesis.

[99] S.M. Nowick, M.E. Dean, D.L. Dill, and M. Horowitz. The design of a high-performance cache controller: a case study in asynchronous synthesis. *INTEGRATION, the VLSI journal*, 15(3):241–262, October 1993.

[100] S.M. Nowick and D.L. Dill. Synthesis of asynchronous state machines using a local clock. In *Proceedings of the IEEE International Conference on Computer Design*, pages 192–197. IEEE Computer Society Press, October 1991.

[101] S.M. Nowick and D.L. Dill. Exact two-level minimization of hazard-free logic with multiple-input changes. In *Proceedings of the IEEE/ACM International Conference on Computer-Aided Design*, pages 626–630. IEEE Computer Society Press, November 1992.

[102] S.M. Nowick, N.K. Jha, and F.-C. Cheng. Synthesis of asynchronous circuits for stuck-at and robust path delay fault testability. In *Proceedings of the Eighth International Conference on VLSI Design (VLSI Design 95)*. IEEE Computer Society Press, January 1995.

[103] S.M. Nowick, K.Y. Yun, and D.L. Dill. Practical asynchronous controller design. In *Proceedings of the IEEE International Conference on Computer Design*, pages 341–345. IEEE Computer Society Press, October 1992.

[104] S.S. Patil. An Asynchronous Logic Array. Technical Report Technical Memorandom 62, Massachusetts Institute of Technology, Project MAC, 1975.

[105] N.C. Paver. The design and implementation of an asynchronous microprocessor. Technical report, University of Manchester, June 1994. Ph.D. Thesis.

[106] N.C. Paver, P. Day, S.B. Furber, J.D. Garside, and J.V. Woods. Register locking in an asynchronous microprocessor. In *Proceedings of the IEEE International Conference on Computer Design*, pages 351–355. IEEE Computer Society Press, October 1992.

[107] J.L. Peterson. *Petri Net Theory and the Modeling of Systems*. Prentice-Hall, Englewood Cliffs, NJ, 1981.

[108] R. Puri and J. Gu. Area efficient synthesis of asynchronous interface circuits. In *Proceedings of the IEEE International Conference on Computer Design*, pages 212–216. IEEE Computer Society Press, October 1994.

[109] M. Rem, J.L.A. van de Snepscheut, and J.T. Udding. Trace theory and the definition of hierarchical components. In Randal Bryant, editor, *Proceedings of the Third Caltech Conference on Very Large Scale Integration*, pages 225–239. CSP, Inc., 1983.

[110] R. D. Rettberg, W. R. Crowther, P. P. Carvey, and R. S. Tomlinson. The Monarch Parallel Processor Hardware Design. *Computer*, 23(4):18–30, April 1990.

[111] C.A. Rey and J. Vaucher. Self-synchronized asynchronous sequential machines. *IEEE Transactions on Computers*, C-23(12):1306–1311, December 1974.

[112] M. Roncken. Partial scan test for asynchronous circuits illustrated on a DCC error corrector. In *Proceedings of the International Symposium on Advanced Research in Asynchronous Circuits and Systems (Async94)*, pages 247–256. IEEE Computer Society Press, November 1994.

[113] Marly Roncken and Ronald Saeijs. Linear Test Times for Delay-Insensitive Circuits: a Compilation Strategy. In S. Furber and M. Edwards, editors, *Proceedings of the IFIP WG 10.5 Working Conference on Asynchronous Design Methodologies, Manchester*, pages 13–27. Elsevier Science Publishers B.V., 1993.

[114] L.Y. Rosenblum and A.V. Yakovlev. Signal graphs: from self-timed to timed ones. In *Proceedings of International Workshop on Timed Petri Nets, Torino, Italy*, pages 199–207. IEEE Computer Society Press, July 1985.

[115] C.-J. Seger. A bounded delay race model. In *Proceedings of the IEEE International Conference on Computer-Aided Design*, pages 130–133. IEEE Computer Society Press, November 1989.

[116] C.L. Seitz. Asynchronous machines exhibiting concurrency. In *Conference Record of the Project MAC Conference on Concurrent Systems and Parallel Computation*, 1970.

[117] C.L. Seitz. *Graph representations for logical machines*. PhD thesis, MIT, Jan 1971.

[118] P. Siegel, G. De Micheli, and D. Dill. Technology mapping for generalized fundamental-mode asynchronous designs. In *30th ACM/IEEE Design Automation Conference*, June 1993. To appear.

[119] R. L. Sites. *Alpha Architecture Reference Manual*. Digital Equipment Corporation, 1992.

[120] J. Sparso and J. Staunstrup. Design and performance analysis of delay insensitive multi-ring structures. In *Proceedings of the Twenty-Sixth Annual Hawaii International Conference on System Sciences*, volume I, pages 349–358. IEEE Computer Society Press, January 1993.

[121] R.F. Sproull, I.E. Sutherland, and C.E. Molnar. The counterflow pipeline processor architecture. *IEEE Design & Test of Computers*, 11(3):48–59, 1994.

[122] K. Stevens. Practical Verification and Synthesis of Low Latency Asynchronous Systems. PhD Thesis, Computer Science Department, University of Calgary, 1994.

[123] K. Stevens, J. Aldwinckle, G. Birtwistle, and Y. Liu. Designing parallel specifications in CCS. In *Proceedings of Canadian Conference on Electrical and Computer Engineering*, Vancouver, 1993.

[124] I.E. Sutherland. Micropipelines. *Communications of the ACM*, 32(6):720–738, June 1989.

[125] M.A. Tapia. Synthesis of asynchronous sequential systems using boolean calculus. In *14th Asilomar Conference on Circuits, Systems and Computers*, pages 205–209, November 1980.

[126] J.H. Tracey. Internal state assignments for asynchronous sequential machines. *IEEE Transactions on Electronic Computers*, EC-15:551–560, August 1966.

[127] J.T. Udding. A formal model for defining and classifying delay-insensitive circuits and systems. *Distributed Computing*, 1(4):197–204, 1986.

[128] S.H. Unger. *Asynchronous Sequential Switching Circuits*. Wiley-Interscience, New York, NY, 1969.

[129] S.H. Unger. Asynchronous sequential switching circuits with unrestricted input changes. *IEEE Transactions on Computers*, C-20(12):1437–1444, December 1971.

[130] S.H. Unger. Self-synchronizing circuits and nonfundamental mode operation. *IEEE Transactions on Computers (Correspondence)*, C-26(3):278–281, March 1977.

[131] S.H. Unger. A building block approach to unclocked systems. In *Proceedings of the Twenty-Sixth Annual Hawaii International Conference on System Sciences*, volume I, pages 339–348. IEEE Computer Society Press, January 1993.

[132] C.H. van Berkel and R.W.J.J. Saeijs. Compilation of communicating processes into delay-insensitive circuits. In *Proceedings of the IEEE International Conference on Computer Design*, pages 157–162. IEEE Computer Society Press, 1988.

48

[133] K. van Berkel. *Handshake Circuits. An asynchronous architecture for VLSI programming*. International Series on Parallel Computation 5. Cambridge University Press, 1993.

[134] K. van Berkel, R. Burgess, J. Kessels, A. Peeters, M. Roncken, and F. Schalij. Asynchronous Circuits for Low Power: a DCC Error Corrector. *IEEE Design & Test*, 11(2):22–32, June 1994.

[135] Kees van Berkel. Beware the isochronic fork. *Integration, the VLSI journal*, 13(2):103–128, 1992.

[136] P. Vanbekbergen, F. Catthoor, G. Goossens, and H. De Man. Optimized synthesis of asynchronous control circuits from graph-theoretic specifications. In *Proceedings of the IEEE International Conference on Computer-Aided Design*, pages 184–187. IEEE Computer Society Press, November 1990.

[137] P. Vanbekbergen, B. Lin, G. Goossens, and H. De Man. A generalized state assignment theory for transformations on signal transition graphs. In *Proceedings of the IEEE/ACM International Conference on Computer-Aided Design*, pages 112–117. IEEE Computer Society, November 1992.

[138] V.I. Varshavsky, M.A. Kishinevsky, V.B. Marakhovsky, V.A. Peschansky, L.Y. Rosenblum, A.R. Taubin, and B.S. Tzirlin. *Self-timed Control of Concurrent Processes*. Kluwer Academic Publishers, 1990. Russian edition: 1986.

[139] Tom Verhoeff. Delay-insensitive codes – an overview. *Distributed Computing*, 3(1):1–8, 1988.

[140] T.E. Williams. Self-timed rings and their application to division. Technical Report CSL-TR-91-482, Computer Systems Laboratory, Stanford University, 1991. Ph.D. Thesis.

[141] T.E. Williams and M.A. Horowitz. A zero-overhead self-timed 54b 160ns CMOS divider. *IEEE Journal of Solid-State Circuits*, 26(11):1651–1661, November 1991.

[142] A.V. Yakovlev. On limitations and extensions of STG model for designing asynchronous control circuits. In *Proceedings of the IEEE International Conference on Computer Design*, pages 396–400. IEEE Computer Society Press, October 1992.

[143] O. Yenersoy. Synthesis of asynchronous machines using mixed-operation mode. *IEEE Transactions on Computers*, C-28(4):325–329, April 1979.

[144] M.L. Yu and P.A. Subrahmanyam. A path-oriented approach for reducing hazards in asynchronous designs. In *Proceedings of the 29th IEEE/ACM Design Automation Conference*, pages 239–244. IEEE Computer Society Press, June 1992.

[145] K.Y. Yun and D.L. Dill. Automatic synthesis of 3D asynchronous finite-state machines. In *Proceedings of the IEEE/ACM International Conference on Computer-Aided Design*. IEEE Computer Society Press, November 1992.

[146] K.Y. Yun and D.L. Dill. Unifying synchronous/asynchronous state machine synthesis. In *Proceedings of the IEEE/ACM International Conference on Computer-Aided Design*, pages 255–260. IEEE Computer Society Press, November 1993.

[147] K.Y. Yun, D.L. Dill, and S.M. Nowick. Practical generalizations of asynchronous state machines. In *The 1993 European Conference on Design Automation*, pages 525–530. IEEE Computer Society Press, February 1993.

Parallel Program and Asynchronous Circuit Design

Jo C. Ebergen, John Segers[1], and Igor Benko

Department of Computer Science, University of Waterloo
Waterloo, Ontario, Canada

Abstract

Asynchronous circuit design is a beautiful application area for any formalism that can reason about parallelism. By means of two small, but challenging, exercises we illustrate the similarities and differences between parallel program and asynchronous circuit design. The exercises are simple to state and have many solutions, which are sometimes surprisingly efficient. They all illustrate many aspects of asynchronous circuit design. For each exercise we present several solutions, which are analysed with respect to delay assumptions, safety, progress, and performance issues. We also mention some open problems.

> *It is quite difficult to think of the code*
> *entirely* in abstracto *without any kind of circuit.*
> Alan M. Turing [34].

1   Introduction

The design and analysis of asynchronous circuits has witnessed a remarkable upsurge in the past five years. Many researchers have claimed or demonstrated that asynchronous circuits have a great potential for speed, low power consumption, robustness, modular design, and ease of design. Some of these properties, like modular design and ease of design, were already demonstrated in the 60's in the Macromodules project [7]. The recent advances are characterized by many novel design and verification techniques [1, 3, 6, 11, 12, 18, 22, 33] or by the new applications of and improvements in classical approaches [4, 8, 9, 20, 25, 29]. The performance analysis of asynchronous circuits is also beginning to draw wide attention [2, 5, 13, 28, 31, 37]. The novel techniques indicate that the design and analysis of asynchronous circuits have much in common with the design and analysis of parallel programs. The purpose of this chapter is to illustrate some techniques in designing asynchronous circuits and in analyzing the performance of these circuits with respect to area complexity, power consumption, and response time. The design of a circuit consists of the derivation of a parallel program for a specification. Wherever possible, we try to give a heuristics for each design decision. The derivations are presented in a format that allows a quick verification of all steps.

Our method for designing asynchronous circuits is based on a simple formalism. Specifications are given by means of guarded commands with input

---

[1] Currently at Department of Mathematics and Computing Science, Eindhoven University of Technology, Eindhoven, The Netherlands.

and output actions, a notation inspired by Dijkstra's guarded commands [10] and Hoare's CSP [15]. The formalization of implementation is given by the definition of decomposition. A decomposition of a specification consists of a network of basic components realizing the specified behavior. The notion of a delay-insensitive circuit plays an important role in our implementations. A delay-insensitive circuit is a special type of an asynchronous circuit, which is informally characterized as a network of basic components implementing a specification such that the correctness of the implementation is insensitive to any delays in wire connections or variations in the response time of the basic components.

The modulo-N counter and the up-down N counter are beautiful examples for illustrating both design and analysis techniques for asynchronous circuits. Both components have simple specifications, but admit a surprising variety of implementations. We give detailed derivations of several implementations for the two counters and a performance analysis of the designs. Our final designs for the counters have a bounded response time, a bounded power consumption, and an area complexity logarithmic in N. All bounds are asymptotically optimal. We start by giving a specification for the modulo-N counter. We explain the rules of the game along the way.

2   The Modulo-N Counter

The goal of the game is to find an efficient decomposition of the modulo-N counter into basic components, for any $N > 1$. The decomposition should be efficient with respect to area complexity, power consumption, and response time.

There are several ways to specify a modulo-N counter. Perhaps the most simple one is to view the modulo-N counter as a component with one input r and two outputs $a0$ and $a1$. After each of the first $N-1$ inputs r, the component may respond with output $a1$, and after the Nth input the component may respond with output $a0$. This behavior then repeats. A formal specification for this component can be formulated as follows.

$$\mathbf{pref} * [\, (r?;\ a1!)^{N-1};\ r?;\ a0! \,]$$

where '?' denotes an input action, '!' denotes an output action, ';' denotes concatenation, 'E^N' denotes N-fold repetition of the command E, '*[ ]' denotes arbitrary repetition of the enclosed command (or Kleene's closure), and $\mathbf{pref}\,E$ denotes the prefix-closure of E. (A more precise explanation of the notation follows in the next section.)

A higher-level specification can be given if we consider the outputs $a0$ and $a1$ as special implementations of sending the values 0 and 1 on channel a. Channel a is then considered as an output channel of type binary. We denote this by $a!: bin$. Channel r can be considered an input channel of type unary, since each signal on channel r can be seen as the communication of a value that is always the same. If r is an input channel of type unary, we denote this by $r? : un$. In this more abstract view, the modulo-N counter can be specified as follows.

$ModC(N : int, r? : un, a! : bin)$

$=$ { by definition }

$|[$ **var** $n : int$ ::
  initially $n = 0$ ::
  pref $*[r?; n := (n + 1) \bmod N;$
             if $n \neq 0$ **then** $a := 1$
             $| \ n = 0$ **then** $a := 0$
             fi ; $a!$
             $]$
$]|$

where int denotes the type integer and an occurrence of $a!$ denotes 'the value of a is sent along its channel.' In order to leave as much freedom as possible in choosing an implementation, in particular in implementing the data types, this is the specification we use for deriving decompositions.

A property of this last specification worth remembering is that the precondition for each output $a!$ is given by

$$(\#r \bmod N = 0) \equiv (a = 0)$$

where $\#r$ is the number of inputs r received thus far. For this reason, a postcondition of a corresponding input $a?$ in another component can given by the same assertion. We use this property frequently in the coming derivations.

When trying to find an implementation for the modulo-N counter, we assume that the environment provides the inputs as specified. For example, after providing input r the environment waits for output a before providing the next input. Under this assumption our implementation should provide outputs as specified.

In traditional circuit design, a modulo-N counter is specified by means of a state transition diagram, where states and transitions are given by means of logic values. Implementations are nearly always based on a binary representation of the count and consist of a series of latches and logic gates. Often, these designs have a so-called ripple-carry delay, which makes the response time dependent on N. Furthermore, if every latch is clocked in each clock period, the power consumption of these designs is at least logarithmic in N. Various designs for the modulo-N counter implemented in this way can be found in almost any textbook on switching theory. We demonstrate that modulo-N counters with bounded response time and bounded power consumption can be designed. Our designs are inspired by ideas developed at Philips Research Laboratories [2, 19].

3 Rules of the Game

In the previous section we briefly stated our goal and gave a specification of the modulo-N counter without much explanation. An attentive reader may have lots of questions. For example, what is the formal meaning of our program notation? What do we mean by 'implementation'? What basic components can we use? How can we implement data types? How do we measure area complexity, power consumption, and response time? In short, what are the rules of the game? We answer these questions briefly in the coming sections. (See [12] for more detail.)

3.1 Commands

The modulo-N counter is specified by a so-called *guarded command*. A guarded command prescribes the communication behavior of a component by listing all sequences of communication actions that may occur between the component and its environment. We first consider a subset of guarded commands called *commands*.

The semantics of a command is given by a *trace structure*, which is a triple $\langle I, O, T \rangle$. Set I is the *input alphabet* and represents all input terminals of the component; O is the *output alphabet* and represents all output terminals of the component; T is the *trace set* and represents all possible communication behaviors between component and environment. Every trace in T is constructed from symbols in $I \cup O$.

For command E, the notations iE, oE, and tE stand for the input alphabet, output alphabet, and trace set of the trace structure represented by E respectively. The alphabet of E is denoted by aE and given by $aE = iE \cup oE$. Equality between commands denotes equality of the trace structures represented by the commands.

The so-called atomic commands are *abort*, *skip*, and $a!, a?$, and a for any symbol a. They denote the following trace structures

$$
\begin{aligned}
abort &= \langle \emptyset, \emptyset, \emptyset \rangle \\
skip &= \langle \emptyset, \emptyset, \{\varepsilon\} \rangle \\
a! &= \langle \emptyset, \{a\}, \{a\} \rangle \\
a? &= \langle \{a\}, \emptyset, \{a\} \rangle \\
a &= \langle \{a\}, \{a\}, \{a\} \rangle
\end{aligned}
$$

The trace set of a command is constructed in a similar way as the language of a regular expression. The constructions for concatenation ';', selection (or union) '|', N-fold repetition '$()^N$', and arbitrary repetition '$*[\ ]$' are defined as usual and omitted here. The operation *weaving* is used to express a parallel composition with synchronization on common symbols. Formally, the weave $E0\|E1$ of trace structures $E0$ and $E1$ is defined by

$$
\begin{aligned}
E0\|E1 = \langle\ & iE0 \cup iE1 \\
, &\ oE0 \cup oE1 \\
, &\ \{t \in (aE0 \cup aE1)^* \mid t{\downarrow}aE0 \in tE0 \ \wedge\ t{\downarrow}aE1 \in tE1\} \\
&\rangle.
\end{aligned}
$$

where $t \downarrow aE0$ stands for the projection of trace t on the alphabet of $E0$. (Recall that $aE0 = iE0 \cup oE0$.) Notice that every trace in the weave $E0\|E1$ must be in accordance with a trace of $E0$, if you only look at symbols from $E0$, and with a trace of $E1$, if you only look at symbols from $E1$. Because the weave $E0\|E1$ consists of all behaviors that are in accordance with behaviors in $E0$ *and* in $E1$, weaving can be considered as the *behavioral conjunction* of $E0$ and $E1$. For this reason, the symbol $\|$ is often pronounced as 'and.' There are two special cases of weaving. If $aE0 \cap aE1 = \emptyset$, then weaving $E0$ and $E1$ amounts to the interleaving of the traces of $E0$ and $E1$. If $aE0 = aE1$, then weaving $E0$ and $E1$ amounts to taking the intersection of the traces of $E0$ and $E1$.

The *prefix-closure* of command E, denoted by $\mathbf{pref}\,E$, is a trace structure with the same alphabets as E and with the trace set that consists of all prefixes of all traces in tE. Trace structure E is called *prefix-closed* if $\mathbf{pref}\,E = E$. The **pref** operation constructs prefix-closed trace structures. Specifications are always given by commands that represent prefix-closed, non-empty trace structures with disjoint input and output alphabets. Trace structure E is called *non-empty* if $tE \neq \emptyset$.) Whenever we speak of *component E* in the following, E represents a prefix-closed, non-empty trace structure with disjoint input and output alphabets. The domain of prefix-closed, non-empty trace structures is one of the simplest semantic domains [15]. Notice that *abort*, $a?$, and $a!$ do not represent non-empty, prefix-closed trace structures and therefore do not specify components. The command *skip* does represent a prefix-closed, non-empty trace structure and therefore specifies a component: a component that has no input or output terminals and doesn't do anything.

The **pref** operator allows us to *unfold* a specification. For example, we have

$\mathbf{pref} *[\, a?; b! \,]$

$=$ { unfolding }

$\mathbf{pref}(a?; *[\, b!; a? \,])$

Finally we mention some laws for *skip* and *abort*

$$
\begin{aligned}
E;\ skip &= E \\
E \| skip &= E \\
E \mid abort &= E \\
t(E;\ abort) &= t(abort) \\
t(E \| abort) &= t(abort)
\end{aligned}
$$

3.2   Some Specifications and their Interpretations

In order to familiarize ourselves with commands, we specify the basic components WIRE, IWIRE, MERGE, TOGGLE, JOIN, and the 2-by-1 JOIN. Their specifications are given in Table 1. As for the priority of the binary operators, the parallel bar ($\|$) has the highest binding power, then the semicolon (;), and finally the choice bar ($\mid$).

Name	Specification	Schematic
WIRE	$\textbf{pref} *[\,a?; b!\,]$	a? ⟶ b!
IWIRE	$\textbf{pref} *[\,b!; a?\,]$	a? ⟶▭⟶ b!
MERGE	$\textbf{pref} *[\,(a? \mid b?)\,; c!\,]$	a? b? ⟶ M ⟶ c!
TOGGLE	$\textbf{pref} *[\,a?; b!; a?; c!\,]$	a? ⟶ (b! , c!)
JOIN	$\textbf{pref} *[\,(a?\|\|b?)\,; c!\,]$	a? b? ⟶ ◯ ⟶ c!
2-by-1 JOIN	$\textbf{pref} *[\,(a0?\|\|b?)\,; c0! \mid (a1?\|\|b?)\,; c1!\,]$	a0? a1? b? ⟶ c0! c1!

Table 1: Some basic components

In the previous section we stipulated that components are specified by means of commands representing prefix-closed, non-empty trace structures with disjoint input and output alphabets. We interpret such a trace structure in a special *mechanistic* way as opposed to a physical interpretation related to a particular implementation.

In order to understand this mechanistic interpretation, it is important to explain the role of input and output first. In our interpretation, an output may be produced by a component as soon as that output is enabled in the specification. Similarly, an input may be produced by the environment as soon as that input is enabled in the specification. The environment can be considered as a collection of other components.

The mechanistic interpretation consists of two conditions: a condition related to safety and a condition related to progress.

- Safety: no input or output is produced when not allowed by the specification.

- Progress: each trace specified may occur.

Because of our interpretation of inputs and outputs, the safety condition not only prescribes what the component may not do, but also what the environment may not do. In other words, a specification is a prescription for both component *and* environment. Because of the environment prescription, we can stipulate the conditions under which correct component behavior must be guaranteed. In this context we can interpret specification E as 'if the environment produces the inputs as prescribed in E, then the component may produce the outputs as prescribed in E.' If the environment violates the prescription, nothing is guaranteed and erroneous behavior may occur. The production of an input or output that violates a specification is also called *computation interference*. This term was first introduced in [32]. In [11], it is called a *choke*. Later, when we consider a network of components as an implementation of a specification, one of our proof obligations is to show that computation interference cannot occur.

While the safety condition prescribes what may not happen, the progress condition prescribes to a certain extent what must happen. The progress condition says that, if the environment produces the inputs as specified in E, then the component must behave such that every trace in tE may occur. Here 'may' should be interpreted as 'is possible, but not guaranteed.' This requirement excludes, for example, implementations of components where some traces cannot occur. On the other hand, this prescription does not require that every trace is guaranteed to occur. The actual occurrence of a trace may depend on nondeterministic choices made during the operation of the implementation: if the appropriate choices are made, then each trace can occur. This progress condition is too weak in a number of cases. On the other hand, it is easy to work with and suffices in many cases. Formulating a general, satisfactory progress condition that is convenient to deal with is still an open problem. We will return to a more formal treatment of these conditions and their problems later when we discuss implementations of specifications.

The abstract mechanistic interpretation allows for several physical implementations, like a mechanical, optical, or electrical one. The usual electrical implementation is that each symbol in the alphabet is associated with a terminal of a circuit. Each occurrence of a symbol in a trace corresponds to a voltage transition at that terminal. There is no distinction between rising and falling transitions: both transitions are denoted by the same symbol. This type of signaling is called *transition signaling* [33]. Outputs are transitions caused by the circuit and inputs are transitions caused by the environment.

With the above mechanistic interpretation in mind, we can explain the behavior of the basic components as follows. A WIRE has one input and one output terminal. Each communication behavior is an alternation of inputs and outputs, starting with an input, if any. (A WIRE can be implemented by a physical wire.) Notice that the safety condition prescribes that the environment is not allowed to provide two inputs in a row, nor is the component allowed to produce initially an output, or to produce two outputs in a row. The progress condition, on the other hand, prevents implementations where an output is guaranteed never to occur.

The IWIRE also has one input and one output terminal. Each communication behavior is an alternation of inputs and outputs, starting with an output, if any.

The MERGE has two input terminals and one output terminal. Each communication behavior is an alternation of inputs and outputs: the environment may choose to produce one of the two inputs (not both), the component may then produce an output. (The MERGE can be implemented by an XOR gate.)

The TOGGLE has one input and two output terminals. Each communication behavior is also an alternation of inputs and outputs starting with an input, if any. After each input has been received, an output may be produced. The outputs are produced at alternating terminals.

The JOIN has two input terminals and one output terminal. Each communication behavior is an alternation of two inputs and an output. After both inputs have been received, the JOIN may produce an output. The JOIN can be implemented by a Muller C-ELEMENT [24]. The Muller C-ELEMENT has a more general specification:

$$\textbf{pref} * [\ a?^2 \mid b?^2 \mid (a?\|b?;\ c!)\ ]$$

Notice that the behaviors of the JOIN are a proper subset of the behaviors of the C-ELEMENT. For example, for the C-ELEMENT the environment is allowed to produce an input a and then immediately withdraw that input. The specification of the JOIN prescribes that the environment is not allowed to withdraw an input. Knowing what the environment may do and what it may not do will be essential later when we give our correctness criteria for an implementation. For these reasons we distinguish between the JOIN and the C-ELEMENT. (The JOIN is also known as the RENDEZVOUS element in the Macromodules project [7].)

The 2-by-1 JOIN has three input and two output terminals. Each communication behavior is an alternation of two inputs followed by an output. Either $a0$ and b are received in parallel and then a $c0$ may be produced, or $a1$ and b are received in parallel and then a $c1$ may be produced. We have listed the 2-by-1 JOIN only. Other n-by-m JOINs for $n, m > 0$ are specified in a similar way. Notice that the standard JOIN is a special n-by-m JOIN, viz., a 1-by-1 JOIN. (n-by-m JOINs for $m, n > 1$, are called DECISION-WAIT modules in the Macromodules Project [7].)

3.3 Guarded Commands

If we extend our command language with variables, channels, and guarded selection we get the language of guarded commands with input and output actions. A variable n of type T is declared by var $n : T$. Type T denotes the set of values n can take. For example, the type binary is defined by

type $bin = \{0, 1\}$

If the initial value of a variable n is val then this can be indicated in the program by

initially $n = val$

The scope of variables is delineated by means of the scoping brackets |[and]|. The formal meaning of delineating the scope of a list of variables V is given by hiding those actions that involve changes to those variables. For example, we have

$$|[\ \mathbf{var} \ n : bin \ ::$$
$$\quad \mathbf{initially} \ n = 0 \ ::$$
$$\quad \mathbf{pref} \ast [\ r?; \ n := (n+1) \bmod 2; \ a!1; \ r?; \ n := (n+1) \bmod 2; \ a!0 \ ]$$
$$]|$$

$$= \quad \{ \ \text{hiding local variable} \ n \ \}$$

$$\mathbf{pref} \ast [\ r?; \ a!1; \ r?; \ a!0 \ ]$$

In any parallel composition components are not allowed to share variables; components can only communicate through message passing via channels. For each component a channel is either an input channel or an output channel. An input channel r of type T is declared by $r? : T$ in the specification of the component. The occurrence of $r?$ in the program text denotes an input action and means 'r receives a value on its channel.' If val is a value of type T, then $r?val$ means 'receive value val on channel r.' An output channel b of type T is declared by $b! : T$. The occurrence of $b!$ in the program text denotes an output action and means 'the value of b is sent along its channel.' Similarly, $a!val$ means 'send value val along channel a.' Local channels in a parallel composition are declared as follows.

$$|[\ \mathbf{chan} \ b : T :: \quad C0(a?, b!) \ || \ C1(b?, c!) \ ]|$$

where components $C0(a?, b!)$ and $C1(b?, c!)$ have been specified elsewhere with channel b of type T. If channels have types associated with them, then the formal meaning of a component is given by a trace structure where symbols are pairs (b, val) with b representing the name of the channel and val representing the value that is communicated. (See for example [28].)

The last extension to our language is guarded selection. A guarded selection has the following form.

$$\mathbf{if} \ B0 \ \mathbf{then} \ S0$$
$$| \quad B1 \ \mathbf{then} \ S1$$
$$\mathbf{fi}$$

Here $B0$ and $B1$ are guards of the commands $S0$ and $S1$ respectively. A guard is a Boolean expression. Informally, the meaning of this guarded command is the selection of the commands for which the guards evaluate to true. If no guard evaluates to true, the command is trace-equivalent to *abort*. (Recall that *abort* is the identity of selection.) So the formal meaning is

$$\mathbf{if} \ B0 \ \mathbf{then} \ S0 \ | \ B1 \ \mathbf{then} \ S1 \ \mathbf{fi}$$

$$= \quad \{ \ \text{by definition} \ \}$$

$$(S0' \ | \ S1')$$

where the alphabets of $S0$ and $S0'$ are the same and

$$t(S0') = \left\{ \begin{array}{ll} t(S0) & \text{if } B0 \text{ holds} \\ t(abort) & \text{otherwise} \end{array} \right.$$

A similar meaning applies to $S1'$.

3.4 Two Implementations for Data Communications

If a channel is of type unary, then implementing a data communication along that channel is straightforward: the channel consists of a single wire connection, and sending a single transition along that connection implements a data communication. What if the data types are not unary? For example how do we implement a data communication along a binary channel? We consider two types of implementations: an implementation using a dual-rail transition encoding and an implementation using a single-rail data-bundling encoding. As an example we use a component with a binary input channel a, binary output channel b, and communication behavior given by

$$\textbf{pref} * [\, a?; \; b := a; \; b! \,]$$

With a dual-rail transition encoding, each bit is implemented using two wires, one wire for the value zero and one wire for the value one. Sending a transition along the '0' wire implements the communication of a zero, and sending a transition along the '1' wire implements the communication of a one. With single-rail data bundling, each bit is implemented using a single wire. The level on that wire indicates which binary value is communicated. These wires are called the data wires. Besides the data wires, which indicate *what* value is communicated, there is one wire that is used to indicate *when* the data is communicated. This wire is called the data-valid wire. Whenever a transition occurs at the connection, the data wires are assumed to have reached a stable value. This assumption imposes a delay constraint between the arrival of the transition on the data-valid wire and the 'validity' of the data wires. This constraint is called the data-bundling constraint. In order to satisfy the data-bundling constraint, delays may have to be inserted in the data-valid wire to compensate for any possible delays in data wires. In Figure 1 these delays are indicated by long ovals. As a convention, we indicate the data-valid wires

(a) Dual-rail transition encoding   (b) Single-rail data bundling

Figure 1: Two data implementations for data types

60

by solid lines. The data wires in the data part are indicated by dashed lines. The data wires and the data-valid wire that belong together are 'bundled' by drawing a circle around them.

As a further illustration, we discuss some implementations of two other components.

$$C(a? : bin, \ b? : un, \ c! : bin)$$

$= \quad$ { by definition }

pref $* [a?||b?; \ c := a; \ c!]$

If we forget the data types, the behavior of component C is the same as that of the JOIN. If we use a dual-rail transition encoding, an implementation consists of a 2-by-1 JOIN. The implementation using single-rail data bundling consists of a JOIN, a delay, and a data wire as given in Figure 2.

    (a) Dual-rail transition encoding   (b) Single-rail data bundling

Figure 2: Two implementations for C

Finally we discuss two implementations for the modulo-2 counter. There are several ways in which the specification of the modulo-2 counter can be written. Here are two of them.

$|[$ **var** $n : bin ::$
    initially $n = 0 ::$
    pref $* [r?; \ n := (n+1) \bmod 2; \ a := n; \ a!]$
$]|$

$= \quad$ { hide variable n }

pref $* [\, r?; \ a!1; \ r?; \ a!0 \,]$

When using dual-rail transition encoding, the modulo-2 counter can be implemented by a TOGGLE. This implementation resembles the last specification. The implementation using single-rail data bundling resembles the first specification. It consists of a transition latch marked L, a delay, and a feedback wire with an inverter. When the transition latch receives an input transition, it latches the input data —which becomes the new output data— and then sends the data-valid signal for the output data. Normally, each set of data wires implementing a variable should have a data-valid wire associated with it to compensate for possible delays in the data wires. We have not done so here

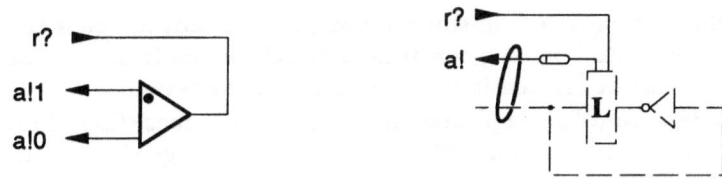

(a) Using dual-rail transition encoding   (b) Using single-rail data bundling

Figure 3: Two implementations for $ModC(2, r?, a!)$

for the feedback wire with the inverter, since the local delays in this wire can easily be compensated for by a delay in the data-valid wire for the output. In the following we apply a similar optimization for local feedback wires.

The generation of the data-valid signals can be seen as just another 'clocking' strategy. Unlike conventional clocking strategies, where the clock signals are generated at regular, fixed intervals and sent to all storage devices, the generation of the data-valid signals depends on the occurrence of a data communication, and the route of a data-valid signal may depend on the value of the data. A convenient component to route the data-valid signal depending on the value of a binary variable is the SELECT. The SELECT can be seen as a demultiplexer for the data valid signal. Its specification is given in Figure 4. Each input t will propagate to either output $q0$ or output $q1$. The level of the

$$\mathbf{pref} * [\ *[\ t?;\ q0!];\ b?;\ *[t?;\ q1!];\ b?\ ]$$

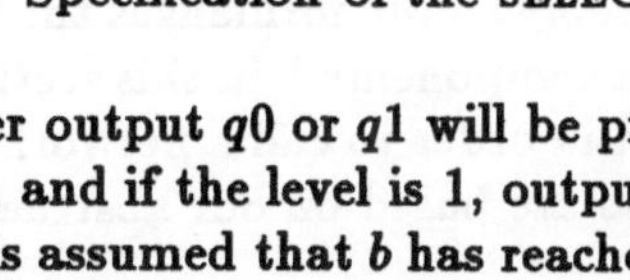

$$t? \quad \boxed{S\ {}^0_1} \quad q0! \quad q1!$$

$$b?$$

Figure 4: Specification of the SELECT

input b will determine whether output $q0$ or $q1$ will be produced. If the level is 0 output $q0$ will be produced, and if the level is 1, output $q1$ will be produced. Initially the level of b is 0. It is assumed that b has reached a stable level before a transition on t occurs. This constraint is in fact a data bundling constraint. For this reason we have encircled the t and b wires in Figure 4.

Implementing data communications using data bundling has been employed extensively in the Macromodules project [7]. One of the advantages of employing data bundling is that conventional logic design can be used for the data part, which often results in smaller circuits. A disadvantage is the implementation of the data-bundling constraints, which results in a worst-case behavior and makes the design sensitive to certain delay variations.

In a data-bundling implementation, the circuit consisting of the data-valid wires is the control circuit. This circuit can be designed separately from the data part. When combining the control part and the data part, the data-bundling constraints are met by inserting appropriate delays in the (data-valid) wires of the control part. Of course, such a method will only work if the control part is a delay-insensitive circuit. That is, the correctness of the circuit for the control part should be insensitive to adding arbitrary delays in the wire connections. Accordingly, delay-insensitive circuits play an important role in the design of data-bundling implementations. In the next sections we explain what exactly is a delay-insensitive circuit.

3.5 Decomposition versus Parallel Composition

Expressing a specification as a parallel composition (i.e., a weave) of smaller specifications is a first step towards finding a network implementation. Unfortunately, it is not the case that each parallel composition can be interpreted as a network of components that implements the specification. The problem is that that our parallel composition expresses a synchronization on common symbols irrespective of whether these symbols are input or output symbols. In general, a component in a network will produce an output as soon as that output is locally enabled. It will not wait for other components, where that same symbol is an input for example, to synchronize on the production of that symbol. There are several things we can do in order to transform a parallel composition into a network implementation and avoid these synchronization failures. One solution is to enforce the proper synchronizations by introducing handshakes for each communication on each channel. This method is proposed by Martin [22] and is also used by Van Berkel [1]. Another solution is to design your parallel composition in such a way that no handshakes are needed: whenever a component can produce an output, the components for which that symbol is an input are waiting for that input to happen. We need to formalize this property more precisely. We first discuss our correctness criteria for a network implementation of a specification.

The definition of *decomposition* formalizes the idea of 'implementing a specification by a network of components.' In this section we discuss four conditions that have to be satisfied in order to call a network a decomposition of a specification. These conditions are based on our abstract mechanistic interpretations of specifications.

Let us briefly recall the mechanistic interpretation of a specification. Our interpretation is based on the distinction between the component and the environment and on two conditions, the safety and the progress condition. Both conditions prescribe the behavior of the component and the environment. The safety condition says that no input or output may be produced when not allowed by the specification. The progress condition says that every trace specified may occur. It is important to distinguish the component's prescriptions and the environment's prescriptions in a specification. When implementing a specification, we try to implement the component's prescriptions by a network of components assuming that the environment's prescriptions are satisfied.

We formulate the conditions for a decomposition of component E into components $E_1, \ldots, E_n$, $n > 0$. The network of components E_1 through E_n is denoted by $(E_1, \ldots, E_n)$. The property that E can be decomposed into the network consisting of E_1 through E_n is denoted by $E \to (E_1, \ldots, E_n)$.

First, we take into account the behavior of the environment with respect to the network $(E_1, \ldots, E_n)$. The environment's prescriptions for the network are given in E. In order to consider the production of an input by this environment as the production of an output by a component, we consider the *reflection* of E, denoted by $\overline{E}$ and defined by

$$\overline{E} = \langle \mathsf{o}E, \mathsf{i}E, \mathsf{t}E \rangle$$

Consequently, $\mathsf{i}\overline{E} = \mathsf{o}E$, $\mathsf{o}\overline{E} = \mathsf{i}E$, and $\mathsf{t}\overline{E} = \mathsf{t}E$. By reflecting E, we interchange the prescriptions for the component and the environment. Instead of considering $\overline{E}$ and network $(E_1, \ldots, E_n)$, we now consider the network $(E_0, \ldots, E_n)$, where $E_0 = \overline{E}$.

In order for E to be decomposable into the network $(E_1, \ldots, E_n)$, four conditions have to hold for $(E_0, \ldots, E_n)$. Two conditions concern the so-called structure of the network and two conditions concern the behavior of the network. The behavioral conditions are related to the safety and the progress condition in our interpretation. We discuss these four conditions below.

- **Closed network**
  In the network $(E_0, \ldots, E_n)$ there are no dangling inputs and outputs: every input is connected to an output and every output is connected to an input.

- **No output interference**
  Outputs of distinct components are not connected to each other. In other words, the output alphabets of the components in the networks are pairwise disjoint. The first two conditions guarantee that each symbol is an output of exactly one component and an input of at least one component. (Notice that an output may be connected to multiple inputs.)

- **No computation interference (Safety)**
  The third condition states that the environment's prescriptions for every specification in the closed network may not be violated. In other words, no output may be produced by any component when not allowed by the specifications for which this symbol is an input. Or, to put it positively, every output that can be produced by a component can be accepted as input by the receiving component. This condition can be checked by constructing all possible network behaviors and verifying that no environment prescription is violated. The trace set T of all possible network

behaviors can be constructed as follows. Initially, $T = \{\varepsilon\}$. We repeatedly enlarge trace set T with extensions of traces in T as indicated below until T can no longer be enlarged. The rule for extending T is as follows. Choose a trace $t \in T$, symbol z, and component E_i, such that after network behavior t, component E_i may produce output z. Since component E_i cannot be prevented from producing output z after behavior t, tz is also a possible network behavior, and we add tz to T. Our third condition states that any possible network behavior must be in accordance with every specification, which can be formulated by

$$T \downarrow \mathbf{a}E_i \subseteq \mathbf{t}E_i \quad \text{for all } i = 0, \ldots, n \tag{1}$$

where T represents the trace set of all networks behaviors. If this condition holds, we say that the network is *free of computation interference*.

- **Completeness with respect to specification (Progress)**
  The fourth condition states that every trace of the specification E may also occur in the network behaviors. This condition is formulated as

$$T \downarrow \mathbf{a}E = \mathbf{t}E \tag{2}$$

where again T represents the trace set of all network behaviors. If this condition holds we say that the network behaviors are *complete with respect to the specification*.

3.6   Some Examples

We consider some examples to familiarize ourselves with this definition of decomposition. We are given component E specified by

$$E = \mathbf{pref} * [\, r?; \, (a!; \, r?) \parallel (sr!; \, sa?); \, a! \,]$$

For the moment we consider all channels to be unary. Notice that for every four communications on r and a there are two communications on sr and sa. For this reason, this component can be called a four-phase-to-two-phase converter. In Figure 5 two decompositions for E are given. Both networks contain a TOGGLE, JOIN, and a MERGE. Network (b) also contains an IWIRE. It is easy to verify that both networks, *including* the environment, form a closed network without output interference. Notice that in networks (a) and (b) there are multiple input connections at terminals x and y respectively. If we construct the network behaviors for network (a), again including the environment, we find

$$T = \mathbf{t} \left(\mathbf{pref} * \left[\, r?; \, x; \, ((a!; \, r?; \, y) \parallel (sr!; \, sa?)); \, z; \, a! \,\right]\right)$$

If we project trace set T on each of the component's alphabets, we obtain a subset of the trace set of the component. That is,

$$T \downarrow \mathbf{a}E_i \subseteq \mathbf{t}E_i$$

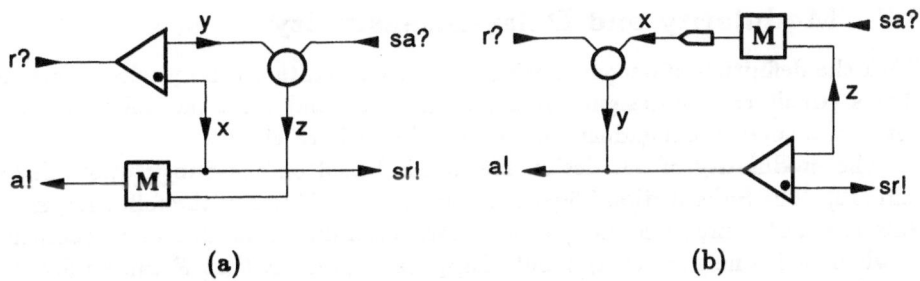

Figure 5: Two decompositions for a four-to-two-phase converter E

where E_i is a specification of a component. We even have $T \downarrow aE = tE$. From these observations we can conclude that network (a) is indeed a decomposition of the four-to-two-phase converter.

The network behaviors for network (b), including the environment, are given by

$$T = \mathbf{t}\left(\mathbf{pref}\left(r?\|x;\ *\Big[\ y;\ ((a!;\ r?)\ \|\ (sr!;\ sa?;\ x));\ y;\ ((a!;\ r?)\ \|\ (z;\ x))\ \Big]\right)\right)$$

If we project T on each of the component's alphabets, we obtain subsets of the trace sets of the components: no specification is violated. Furthermore, we have that $T \downarrow aE = tE$. (Notice that by deleting x, y, and z and after some folding we get specification E again.) Consequently, network (b) is also a decomposition of the four-to-two-phase converter.

As an example where the safety condition is violated, we can take network (a) with the WIRE between terminal x and sr replaced by an IWIRE. Since initially this network can produce sr, the set of network behaviors would now include trace sr. According to the specification E, sr may not occur initially, so the safety condition is violated.

As an example where the progress condition is violated, we take network (b) with the IWIRE replaced by a WIRE. The network behaviors are then given by $T = \mathbf{t}\,(\mathbf{pref}\ r?)$. After the initial input r, nothing will ever be produced. Obviously, not every trace from the specification E may occur, so the progress condition is violated.

The above decompositions only apply if the channels are unary channels. If the channels are not unary, we may get a different decomposition depending on how we implement data communications. If data communications are implemented using a single-rail data bundling scheme, these decompositions may be used as a control circuit for directing the data-valid signals. When the control part is combined with the data part, delays may have to be inserted in some of the wire connections. The important question then is whether these delay insertions will affect the correctness of the decomposition. In other words, do these networks represent delay-insensitive circuits? This question is answered in the next section.

3.7 Modularity and Delay-Insensitivity

With the definition of decomposition we can now state precisely what it means that a circuit can be designed in a modular way and what the differences are between a speed-independent and a delay-insensitive circuit.

The modularity of our design method is based on the Substitution Theorem [12]. The Substitution Theorem applies to problems of the following kind. Suppose that component E_0 can be decomposed into a number of components of which F is one such component. Suppose, moreover, that F can be decomposed further into a number of components. Under what conditions can the decomposition of F be substituted for F in the decomposition of E_0?

Theorem 1 (Substitution Theorem) *For components E_0, E_1, E_2, E_3, and F we have*

$$
\begin{aligned}
\text{if} \quad & E_0 \ \rightarrow \ (\ E_1, \ F\ ) \\
\text{and} \quad & F \ \rightarrow \ (\ E_2, \ E_3\ ) \\
\text{then} \quad & E_0 \ \rightarrow \ (\ E_1, \ E_2, \ E_3\ ),
\end{aligned}
$$

if the following condition is satisfied.

$$(\mathbf{a}E_0 \cup \mathbf{a}E_1) \cap (\mathbf{a}E_2 \cup \mathbf{a}E_3) = \mathbf{a}F. \tag{3}$$

Condition (3) states that the only symbols that the decompositions of E_0 and F have in common are symbols from $\mathbf{a}F$. It is essentially a trivial condition, since, by an appropriate renaming of the internal symbols in the decomposition of F, this condition can always be satisfied. The internal symbols of the decomposition of F are given by $(\mathbf{a}E_2 \cup \mathbf{a}E_3) \backslash \mathbf{a}F$, where '$\backslash$' means set deletion.

The theorem above considers decompositions into two components only. The generalization of this theorem to decompositions into more than two components is straightforward.

Some of the main attractions that are often mentioned in connection with speed-independent and delay-insensitive circuit design are the 'modular design approach,' the 'hierarchical design approach,' 'design by stepwise refinement,' or the 'possibility for incremental improvements.' All these characteristics refer to the same property in our formalism and are symbolized in the Substitution Theorem.

If a network is a decomposition of a specification, it represents a *speed-independent circuit*. Therefore, a decomposition is also called an *SI decomposition*. A speed-independent circuit, however, is not necessarily a delay-insensitive circuit. The reason is that decompositions are invariant under variations in response times of components, but not under the insertion of communication delays in the connections. If the correctness of the circuit is invariant under the insertion of communication delays in the connections as well, then we call such a circuit a *delay-insensitive circuit*. While a speed-independent circuit is formally described by means of (SI) decomposition, a delay-insensitive circuit is formally described by means of *DI decomposition*. A DI decomposition is a decomposition in which all communication delays between the components are taken into account. Formally, these communication delays are represented by WIRES.

We say that the network $(E_1, \ldots, E_n)$ forms a DI decomposition of component E, denoted by $E \xrightarrow{DI} (E_1, \ldots, E_n)$ if and only if

$$E \to (\; ren(E_1),\; Wires(E_1), \ldots, ren(E_n),\; Wires(E_n)\;).$$

where $ren(E_i)$ is a renaming of E_i to a 'localized' version and $Wires(E_i)$ is the collection of WIRE components connecting $ren(E_i)$ with its original terminals aE_i.

In general, DI decompositions are more difficult to derive and verify than SI decompositions, because of all the connection WIRES. By the Substitution Theorem, it follows that a SI decomposition is a DI decomposition, if all constituent components are DI components. A component E is called a DI component, if

$$E \to (ren(E),\; Wires(E)).$$

The DI property formalizes that the communication behavior between component and environment is insensitive to the insertion of communication delays. In other words, specification E is invariant under any extension with WIRES at its input or output terminals. All basic components of Figure 1 are DI components. Consequently, the decompositions of Figure 5 are DI decompositions. The C-ELEMENT and the SELECT, however, are not DI components.

The idea of formalizing delay-insensitivity using a characterization of a DI component originates from Molnar [23]. Udding was the first to give a rigorous formulation of the DI property in terms of trace structures [35]. More information on SI decomposition, DI decomposition, and DI components, can be found in [12].

3.8 Limitations of Decomposition

Our correctness conditions for decomposition have certain limitations, in particular the progress condition. The safety condition guarantees that whenever an output can be produced, the production causes no problems. It does not guarantee that every output *will* be produced. Therefore we need a progress condition. Our progress condition excludes, for example, decompositions into so-called 'accept-everything-do-nothing' modules: components that accept every possible input but never produce any output. On the other hand, although the progress condition requires that each trace in tE *may* occur in the network behaviors, in general it does not require that some traces, or outputs, are guaranteed to occur. The occurrence of a trace, or output, may depend on nondeterministic choices made by some components. Figure 6 illustrates this property by means of two decompositions of a WIRE. The component depicted by the left half of a circle is a SELECTOR. Its specification is given by

$$\mathbf{pref} * [\; a?;\; (b!|c!)\;]$$

After each receipt of input a, either output b or output c may be produced. The output is chosen nondeterministically. The component depicted by an open

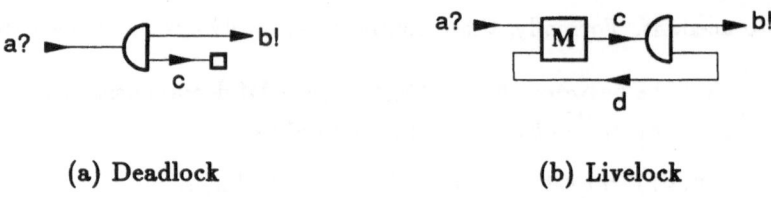

(a) Deadlock (b) Livelock

Figure 6: Examples of deadlock and livelock

square is a SINK. It only accepts inputs. Decompositions (a) and (b) satisfy all four conditions for decomposition. Decomposition (a), however, exhibits a behavior that could be characterized by deadlock: once output c is chosen by the SELECTOR, output b will never be produced. Decomposition (b) exhibits a behavior that could be characterized as livelock: if the SELECTOR makes enough 'bad' choices, there can be an unbounded number of internal communications on d and c, and output b may not be produced. Obviously, if we design an implementation, we do not want deadlock or livelock to occur. So, in formulating a better progress condition we want to exclude implementations with deadlock and livelock. This is easier said than done. Do there exist 'proper' definitions for deadlock and livelock (or of progress) in the context of our formalism? If so, what are they? If not, how much should our formalism be extended to accommodate proper definitions? For example, an obvious definition for livelock might be that in a trace an unbounded number of internal symbols can occur between two external symbols. With this definition in mind consider the decomposition of a 'one-shot' WIRE, pref(a?; b!), in Figure 7. If we look to

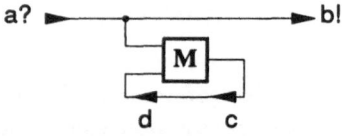

Figure 7: Livelock?

the traces of this decomposition, then after input a an unbounded number of internal symbols c and d can occur before output b is produced. According to our tentative definition, we should conclude that livelock can occur. On the other hand, from experience we know that output b will eventually occur, if the WIRES are implemented by physical wires. So in our context, this tentative definition may not be a 'proper' definition of livelock. Finding definitions of deadlock and livelock (or of progress) that are simple enough to state, have a proper justification in the context of speed-independent circuit design, are easy to verify, and satisfy a property like the Substitution Theorem is still an open problem.

4 How to Play the Game

Knowing what the rules of the game are, there can be many ways of playing the game. For instance, we can simply apply a trial-and-error process: propose a decomposition and verify all conditions. If the conditions are not satisfied, we try another decomposition. Such a style is rather unsatisfactory, since it does not give much guidance in how one could derive a decomposition in a systematic way.

We explore a different way. It is based on a few simple principles. We first try to obtain a parallel decomposition of our specification E. For example, we try to express E as

$$E = |[\text{ chan } V :: \ F \parallel G \ ]|$$

by using techniques from parallel program design. The obvious candidate for a decomposition then is

$$E \to (F , G)$$

The conditions for decomposition can be verified easily by inspecting the syntax of E, F, and G. For example, the conditions for a closed network and absence of output interference are easily verified by checking the alphabets. Absence of computation interference can be verified by checking that as soon as an output can be produced by a component locally, then the receiving components are waiting for that output to happen. With a little care in specifying the components F and G, this condition can often be satisfied. Finally, we mention without proof that the progress condition (that is, completeness of the network behaviors) is automatically satisfied if we already have $E = |[\text{ chan } V :: F \parallel G]|$.

5 The Modulo-N Counter Continued

5.1 A First Decomposition: Divide and Conquer

How can we decompose the modulo-N counter into smaller components? An obvious choice is to try a divide-and-conquer approach: decompose the modulo-$2N$ counter into a modulo-N counter and a 'small' subcomponent. 'Small' can be interpreted as the specification has a small number of states, which is independent of N. We derive two such decompositions. The first decomposition is based on the following idea. Keep track of two counts represented by n and k, where n is incremented modulo N each time an input r occurs and k is incremented modulo 2 each time n reaches 0. Initially, both n and k are 0. While incrementing n and k, we maintain the invariant

$$\#r \bmod 2N = kN + n$$

In other words, $\#r \bmod 2N = 0$ is equivalent to $n = 0$ and $k = 0$. This observation leads to the following first derivation step. For $2N > 0$, we have

$$ModC(2N, r?, a!)$$

$= \quad \{ \#r \bmod 2N = kN + n \}$

$\quad |[\ \textbf{var}\ n : int, k : bin\ ::$
$\quad\quad \textbf{initially}\ n = 0, k = 0\ ::$
$\quad\quad \textbf{pref} * [\ r?;\ n := (n + 1) \bmod N$
$\quad\quad\quad\quad ;\ \textbf{if}\ n = 0\ \textbf{then}\ k := (k + 1) \bmod 2\ |\ n \neq 0\ \textbf{then}\ skip\ \textbf{fi}$
$\quad\quad\quad\quad ;\ \textbf{if}\ n \neq 0 \lor k \neq 0\quad \textbf{then}\ a := 1$
$\quad\quad\quad\quad |\quad n = 0 \land k = 0\quad \textbf{then}\ a := 0$
$\quad\quad\quad\quad \textbf{fi}\ ;\ a!\ ]$
$\quad]|$

Incrementing n modulo N and indicating whether $n \neq 0$ or $n = 0$ can be done by a modulo-N counter. Consequently, in our second step we introduce a modulo-N counter and replace each statement $n := (n+1) \bmod N$ by the communication actions $sr!$; $sa?$. Since the postcondition for each communication on sa is given by

$$(n = 0) \equiv (sa = 0)$$

where $n = \#sr \bmod N$, we can replace the conditions $n \neq 0$ and $n = 0$ by $sa \neq 0$ and $sa = 0$ respectively. These observations then lead to the following parallel composition.

$$ModC(2N, r?, a!)$$

$= \quad \{ \text{def. of } ModC(N, sr?, sa!) \}$

$\quad |[\ \textbf{chan}\ sr : un, sa : bin\ ::$
$\quad\quad CELL0(r?, a!, sr!, sa?)\ \|\ ModC(N, sr?, sa!)$
$\quad]|$

where $CELL0$ is defined by

$$CELL0(r? : un, a! : bin, sr! : un, sa? : bin)$$

$= \quad \{ \text{by definition} \}$

$\quad |[\ \textbf{var}\ k : bin\ ::$
$\quad\quad \textbf{initially}\ sa = 0, k = 0\ ::$
$\quad\quad \textbf{pref} * [\ r?;\ sr!;\ sa?$
$\quad\quad\quad\quad ;\ \textbf{if}\ sa = 0\ \textbf{then}\ k := (k + 1) \bmod 2\ |\ sa \neq 0\ \textbf{then}\ skip\ \textbf{fi}$
$\quad\quad\quad\quad ;\ \textbf{if}\ sa \neq 0 \lor k \neq 0\quad \textbf{then}\ a := 1$
$\quad\quad\quad\quad |\quad sa = 0 \land k = 0\quad \textbf{then}\ a := 0$
$\quad\quad\quad\quad \textbf{fi}\ ;\ a!\ ]$
$\quad]|$

The parallel composition above is a candidate for a network implementation of the modulo-$2N$ counter. We have to check if the conditions for decomposition are satisfied. First, we observe that the network of $CELL0$, the modulo-N counter, and the environment is closed and has no output interference. Second, we observe that as soon as a component or the environment can perform an output action, the corresponding input action is also enabled: no computation interference can occur in the communications between $CELL0$, the modulo-N counter, and the environment. We may therefore conclude the following decomposition

$$ModC(2N, r?, a!)$$

$\rightarrow$ \quad \{ def. of decomposition \}

$$(CELL0(r?, a!, sr!, sa?) , \quad ModC(N, sr?, sa!))$$

We briefly present two implementations for $CELL0$. We observe that the guarded command for $CELL0$ can also be written as

pref*[$r?$; $sr!$; ($sa?1$; $a!1$

$\qquad\qquad$ | $sa?0$; $k := (k + 1)$ mod 2;

$\qquad\qquad\qquad$ if $k = 1$ then $a!1$ | $k = 0$ then $a!0$ fi)

$\qquad$]

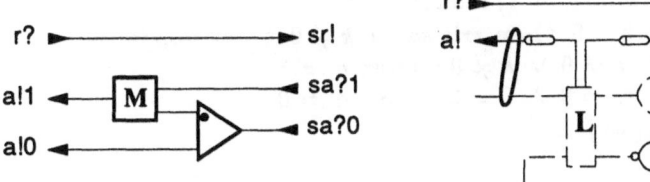

(a) Using dual-rail transition encoding \qquad (b) Using single-rail data bundling

Figure 8: Implementations of $CELL0$

Recalling that counting modulo-2 can be done by a TOGGLE, it is now not difficult to verify that the network given in Figure 8(a) represents an implementation of $CELL0$ using a dual-rail transition encoding. An implementation using single-rail data bundling is given in Figure 8(b).

5.2 A Second Decomposition

In the first step of our previous decomposition, we may choose to increment k modulo-2 after each input r and increment n modulo N each time k reaches 0. Now the invariant is given by

$$\#r \bmod 2N = 2n + k$$

Based on this observation we obtain the following first derivation step.

$ModC(2N, r?, a!)$

$=$ $\{ \#r \bmod 2N = 2n + k \}$

$\|[$ **var** $n : int, k : bin$ ::
     initially $n = 0, k = 0$ ::
     pref $* [r?;\ k := (k + 1) \bmod 2$
         ; **if** $k = 0$ **then** $n := (n + 1) \bmod N$ $|$ $k \neq 0$ **then** $skip$ **fi**
         ; **if** $n \neq 0 \lor k \neq 0$ **then** $a := 1$
         $|$ $n = 0 \land k = 0$ **then** $a := 0$
         fi $;\ a!]$
$]|$

The next steps are now similar to the previous section. We introduce a modulo-N counter and replace the statement $n := (n + 1) \bmod N$ by $sr!;\ sa?$. The conditions $n \neq 0$ and $n = 0$ are replaced by $sa \neq 0$ and $sa = 0$ respectively. If we define $CELL1$ as follows

$CELL1(r? : un,\ a! : bin,\ sr! : un,\ sa? : bin)$

$=$ $\{$ by definition $\}$

$\|[$ **var** $k : bin$ ::
     initially $sa = 0,\ k = 0$ ::
     pref $* [r?;\ k := (k + 1) \bmod 2$
         ; **if** $k = 0$ **then** $sr!;\ sa?$ $|$ $k \neq 0$ **then** $skip$ **fi**
         ; **if** $sa \neq 0 \lor k \neq 0$ **then** $a := 1$
         $|$ $sa = 0 \land k = 0$ **then** $a := 0$
         fi $;\ a!]$
$]|$

we obtain the parallel composition and subsequent network decomposition

$ModC(2N, r?, a!)$

$=$ $\{$ def. of **weave** $\}$

$\|[$ **chan** $sr : un,\ sa : bin$ :: $CELL1(r?, a!, sr!, sa?) \| ModC(N, sr?, sa!)]|$

$\rightarrow$ $\{$ def. of decomposition $\}$

$(CELL1(r?, a!, sr!, sa?)\ ,\ ModC(N, sr?, sa!))$

Let us briefly look at some implementations for $CELL1$. First, we observe that $CELL1$, after some simplification, can also be written as

pref $* [r?;\ a!1;\ r?;\ sr!;\ (sa?0;\ a!0\ |\ sa?1;\ a!1)]$

Apparently, we again need a modulo-2 counter to record whether the input $r?$ has to propagate to output $a!1$ or to output $sr!$. This observation then quickly leads to the implementations in Figure 9.

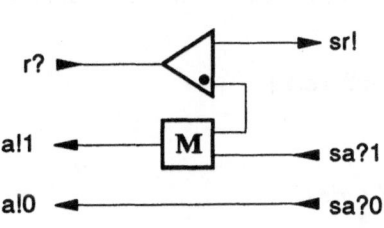
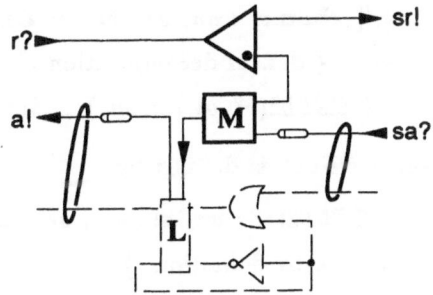

(a) Using dual-rail transition encoding   (b) Using single-rail data bundling

Figure 9: Implementations for $CELL1$

5.3 What about odd N?

So far we have presented two decompositions for even N. What about odd N? Although there are some decompositions that apply to odd N only, we present a decomposition that applies to any $N > 0$. We try to decompose a modulo-$(N + 1)$ counter into a modulo-N counter and a small subcomponent. Again we keep track of two integers n and k, where we maintain the invariant

$$\#r \bmod (N + 1) = n + k$$

Initially $k = 0$ and $n = 0$. If $k = 0$ and an r is received, then k is set to 1. If $k = 1$, then for each r received, n is incremented by one modulo N. When n becomes 0, k is also set to zero. This leads to the following derivation step.

$$ModC(N + 1, r?, a!)$$

$$= \quad \{ \ \#r \bmod (N + 1) = n + k \ \}$$

$$|[\ \textbf{var } n : int, k : bin ::$$
$$\quad \textbf{initially } n = 0, k = 0 ::$$
$$\quad \textbf{pref} * [\ r?; \ \textbf{if } k = 1 \ \textbf{then } n := (n + 1) \bmod N \ | \ k = 0 \ \textbf{then } skip \ \textbf{fi}$$
$$\quad \quad ; \ \textbf{if } k = 0 \ \vee \ n \neq 0 \ \ \textbf{then } k := 1; \ a := 1$$
$$\quad \quad \ | \ \ k = 1 \ \wedge \ n = 0 \ \ \textbf{then } k := 0; \ a := 0$$
$$\quad \quad \textbf{fi} \ ; \ a! \ ]$$
$$]|$$

Similar to the previous decompositions we introduce a modulo-N counter for incrementing n modulo N and indicating whether $n \neq 0$ or $n = 0$. We then obtain the following parallel composition and network decomposition.

$$ModC(N + 1, r?, a!)$$

$$= \quad \{ \ \text{intro. of mod-}N \text{ counter, def. of parallel composition} \ \}$$

$$|[\ \textbf{chan} \ sr : un, \ sa : bin \ :: \ CELL2(r?, a!, sr!, sa?) \parallel ModC(N, sr?, sa!) \ ]|$$

$\rightarrow \quad \{ \text{ def. of decomposition } \}$

$$(\ CELL2(r?, a!, sr!, sa?) \ , \ ModC(N, sr?, sa!) \ )$$

where $CELL2$ is defined by

$$CELL2(r? : un, \ a! : bin, \ sr! : un, \ sa? : bin)$$

$= \quad \{ \text{ by definition } \}$

$|[\ \textbf{var} \ k : bin ::$
   $\textbf{initially} \ sa = 0, \ k = 0 ::$
   $\textbf{pref} * [\ r?; \ \textbf{if} \ k = 1 \ \textbf{then} \ sr!; \ sa? \ | \ k = 0 \ \textbf{then} \ skip \ \textbf{fi}$
      $; \ \textbf{if} \ k = 0 \ \vee \ sa \neq 0 \ \ \textbf{then} \ k := 1; \ a := 1$
      $| \ \ k = 1 \ \wedge \ sa = 0 \ \ \textbf{then} \ k := 0; \ a := 0$
      $\textbf{fi} \ ; \ a! \ ]$
$]|$

We give two implementations for $CELL2$ without proof. In the case that binary channels are implemented using a dual-rail transition encoding, $CELL2$ can be implemented as given in Figure 10(a). (Notice that the 2-by-1 JOIN is used for implementing the binary variable k. An IWIRE is used for the proper initialization.) In the case that single-rail data bundling is used for implementing binary channels, $CELL2$ can be implemented as given in Figure 10(b).

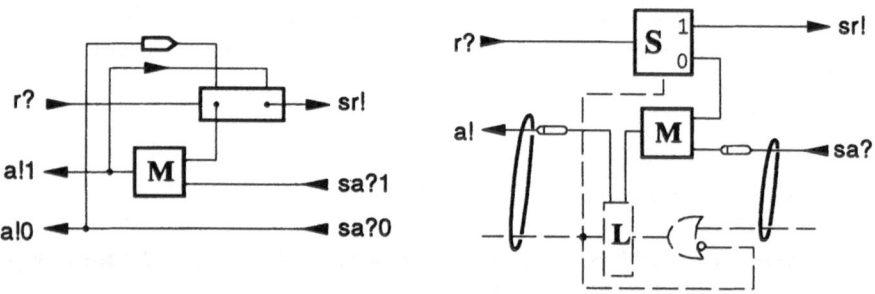

(a) Using dual-rail transition encoding   (b) Using single-rail data bundling

Figure 10: Implementations for $CELL2$

5.4   What about Parallelism?

All our decompositions thus far have a sequential behavior in the sense that all communication actions (and even internal actions) are totally ordered. As such, our programs are not much different from a normal sequential program with procedure calls. How can we derive decompositions that exhibit parallel behavior? For example, would it be possible to introduce some parallel behavior in our sequential decompositions without invalidating their correctness? There are several ways to do this. We discuss two.

Observe our specification of the modulo-N counter once more

$ModC(N : int, r? : un, a! : bin)$

$=$ { by definition }

$|[$ **var** $n : int$ ::
    initially $n = 0$ ::
    pref $* [r?; \; n := (n + 1) \bmod N;$
            if $n \neq 0$ **then** $a := 1 \mid n = 0$ **then** $a := 0$ **fi** ; $a!$
            ]
$]|$

Notice that the value to be sent on channel a can be computed ahead of time, possibly before r arrives. Only when r arrives the value of a is sent along channel a, and, in parallel, the computation of the next value of a is initiated. In order to compute the value of a before r arrives, we can put component P in front of $ModC(N, sr?, sa!)$.

$P(r? : un, a! : bin, sr! : un, sa? : bin)$

$=$ { by definition }

    pref$(r? \parallel (sr!; \; sa?); \; *[a := sa; \; ((a!; r?) \parallel (sr!; sa?))])$

Initially, in parallel with receiving input r, the first value of a modulo-N counter is requested and received via channels sr and sa respectively. After both input r and the first value on sa have been received, the value of sa can be sent on channel a. In parallel with sending a and receiving the next request r, the next value from the modulo-N counter is requested and received on channels sr and sa, respectively. This behavior then repeats. It is not difficult to see that the modulo-N counter can be decomposed into a renaming of itself and component P.

$ModC(N, r?, a!)$

$\rightarrow$ { def. of decomposition }

    $(P(r?, a!, sr!, sa?) , \; ModC(N, sr?, sa!))$

Since component P can be put in front of every modulo-N counter, it can also be put in front of every $CELL$ without changing the correctness of the decomposition. In this way we can obtain decompositions where many communications may take place in parallel. Notice, however, that the communication behavior on the pair of channels between any two adjacent cells remains invariant when we insert component P between the cells: after the insertion, actions from different pairs of channels can occur in parallel, but actions from the same pair still occur in the same sequential order and the same values are communicated.

If we decide to implement binary channels by means of dual-rail transition encoding, then component P can be implemented by a 2-by-1 JOIN, a MERGE, and an IWIRE. See Figure 11(a). If we decide to implement binary channels by means of single-rail data bundling, then component P can be implemented by a transition latch, a JOIN, and an IWIRE. See Figure 11(b).

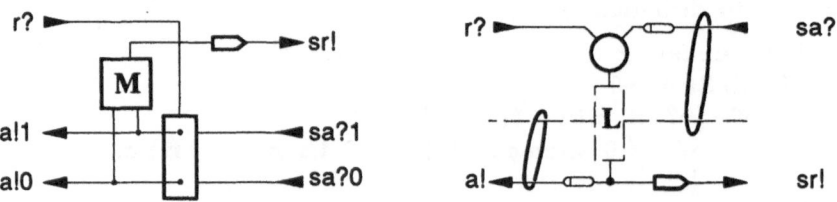

(a) Using dual-rail transition encoding   (b) Using single-rail data bundling

Figure 11: Implementations for component P

Another way to introduce parallelism is to consider the specification of each *CELL* in isolation and to try changing or reordering the statements in the guarded command such that communications on the channels r and a can be done in parallel with the communications on channels sr and sa. In doing so we should not destroy the correctness of the decomposition. In other words, if *PCELL* is the specification obtained after changing *CELL*, then *PCELL* should still satisfy the decompositions in which *CELL* is used. In order not to destroy the correctness of the decomposition, we keep the communication behavior on the channels r and a invariant and, for reasons of symmetry, also the communication behavior on the channels sr and sa. Communications from different pairs of channels, however, may overlap.

Based on these observations we try to specify the communication behavior of *PCELLi* as follows.

> $PCELLi(r? : un,\ a! : bin,\ sr! : un,\ sa? : bin)$
>
> $=$ { by definition }
>
> |[**var** $k : bin$ ::
>   initially $sa = 0,\ k = 0$ ::
>   **pref** (**if** $B.i$ **then** $r?$ || $(sr!;\ sa?)$ | $\neg B.i$ **then** $r?$ **fi** ; $S.i$;
> $*[$ **if** $B.i$ **then** $(a!;\ r?)$ || $(sr!;\ sa?)$ | $\neg B.i$ **then** $(a!; r?)$ **fi** ; $S.i$])
>   ]|

where $S.i$ represents the statement that calculates the next value of a (possibly using some local variables) and $B.i$ is a guard. Notice that in each repetition step there is one pair of communications with the environment through a and r, and possibly one pair of communications with the subcomponent through sr and sa. The occurrence of a communication with the subcomponent depends on the value of $B.i$. We try to find appropriate values for $B.i$ and $S.i$ by changing the specification of $CELLi$. The specification for $PCELLi$ should still satisfy the decomposition for the modulo-N counter in which $CELLi$ was used.

As an example we consider $CELL1$ once more.

$$CELL1(r? : un,\ a! : bin,\ sr! : un,\ sa? : bin)$$

$=$ { by definition }

$|[$ **var** $k : bin ::$
   initially $sa = 0,\ k = 0$ $::$
   pref $*\ [\ r?;\ k := (k + 1)$ mod 2
        $;$ **if** $k = 0$ **then** $sr!;\ sa?\ |\ k \neq 0$ **then** $skip$ **fi**
        $;$ **if** $sa \neq 0 \vee k \neq 0$ **then** $a := 1$
        $|$ $sa = 0 \wedge k = 0$ **then** $a := 0$
        fi $;\ a!\ ]$
$]|$

We first switch $k := (k + 1)$ mod 2 with the succeeding **if** ...**fi**, thereby also changing the guards of the selection.

pref $*\ [\ r?;$ **if** $k \neq 0$ **then** $sr!;\ sa?\ |\ k = 0$ **then** $skip$ **fi**
        $;\ k := (k + 1)$ mod 2
        $;$ **if** $sa \neq 0 \vee k \neq 0$ **then** $a := 1$
        $|$ $sa = 0 \wedge k = 0$ **then** $a := 0$
        fi $;\ a!\ ]$

After some unfolding and reordering we get a specification of the desired form, where

$B.1 = (k \neq 0)$

$S.1 = (k := (k + 1)$ mod $2;$
        if $sa \neq 0 \vee k \neq 0$ **then** $a := 1$
        $|$ $sa = 0 \wedge k = 0$ **then** $a := 0$
        fi $)$

The other cells only require some unfolding and reordering of $a!;\ r?$ to obtain the desired form. We just give the outcome of this exercise here. For $PCELL0$ we get

$B.0 = true$

$S.0 =$ **if** $sa = 0$ **then** $k := (k + 1)$ mod $2\ |\ sa \neq 0$ **then** $skip$ **fi**
       $;$ **if** $sa \neq 0 \vee k \neq 0$ **then** $a := 1$
       $|$ $sa = 0 \wedge k = 0$ **then** $a := 0$
       fi $;$

For $PCELL2$ we get

$B.2 = (k = 1)$

$$S.2 = (\ \textbf{if}\ k = 0\ \lor\ sa \neq 0\ \textbf{then}\ k := 1;\ a := 1$$
$$|\ k = 1\ \land\ sa = 0\ \textbf{then}\ k := 0;\ a := 0$$
$$\textbf{fi}\ )$$

For each of the cells $PCELLi$, $0 \leq i \leq 2$ we can try to find an implementation using a dual-rail transition encoding for the binary channels. It turns out that $PCELLi$ can be decomposed into P and $CELLi$, for $0 \leq i \leq 2$. In other words, we can put the implementation of cell P in front of the implementation for $CELLi$, and we get an implementation for $PCELLi$. These are not the only implementations for $PCELLi$, however. For example, for $PCELL1$ and $PCELL2$ some smaller implementations can be obtained. See Figures 12 and 13. The verification of these implementations is left to the reader. Notice that

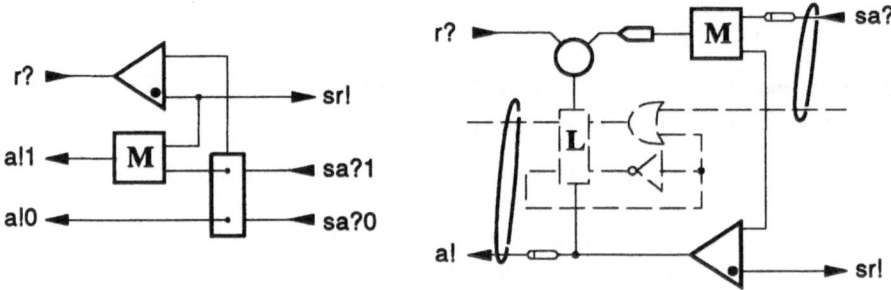

(a) Using dual-rail transition encoding   (b) Using single-rail data bundling

Figure 12: Implementations for $PCELL1$

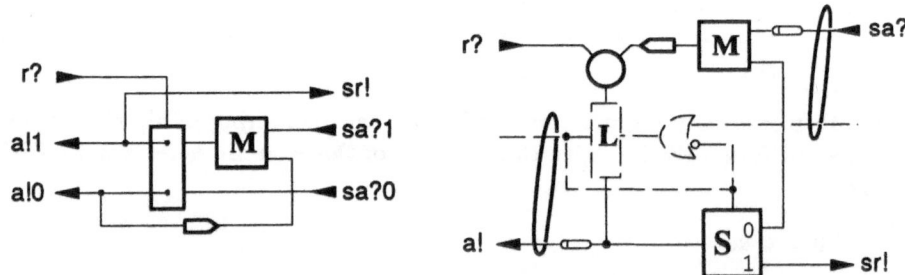

(a) Using dual-rail transition encoding   (b) Using single-rail data bundling

Figure 13: Implementations for $PCELL2$

the combinational logic in Figures 12(b) and 13(b) are straightforward implementations of the statements $S.1$ and $S.2$. Furthermore, the implementations for the control part correspond to the network given in Figure 5(b). The condition $B.i$ is the input to the SELECT that controls the communications with the subcounter. Notice also that if the data input to the SELECT alternates in value, as is the case with $S.1$, then the SELECT boils down to a TOGGLE.

6   Performance Analysis

By now we have many ways to decompose a modulo-N counter. How do we compare these decompositions? Do there exist some measures for our parallel programs which are similar to the running time or space complexity of an algorithm? Are 'time' and 'space' the only interesting performance aspects? How do we measure each performance aspect?

We consider three performance criteria: area complexity, power consumption, and response time. For each performance criterion we give a performance measure. Our estimates for each of these performance measures satisfy four properties. First, for all measures we only look to an order-of-growth estimate; no estimates are given in terms of square millimeters, microwatts, or nanoseconds. Second, our estimates are based on the program texts: they do not rely on a specific implementation of the basic cells. Third, certain conditions apply in order for our estimates to be accurate first-order approximations. Some of these conditions are discussed later. Fourth, calculating these estimates can be done on the back of an envelope. We emphasize once more that our performance estimates are just first-order approximations. We should always be aware that the hidden constants and second order terms can vary heavily among implementations. A more detailed performance estimate will require further knowledge of the particular implementation.

As usual in the analysis of algorithms, we use the notations $\mathcal{O}, \Omega$, and Θ to denote an upper bound, lower bound, and tight bound for the order of growth of a function.

6.1   Area Complexity

Our first performance criterion is area complexity. The area complexity is a rough estimate for the area occupied by a physical implementation such as an integrated circuit. As a measure for area complexity we take the number of 'basic' components in the decomposition. Here, a basic component can be any component as long as the number of states is bounded by a predetermined constant. (Notice that for a given constant there can only be a bounded number of basic components.) For regular implementations consisting of a linear array of basic components, like our modulo-N counter implementations, our measure gives a good first-order approximation. The accuracy of the estimate may change, however, when the connections among the basic components become more complex.

Before we study the area complexity of some decompositions, let us make a list of the decompositions we have so far. Let $ModC(N)$ and $sModC(N)$ be abbreviations for $ModC(N, r?, a!)$ and $ModC(N, sr?, sa!)$ respectively. The cells have the usual input and output channels. We take $N > 1$.

$$ModC(2N) \rightarrow (CELL0 , ModC(N)) \tag{0}$$

$$ModC(2N) \rightarrow (CELL1 , sModC(N)) \tag{1}$$

$$ModC(N+1) \rightarrow (CELL2 , sModC(N)) \tag{2}$$

$$ModC(2N) \rightarrow (PCELL0 , sModC(N)) \tag{3}$$

$$ModC(2N) \rightarrow (PCELL1 , sModC(N)) \tag{4}$$

$$ModC(N+1) \rightarrow (PCELL2 , sModC(N)) \tag{5}$$

$$ModC(N) \rightarrow (P , sModC(N)) \tag{6}$$

First, we remark that the number of states of each cell is independent of N and therefore bounded. Accordingly, we can consider these cells as 'basic' components in our area complexity analysis. Second, we remark that the decompositions (3)-(5) are similar to decompositions (0)-(2), as far as area complexity is concerned. Finally, we observe that it does not make sense to use decomposition step (6) to obtain an efficient area complexity. For these reasons, we only concentrate on the first three decompositions.

With decomposition step (2) we can decompose any modulo-N counter for $N > 1$ into basic components $CELL2$ and a modulo-2 counter. The area complexity of such a decomposition is linear in N, that is $\Theta(N)$. If we use decomposition step (0) or (1) when N is even and decomposition step (2) when N is odd, we obtain a decomposition into basic components $CELL0$, $CELL1$, $CELL2$, and a modulo-2 counter with a logarithmic area complexity, that is, the area complexity is $\Theta(\log N)$. (Notice that at least every other decomposition step is of the form (0) or (1), which gives the logarithmic complexity.)

Is there a decomposition into basic components that has an area complexity that grows less than logarithmic in N? In fact, what is the lower bound of the area complexity of the modulo-N counter taken over all decompositions into basic components? It turns out that we cannot do any better than logarithmic in N. Here is an argument why. Consider a network with k basic components, and each component has at most q states for a given constant q. This network can implement a specification with at most $O(q^k)$ states. The modulo-N counter has $\Theta(N)$ states. Consequently, any decomposition into basic components of the modulo-N counter has at least $\Omega(\log N)$ basic components. A similar reasoning can be applied to any specification to obtain a lower bound for the area complexity.

6.2 Power Consumption

Our second performance measure is power consumption. In physical terms, the power consumption of an integrated circuit is the energy dissipated per time unit. Since in our abstract approach there is no time metric, we consider the energy dissipated per external action. As a measure for the energy we take the total number of communication actions in a behavior. Furthermore, in this note we are not interested in incidental peaks in the power consumption: we only consider the power consumption over the long term. That is, we amortize all communication actions over the external actions. Finally, since the power consumption may depend on what external actions are performed by the environment, we assume a worst-case environment for our power consumption analysis. A worst-case environment is an environment that communicates with the implementation in such a way that the total number of communication actions is maximized over the long term. For these reasons, we take as a measure for the power consumption the total number of communication actions amortized over the external communication actions for a worst-case environment. In order for this measure to be a good first-order approximation a number of conditions must be satisfied. Some of these conditions are discussed below.

The power consumption of a circuit is determined by the dynamic and static power consumption. The dynamic power consumption is dominated by the charging and discharging of capacitances. The static power consumption in CMOS circuits is due to leakage current. Our measure, which is based on counting communication actions, is intended to be an estimate of the dynamic power consumption. In order for this 'communication count' to be a good first-order approximation for the power consumption, the static power consumption should be negligible to the dynamic power consumption. This condition can be met in a CMOS implementation, if the frequency at which the communication actions occur is high enough. If the frequency becomes too low, static power consumption is no longer negligible [36].

The second condition is that all voltage transitions in the implementation require about the same amount of energy. The amount of energy needed for a transition depends on the load capacitance. Load capacitances can vary orders of magnitudes in an integrated circuit. For example, the load capacitance of a long wire with a large fan-out is much higher than the load capacitance of a short wire with a fan-out of only one. In particular, in an implementation with an irregular structure and large differences in capacitances, our estimate becomes inaccurate.

The third condition is that the power consumption of (CMOS implementations of) our basic components is proportional to the number of external communication actions performed on it. This assumption requires, for example, that the implementations of our basic components do not exhibit any livelock or metastable behavior.

If one of the conditions above is not satisfied, then the power consumption can only be higher. In this respect, our measure will always give a lower bound for the order of growth of the power consumption.

Let us return to defining the power consumption of a decomposition in terms of the 'communication count.' For a decomposition into basic components, the *power consumption of a behavior t* is defined as the number of communication actions in t amortized over the number of external communication actions in t, where t is 'long enough'. We take the amortized power consumption of a behavior in order to spread out evenly the cost of all communication actions over the external communication actions. For example, initialization effects can be spread out in this way over many external communication actions. If we

defined the power consumption of a behavior simply as the number of all communication actions divided by the external communication actions, we could
get extremely high power consumptions during initialization or could even get
division by zero. The *power consumption of the decomposition* is defined as the
maximum of the power consumptions over all behaviors of the decomposition.
We say that the decomposition has bounded power consumption if its power
consumption is bounded from above by a constant. These definitions have been
inspired by [2].

With these definitions we can calculate the power consumption for various
decompositions. As a first example, we take the decomposition where only
decomposition step (2) (i.e., *CELL2*) is used. Here are some behaviors of that
decomposition

$$r_0 a_0, \; r_0 r_1 a_1 a_0, \; r_0 r_1 r_2 a_2 a_1 a_0, \ldots$$

where r_0 represents a communication on channel r between the environment
and cell 0, r_1 represents a communication on channel r between cell 0 and cell
1, etc. A similar meaning applies to a_i. See Figure 14. For the decomposition

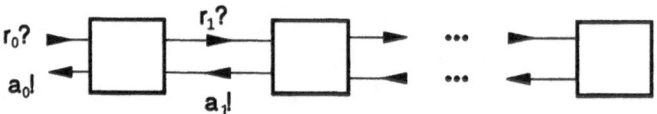

Figure 14: A linear array of cells

of a modulo-N counter, there are $N-2$ components *CELL2* and one modulo-2
counter as end cell. The first external communication on channel r propagates
to cell 0 and then returns via channel a, the second external communication
propagates to cell 1 and then returns via the a channels, and so on. The $N-1$st
and N-th external communication propagate to the end cell and then return via
the a channels. These behaviors then repeat. So for every part of a behavior
where $2N$ external communication actions (N on channel r and N on channel
a) occur, there will be a total of

$$(\sum_{i=1}^{N-2} 2i) + 4(N-1) = (N-1)(N+2)$$

external and internal communication actions. So the power consumption amortized over the external communication actions for this part of the behavior is
$\Theta(N)$. Since every behavior consists of a repetition of these part behaviors,
every behavior has a power consumption of $\Theta(N)$. Consequently, the complete
decomposition also has a power consumption of $\Theta(N)$.

Let us calculate the power consumption of a decomposition of the modulo-
N counter where only decomposition step (0) is used. In such a decomposition

each external action on the r channel propagates through the entire array of $\Theta(\log N)$ cells to the end cell and then back on the a channels. Consequently, every behavior has a power consumption of $\Theta(\log N)$, and therefore this decomposition also has a power consumption of $\Theta(\log N)$.

What would the power consumption be of a decomposition where only step (1) is used? Notice that in this decomposition each cell propagates every other input r further in the linear array of cells. So if in a communication behavior $2k$ external communication actions take place, at most $2k/2$ communication actions take place between the first and second cell, at most $2k/4$ communication actions take place between the second and third cell, and so on. Consequently, each communication behavior with $2k$ external communication actions has a power consumption of at most

$$(\sum_{i=0}^{\infty} 2k/2^i)/2k \leq 2$$

From this observation we may conclude that every behavior has a bounded power consumption, and therefore the decomposition also has a bounded power consumption.

What would be the power consumption of a decomposition using step (1) for even N and step (2) for odd N? It turns out that such a decomposition also has bounded power consumption. The argument why this is so is only a slight elaboration of the argument used in the previous analysis. First we observe that for $CELL2$ after any behavior the number of communication actions on channels sr and sa is at most the number of communication actions on channels r and a. Second, we observe that at most every other cell in the decomposition is of type $CELL2$. This means that the total number of communication actions in the complete decomposition is at most double that of a decomposition where only $CELL1$ is used. Since a decomposition where only $CELL1$ is used has a bounded power consumption, a decomposition where $CELL1$ is used for even N and $CELL2$ for odd N also has a bounded power consumption.

Can we improve any of these bounds if we use any of the steps (3)-(5) instead of (0)-(2) respectively? It turns out that none of these bounds improves. Notice that by using any of the components $PCELLi$ instead of $CELLi$, $0 \leq i \leq 3$, many actions can take place in parallel, but the communication behavior between any two neighboring cells remains almost invariant for a given external behavior. The only difference is that some extra internal communications take place for computing some responses ahead of time. The number of extra internal actions is at most $O(L)$, where L is the length of the array. Since we only consider long-term behaviors where the number of external actions $k \gg L$, these extra communication actions are negligible in calculating the power consumption. For these reasons the results of the analyses of the 'sequential' cells are still valid when using the 'parallel' cells.

6.3 Response Time

Our third performance measure is the response time of a decomposition. The response time is the delay between the receipt of an input and the production of the succeeding output. The response time is always measured from the time the *last* input arrives that enables the production of that output to the actual production of that output. We are particularly interested in the worst-case response time.

In calculating the response time of a decomposition, we assume that the response times of our basic components are bounded from above and below by fixed constants. Consequently, all implementations of the basic components may not exhibit livelock, deadlock, or metastable behavior, since then there is no guaranteed upper bound. We do not require that the response times of basic components are constant. Response times may vary arbitrarily between lower and upper bound. For example, response times may depend on the actual values of the data received, like addition may depend on the actual values added. Response times may also vary over different instances of the same basic component or may vary over time. These delay assumptions are more general than the assumptions made in [28], for example, and may sometimes lead to results that are too pessimistic. More detailed performance analysis and optimization techniques are given in [5, 16]. Techniques for analyzing the throughput and latency of micropipelines are proposed in [31, 37].

Let us see if we can calculate the response times of the various decompositions for the modulo-N counter. If we only consider the steps (0)-(2), then the decompositions do not exhibit any parallel behavior. Let us first consider a decomposition that uses only step (2). In the worst case an input r propagates through all $N-2$ cells to the end cell and then back. Consequently, this decomposition has a response time that grows linearly with N. What is the response time of a decomposition using only step (0)? In such a decomposition every input propagates through all $\Theta(\log N)$ cells and then back. Consequently, the decomposition has a response time of $\Theta(\log N)$. A decomposition using only step (1) also has a response time of $\Theta(\log N)$, since in the worst case an input r propagates through all $\Theta(\log N)$ cells and then back.

Response time analysis is perhaps the most difficult analysis to perform, in particular when there is a high degree of parallelism. In order to facilitate the analyses of decompositions using steps (3)-(6), we present some theorems on response times for linear arrays of cells that exhibit parallel behavior.

We consider a linear array of cells as given in Figure 15. We assume that

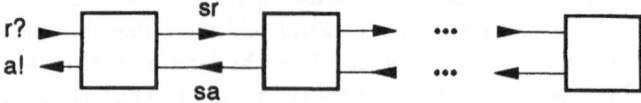

Figure 15: A linear array of cells

each cell in the linear array has a communication behavior given by

pref(*INIT*; *[**if** B **then** $(a!; r?)\|(sr!; sa?) \mid \text{not}(B)$ **then** $(a!; r?)$ **fi** ; S]) (7)

where S represents the computation of the values to be communicated on a and sr, and *INIT* is some initial behavior that can be represented by a proper 'postfix' of the behavior in the repetition. For example, $INIT = r?;$ S or $INIT = (r? \parallel (sr!; sa?));$ S. The guard B depends on the values that are communicated and possibly the values of the local variables. Since we are only interested in when which communication action can occur, we have abstracted as much as possible from the computation of the values that are communicated. There are two special cases for the guards in which we are interested. One case is where $B = true$ in each repetition step; so in every repetition step there is a pair of communications on a and r and on sr and sa. The other case is where at most every other repetition step $B = true$. More specifically, if B_i represents the value of guard B in repetition step $i, i \geq 0$, then we have

$$B_i \Rightarrow \text{not}(B_{i+1}) \quad \text{for all } i \geq 0$$

In other words, if in a certain repetition step there is a pair of communications on sr and sa, then in the next repetition step there are no communications on sr and sa.

Each cell has two outputs: a and sr. For each of these outputs, each of the inputs can be the last input to arrive that enables the production of that output. Accordingly, the response times for a cell are denoted by

$$\tau(r?; a!), \ \tau(sa?; a!), \ \tau(sa?; sr!), \ \tau(r?; sr!)$$

The response time $\tau(r?; a!)$ is the time it takes to produce output $a!$ after input $r?$ has been received, where we assume that input $r?$ is the last input that enables output $a!$. The other response times are described similarly. The response times of the cells may vary. For example, they may vary over different instances of a cell, over time, or they may depend on the state of a cell. We assume, however, that all response times τ of each cell have upper bound Δ and lower bound δ, that is

$$\delta \leq \tau \leq \Delta$$

Furthermore, we assume that also the environment of the linear array of cells takes at least time δ to produce the next request after an acknowledgement. Finally, the end cell has a behavior given by

$$\textbf{pref} * [\ r?; \ a! \ ]$$

The response time of this end cell is given by $\tau_L(r?; a!)$. We assume that the response time of the end cell has lower bound d and upper bound D, that is

$$d \leq \tau_L \leq D$$

Let R stand for the response time of the decomposition. Since we have chosen to calculate the worst-case response time, R is the maximum of all $R.i$, $i \geq 0$, where $R.i$ is the maximum delay between occurrence i of input r and output a in the decomposition. We have the following theorems.

Theorem 2 *If for all cells in the decomposition $B = true$ in every repetition step, then*

$$R \leq D + (L+1)(\Delta - \delta)$$

where L is the number of non-end cells in the linear array and $D \geq \Delta$.

With the proper distribution of delays, this upper bound can indeed be attained. The delays then should be independent of each other. Recall furthermore that delays may vary in 'time and space,' that is, in occurrence of an output and in instance of a component. The freedom of varying delays independent of each other and in time and space may give a too pessimistic bound for some applications. If more information is available about the possible delay distributions, then tighter upper bounds may be obtained.

Theorem 3 *If for all cells in the decomposition $B = true$ in at most every other repetition step, and D and L satisfy*

$$\Delta \leq D \leq 2^{L+1}\delta - 2L\Delta + C0$$

then

$$R \leq C1$$

where L is the number of non-end cells in the linear array and $C0$ and $C1$ are constants depending on Δ and δ only.

Notice that here the upper bound for the response time R is independent of L, the number of cells in the array, and D, the maximum response time of the last cell. The response time R only depends on the values of δ and Δ. In other words, the response time is bounded under any delay distribution (in time and space) and for any length of the array. The requirement for D and L is easy to satisfy. For large L, D may be chosen very large such that still a bounded response time is guaranteed. In fact, D may increase exponentially with L. This property can be exploited by taking, for large L, an end cell that is very slow, but has a low power consumption, for example. (A very rough upperbound for both $C0$ and $C1$ is $5\Delta \log(2\Delta/\delta)$.)

What is the response time for a decomposition using only $PCELL0$? Such a decomposition has $L = O(\log N)$ cells, and each cell, except the end cell, is of the form (7). Furthermore, in each repetition step there is a pair of communications with both neighbors. Consequently, Theorem 2 applies. From Theorem 2 we then derive that the decomposition has a response time of at most $O(\log N)$, if $\Delta \neq \delta$. If $\Delta = \delta$, then the response time is bounded by the response time of the end cell. (It can be proven that the upper bound of $O(\log N)$ can be attained, if the distribution of the delays is such that, for example, 1's propagate twice as fast as 0's through the cells.)

In order to calculate the response time of a decomposition using only $PCELL1$ we can apply Theorem 3. Notice that $PCELL1$ communicates on channels sr and sa in at most every other repetition step. We assume that the end cell has a response time of at most Δ, that is, $D \leq \Delta$. By Theorem 3 it now follows that this decomposition has a bounded response time.

For a decomposition using $PCELL1$ for even N and $CELL2$ for odd N, we can also use Theorem 3 to conclude a bounded response time. This conclusion follows from looking at the composite behavior of $PCELL1$ followed by $CELL2$. Without proof we mention that this behavior is also of the form (7) and has the property that at most every other repetition step there is pair of communications on channels sr and sa.

There are many other combinations of decomposition steps we haven't analyzed. For example, what is the response time of a decomposition using $PCELL2$ only? What is the response time of a decomposition using $PCELL0$ and $PCELL2$? Or, of a decomposition using $PCELL1$ and $PCELL2$? None of the two theorems directly applies to these decompositions. For most of these combinations the analysis is non-trivial. Having more general theorems than the two we presented would be helpful to calculate response times of such decompositions.

7 The Up-Down N Counter

7.1 Specification

An up-down counter is a component on which two operations can be performed: an 'up', which is an increment by one, and a 'down', which is a decrement by one. For each of those operations, one of three replies will be sent back to the environment. The reply depends on the count and the range of the counter. The count of the counter is the number of ups minus the number of downs. The range of the up-down N counter is $[0..N]$, where $N > 0$. Initially, the count of the counter is 0. If after a down, the count of the up-down N counter reaches 0, then the reply will be 'empty'. If after an up, the count of the up-down N counter reaches N, then the reply will be 'full'. Otherwise, the reply will be 'ack'. In order for the count of the up-down N counter to stay in the range $[0..N]$, we assume that the environment will not attempt a down when the count is 0 and will not attempt an up when the count is N. Notice that the environment is informed after each operation whether the counter is empty, full, or neither. We stipulate that initially the counter is empty.

Here is a specification of the up-down N counter using guarded commands. We take the following type definitions throughout these notes.

type $ud = \{up, down\}$
type $efa = \{empty, full, ack\}$

The specification reads

$UDC(N : int,\ r? : ud,\ a! : efa)$

$=$ { by definition }

```
|[ var n : [0..N] ::
   initially n = 0 ::
   pref * [ r?;
              if r = up ∧ n = N − 1 then a := full; n := n + 1
              |  r = up ∧ n < N − 1 then a := ack; n := n + 1
              |  r = down ∧ n = 1   then a := empty; n := n − 1
              |  r = down ∧ n > 1   then a := ack; n := n − 1
           fi ; a!
           ]
]|
```

For the repetition we have the invariant

$$n = (\#r?up - \#r?down)$$

Notice that the two alternatives

$$r = down \ \wedge \ n = 0$$
$$r = up \ \wedge \ n = N$$

do not occur in the guarded selection, since we assume that the environment does not attempt a down when the count is 0 and not attempt an up when the count is N. (By definition of the guarded selection, the if ..fi statement amounts to an *abort* in states where these alternatives would apply.)

The important property we should remember is the precondition for every output a:

$$(a = full) \ \equiv \ (n = N) \quad \wedge \quad (a = empty) \ \equiv \ (n = 0)$$

Here is one special case of the up-down N counter. For $N = 1$ we have

pref $*$ [r?up; a!full; r?down; a!empty]

Given the specification of the up-down N counter we are asked to find an efficient asynchronous implementation for this component for any $N > 0$. Try it, before reading any further.

7.2   Related Work and Results

There are many implementations of up-down counters commercially available. (See, for example, the data books of several manufacturers.) There are also many applications for an up-down counter. For example, they can be used in any application that keeps track of a count within a range $[0..N]$, and where the only operations on the count are increments and decrements by one and testing whether the value of the count is 0 or N. Applications that come to mind immediately are semaphores, bounded stacks, and bounded queues.

There is also quite a rich literature on all sorts of counters. Designing synchronous implementations (as opposed to asynchronous implementations) of an up-down counter is usually considered a standard exercise in almost every textbook on digital design. In [26] one can find designs for many types of counters, including up-down counters. In [14] a synchronous implementation of an up-down counter is discussed, where the output is available after a constant number of clock cycles. This design, however, does not detect whether the counter is full. The idea of this implementation is the same as the idea underlying our first design. In [17] an up-down $2N$ counter is implemented by N identical modules. The output is also available after a constant number of clock cycles under the assumption that the inputs can be broadcast to all modules in a constant amount of time. In our designs no broadcast is needed. In [27], which is based on [14], several up-down counter designs are presented. The counters are slightly different in the sense that they behave like modulo-N counters when an increment occurs in the full state.

Most counters described in the literature are counters that report the value of the count in some radix representation after each operation. For the up-down counter, however, there is no need to know the value of the count after each operation. The only information that is needed is whether the count is equal to one of the two boundary values, and, if so, which one. It is, however, possible to implement an up-down counter that is based on a counter that reports the value of the count after each operation. From this value the response *full*, *empty*, or *ack* can be calculated. This calculation must consider all digits in the radix representation. If you want to obtain a bounded response time, it is hard to imagine that this could be achieved with such a design.

Most published implementations we have found are synchronous implementations and are based on some sort of binary representation of the count. As a consequence the clock frequency depends on the counter size N. For most designs this is due to the carry and borrow propagation during increments and decrements [21]. These carry and borrow propagations give a response time that is at least logarithmic in N. Furthermore, every synchronous implementation based on a binary representation of the count has to clock in each period about $\log N$ storage devices, making the power consumption at least proportional to $\log N$.

In conclusion, we have found that all published designs are clocked designs, have a logarithmic area complexity, usually do not give a response time analysis, do not give a power consumption analysis, and only apply to very few values of N (usually only for values $2^k - 1, k > 0$). We present several designs for the up-down N counter that have an area complexity of $\log N$, but unlike all previously known designs, are unclocked, have a bounded response time, have a bounded power consumption, and apply to any $N > 0$. These bounds are asymptotically optimal.

7.3 A First Decomposition

The derivation of our first decomposition is based on the usual binary representation of numbers. Suppose the count is represented by k bits, $k > 0$. An increment can be implemented by adding 1 to the least significant bit modulo 2, where a carry may propagate possibly all the way to the most significant bit. Similarly, a decrement is implemented by subtracting 1 from the least significant bit modulo 2, where a borrow may propagate possibly all the way to the most significant bit. Keeping track of whether the counter is full or empty can be done while returning to the least significant bit from a carry or borrow propagation. Assuming that the counter is full when all bits are 1 and empty when all bits are 0, then we can have the following scenario. Each cell records whether the rest of the (more significant) bits are all 1 (subcounter full), all 0 (subcounter empty), or neither all 1 nor all 0. After an increment to a cell, the response is 'full' if the subcounter is full and the new value of the bit is 1. Similarly, after a decrement to a cell, the response is 'empty' if the subcounter is empty and the new value of bit is 0. In all other cases the response is 'ack.'

In order to give a formal derivation, we show that a $2N + 1$ counter can be decomposed into an N counter and a cell. Let $N > 0$. We first give the specification of the $2N + 1$ counter. The specification for the cell will arise in the derivation.

$$UDC(2N + 1 : int, \ r? : ud, \ a! : efa)$$
$$= \quad \{ \text{ by definition } \}$$

$$
\begin{aligned}
&|[\ \textbf{var } nn : [0..(2N + 1)] :: \\
&\quad \textbf{initially } nn = 0 :: \\
&\quad \textbf{pref} * [\ r?; \\
&\qquad\qquad \textbf{if } r = up \ \wedge \ nn = 2N \quad \textbf{then } a := \textit{full}; \ nn := nn + 1 \\
&\qquad\qquad | \ \ r = up \ \wedge \ nn < 2N \quad \textbf{then } a := \textit{ack}; \ nn := nn + 1 \\
&\qquad\qquad | \ \ r = down \ \wedge \ nn = 1 \quad \textbf{then } a := \textit{empty}; \ nn := nn - 1 \\
&\qquad\qquad | \ \ r = down \ \wedge \ nn > 1 \quad \textbf{then } a := \textit{ack}; \ nn := nn - 1 \\
&\qquad\qquad \textbf{fi} \ ; \ a! \\
&\qquad\quad] \\
&]|
\end{aligned}
$$

In our first step we introduce binary variable k, representing the value of the bit in the cell, and variable n, representing the count of the subcounter. The variables k and n are related to nn by the invariant

$$P : \ nn = 2 * n + k \ \ \wedge \ \ 0 \le n \le N \ \ \wedge \ \ 0 \le k \le 1$$

From this invariant we derive the following equivalences.

$$
\begin{aligned}
nn = 2N &\equiv (n = N \ \wedge \ k = 0) \\
nn < 2N &\equiv n \ne N \\
nn = 1 &\equiv (n = 0 \ \wedge \ k = 1) \\
nn > 1 &\equiv n \ne 0
\end{aligned}
$$

These observations then lead to the following program for $UDC(2N + 1, r, a)$.

$$
\begin{aligned}
&|[\ \textbf{var } nn : [0..(2N + 1)], n : [0..N], \ k : bin :: \\
&\quad \textbf{initially } n = 0, k = 0 :: \\
&\quad \textbf{pref} * [\ r? \\
&\qquad ; \textbf{if } r = up \ \wedge \ n = N \ \wedge \ k = 0 \quad \textbf{then } a := \textit{full}; \ nn := nn + 1
\end{aligned}
$$

$$\begin{array}{lll}
| & r = up \ \wedge \ n \neq N & \textbf{then } a := ack; \ nn := nn + 1 \\
| & r = down \ \wedge \ n = 0 \ \wedge \ k = 1 & \textbf{then } a := empty; \ nn := nn - 1 \\
| & r = down \ \wedge \ n \neq 0 & \textbf{then } a := ack; \ nn := nn - 1
\end{array}$$
$\quad$ **fi**
$\quad$; **if** $r = up \ \wedge \ k = 1 \quad$ **then** $n := n + 1$
$\quad\quad$ | $\ r = down \ \wedge \ k = 0$ **then** $n := n - 1$
$\quad\quad$ | $\ (r = up \ \wedge \ k = 0)$ or $(r = down \ \wedge \ k = 1)$ **then** *skip*
$\quad$ **fi**
$\quad$; $k := (k + 1) \bmod 2$
$\quad$; $a!$
$\quad$]
]|

The alternative $r = up \ \wedge \ k = 1$ corresponds to a carry propagation, here represented by $n := n + 1$, and the alternative $r = down \ \wedge \ k = 0$ corresponds to a borrow propagation, here represented by $n := n - 1$. It is not difficult to see that $nn = 2 * n + k \ \wedge \ 0 \leq k \leq 1$ are indeed invariants of this program. Let us check that $0 \leq n \leq N$ is an invariant of the program as well. Because of the semantics of guarded selection, a postcondition for the first guarded selection, which is the precondition for the second guarded selection, is

$$(r = up \ \wedge \ k = 1 \Rightarrow n < N) \quad \wedge \quad (r = down \ \wedge \ k = 0 \Rightarrow n > 0)$$

In other words, an increment to n is not done when $n = N$ and a decrement to n is not done when $n = 0$. From this observation we may conclude that $0 \leq n \leq N$ is an invariant of the repetition.

After the first derivation step, we can make a couple of observations. First, the variable nn is a ghost variable: it is never inspected and is only used for the correctness proof of the derivation step. Consequently, we can remove all statements involving nn from the program. Second, we observe that the only operations on variable n are increments, decrements, and tests whether $n = 0$ or $n = N$. These operations can be performed by an up-down N counter. For this purpose we introduce an up-down N counter with input channel sr and output channel sa, and whenever an increment to n is done we replace this statement by $sr!up; sa?$, whenever a decrement to n is done we replace this statement by $sr!down; sa?$. After every communication on channel sa we can then assert

$$(sa = full) \equiv (n = N) \ \wedge \ (sa = empty) \equiv (n = 0)$$

where $n = (\#sr?up - \#sr?down)$. These observations then lead to the following parallel composition.

$\quad UDC(2N + 1, r?, a!)$

$=\quad$ { def. of $UDC(N, sr?, sa!)$, see above }

$\quad$ |[**chan** $sr : ud, \ sa : efa \ ::$
$\quad\quad CELL0(r?, a!, sr!, sa?) \ \| \ UDC(N, sr?, sa!)$
$\quad$]|

where *CELL0* is defined by

```
|[ var k : bin ::
   initially sa = empty, k = 0 ::
   pref * [ r?
            ; if r = up ∧ sa = full ∧ k = 0        then a := full
              |  r = up ∧ sa ≠ full                 then a := ack
              |  r = down ∧ sa = empty ∧ k = 1  then a := empty
              |  r = down ∧ sa ≠ empty            then a := ack
              fi
            ; if r = up ∧ k = 1  then sr!up; sa?
              |  r = down ∧ k = 0 then sr!down; sa?
              |  (r = up ∧ k = 0) or (r = down ∧ k = 1)  then skip
              fi
            ; k := (k + 1) mod 2
            ; a!
            ]
]|
```

Notice that *CELL0* will not attempt an *sr!down* when the subcounter is empty, nor will *CELL0* attempt an *sr!up* when the subcounter is full. This property follows immediately from the invariant $0 \le n \le N$ of the previous program.

Finally we observe that all conditions for decomposition are satisfied. In particular, no computation interference can occur: in *CELL0* every output *sr!* is immediately followed by input *sa?* and every output *a!* is immediately followed by input *r?*. Accordingly, we can write

$UDC(2N + 1, r?, a!)$

→   { def. of decomposition }

$(CELL0(r?, a!, sr!, sa?) , UDC(N, sr?, sa!))$

7.4 What about even N?

In the previous section we derived a decomposition that applies to odd N only. If we need to have a decomposition that applies to all N, we also need to find a decomposition that applies to even N. Our next decomposition is a generalization of our first decomposition and is not restricted to even N only.

The previous decomposition was based on the unique binary representation of each number. The invariant we used there was

$P: \ nn = 2*n + k \ \land \ 0 \le n \le N \ \land \ 0 \le k \le 1$

where nn represents the count of the $2N + 1$ counter and n represents the count of the N counter. What would happen if we enlarged the range of k? For example, what would change in our derivation if we had $0 \le k \le K$? It turns out that we need to change our derivation only slightly. Let $K > 0$. Our new invariant is

$$nn = 2 * n + k \quad \wedge \quad 0 \leq n \leq N \quad \wedge \quad 0 \leq k \leq K$$

From this invariant we derive the following equivalences.

$$(nn = 2N + K - 1) \equiv (n = N \wedge k = K - 1)$$
$$(nn < 2N + K - 1) \equiv (n \neq N \vee k < K - 1)$$
$$nn = 1 \equiv (n = 0 \wedge k = 1)$$
$$nn > 1 \equiv (n \neq 0 \vee k > 1)$$

These observations then lead to the following program.

$$UDC(2N + K, r?, a!)$$

=   { for def. of $CELL2$ see below }

$$(\ CELL2(K : int, \ r?, \ a!, \ sr!, \ sa?) \ , \ UDC(N, \ sr?, \ sa!) \ )$$

where $CELL2$ is defined by

$$CELL2(K : int, \ r? : ud, \ a! : efa, \ sr! : ud, \ sa? : efa)$$

=   { by definition }

```
|[ var k : [0..K] ::
   initially sa = empty, k = 0 ::
   pref * [ r?
        ;if r = up ∧ sa = full ∧ k = K − 1    then a := full
         | r = up ∧ (sa ≠ full ∨ k < K − 1)   then a := ack
         | r = down ∧ sa = empty ∧ k = 1      then a := empty
         | r = down ∧ (sa ≠ empty ∨ k > 1)    then a := ack
        fi
        ;if r = up ∧ k = K    then sr!up; sa?
         | r = down ∧ k = 0   then sr!down; sa?
         | (r = up ∧ k ≠ K) or (r = down ∧ k ≠ 0)   then skip
        fi
        ;if r = up ∧ k ≠ K     then k := k + 1
         | r = up ∧ k = K      then k := K − 1
         | r = down ∧ k ≠ 0    then k := k − 1
         | r = down ∧ k = 0    then k := 1
        fi
        ; a!
        ]
]|
```

Now we have many decompositions that apply for even N. For example, we can use $CELL2$ for $K = 2$ or $K = 4$, when N is even. For odd N we also have many decompositions. We can use $CELL2$ for $K = 1$ or $K = 3$, when N is odd. If we are restricted to using $CELL2$ for values of $K = 2$ and $K = 3$, then any

N counter for $N \geq 1$ can be decomposed using such cells, if we have at least as end cells a 1 counter, a 2 counter, and a 3 counter. Notice that for $N \geq 1$, we have $2N + 2 \geq 4$ and $2N + 3 \geq 5$.

For $K = 1$, we have the usual binary number system, where each number has a unique representation. For $K > 1$ we use a redundant binary number system, where the digits can range from 0 to K inclusive. For example, for $K = 2$ the number 4 can be represented as 100 and as 012 (least significant digit to the right). This redundancy will pay off later when we examine the power consumption and response time.

7.5   What about Parallelism?

The next step in our derivation of an optimal design is the introduction of some parallelism. We try to do so by allowing as much freedom as possible in the ordering of the communication actions in a cell with the restrictions that the communication behavior on channels r and a remains invariant and the communication behavior on channels sr and sa remains invariant.

We first introduce some parallelism in the specification for $CELL0$. Let us consider the specification for $CELL0$ once more.

$CELL0(r? : ud,\ a! : efa,\ sr! : ud,\ sa? : efa)$

$=$ { by definition }

$|[$ **var** $k : bin ::$
   initially $sa = empty,\ k = 0 ::$
   pref $* [\ r?$
            $; \textbf{if}\ r = up\ \wedge\ sa = full\ \wedge\ k = 0$ **then** $a := full$
            $\mid\ r = up\ \wedge\ sa \neq full$ **then** $a := ack$
            $\mid\ r = down\ \wedge\ sa = empty\ \wedge\ k = 1$ **then** $a := empty$
            $\mid\ r = down\ \wedge\ sa \neq empty$ **then** $a := ack$
          fi
          $; \textbf{if}\ r = up\ \wedge\ k = 1$ **then** $sr!up;\ sa?$
            $\mid\ r = down\ \wedge\ k = 0$ **then** $sr!down;\ sa?$
            $\mid\ (r = up\ \wedge\ k = 0)$ or $(r = down\ \wedge\ k = 1)$ **then** $skip$
          fi
          $;\ k := (k + 1) \bmod 2$
          $;\ a!$
          $]$
$]|$

Observe that although the value of a is calculated immediately after r has been received, it is sent only after there has been a possible communication on the channels sr and sa. Immediately after sending a the next request on channel r can be received, in order to avoid computation interference. Is it possible to send a (and receive the next request on r) in parallel with sending sr (and receiving sa)? Yes, that can be done. If we unfold the repetition a bit and reorder some statements, we get the following program for $PCELL0$

$PCELL0(r? : ud,\ a! : efa,\ sr! : ud,\ sa? : efa)$

$=$ { by definition }

|[**var** $k : bin$ ::
  initially $sa = empty,\ k = 0$ ::
  pref ($r?;\ S_a$;
        *[**if** $r = up \wedge\ k = 1$ **then** $(a!;\ r?) \parallel S_k \parallel (sr!up;\ sa?)$
          | $r = down \wedge\ k = 0$ **then** $(a!;\ r?) \parallel S_k \parallel (sr!down;\ sa?)$
          | $(r = up \wedge\ k = 0)$ **or**
            $(r = down \wedge\ k = 1)$ **then** $(a!;\ r?) \parallel S_k$
        fi
        ; S_a
        ])
 ]|

where $S_k = k := (k + 1) \bmod 2$ and S_a is the statement that calculates the next value for a given by

S_a

$=$ { by definition }

if $r = up \wedge\ sa = full \wedge\ k = 0$ **then** $a := full$
| $r = up \wedge\ sa \neq full$ **then** $a := ack$
| $r = down \wedge\ sa = empty \wedge\ k = 1$ **then** $a := empty$
| $r = down \wedge\ sa \neq empty$ **then** $a := ack$
fi

It is not difficult to verify that the communication behavior of $CELL0$ and $PCELL0$ is the same if we look to the channels r and a only. Similarly, the communication behavior of $CELL0$ and $PCELL0$ is the same if we look to the channels sr and sa only.

The same exercise can be performed on $CELL2$. If we name the resulting specification $PCELL2$, we get

$PCELL2(K : int,\ r? : ud,\ a! : efa,\ sr! : ud,\ sa? : efa)$

$=$ { by definition }

|[**var** $k : [0..K]$ ::
  initially $sa = empty,\ k = 0$ ::
  pref ($r?;\ S_a$;
        *[**if** $r = up \wedge\ k = K$ **then** $(a!;\ r?) \parallel (k := K - 1) \parallel (sr!up;\ sa?)$
          | $r = down \wedge\ k = 0$ **then** $(a!;\ r?) \parallel (k := 1) \parallel (sr!down;\ sa?)$
          | $r = up \wedge\ k \neq K$ **then** $(a!;\ r?) \parallel (k := k + 1)$
          | $r = down \wedge\ k \neq 0$ **then** $(a!;\ r?) \parallel (k := k - 1)$
        fi
        ; S_a
        ])
 ]|

where S_a is the appropriate statement calculating the next value of a.

7.6   An Implementation

As an example of an implementation, we briefly present one for $PCELL0$ using data bundling. As an encoding for the data type ud, we use a single wire and take by definition $up = 1$ and $down = 0$. As an encoding for the type efa we take two wires called f and e. Each value of type efa is then encoded by a pair of binary values (f, e) in the following way.

$full = (1, 0)$
$empty = (0, 1)$
$ack = (0, 0)$

The value $(1,1)$ is not used. These encodings then quickly lead to the implementation of Figure 16 for the up-down 1 counter.

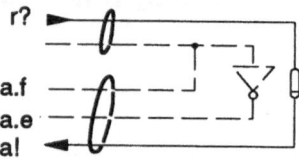

Figure 16: An implementation for $UDC(1)$ using data bundling

From the program for $PCELL0$, we derive that the 'new' values for $a = (a.f, a.e)$ and k can be calculated as follows from the inputs r and $sa = (sa.f, sa.e)$.

$$a.f \quad := \quad r \wedge sa.f \wedge \neg k$$
$$a.e \quad := \quad \neg r \wedge sa.e \wedge k$$
$$k \quad := \quad \neg k$$

The 'new' value for sr is calculated as follows. If r and k are both zero or both one, then sr becomes zero or one respectively. Otherwise, sr retains its 'old' value. In other words, the 'new' value for sr is the majority of r, k, and the old value of sr. Furthermore, a communication with the subcounter takes place only when r and k are both zero or both one. Therefore, we introduce a binary variable $prop$ (of $propagate$) to direct the communications with the subcounter. The values for sr and $prop$ are calculated as follows.

$$sr \quad := \quad maj(r, k, sr)$$
$$prop \quad := \quad (r \wedge k) \vee (\neg r \wedge \neg k)$$

where $prop = 1$ if there is a communication on the channel sr, and $prop = 0$ if there is no communication on the channel sr.

The control part of the implementation consists of a JOIN, a MERGE, a SELECT, and some WIRES. The control part is indicated with solid lines in Figure 17. The data part is indicated by dashed lines in Figure 17.

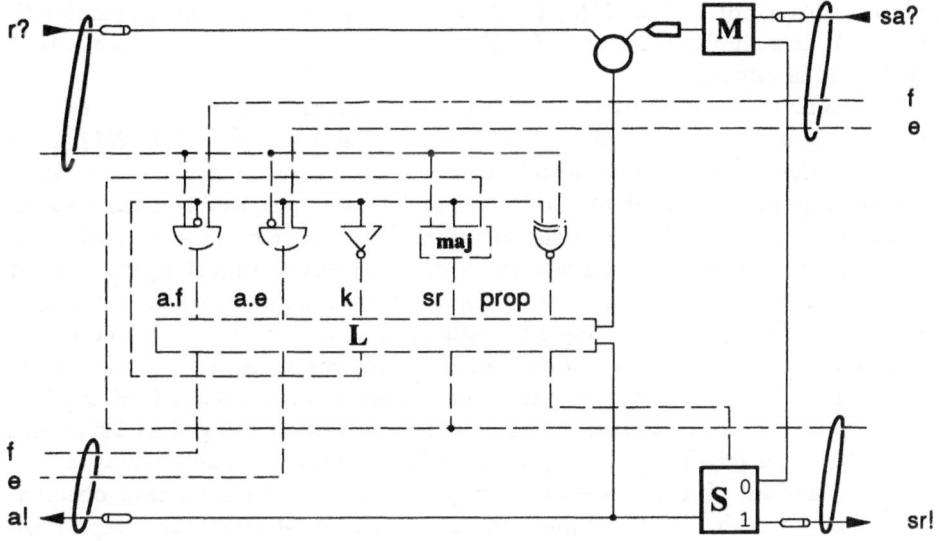

Figure 17: An implementation of *PCELL0* using data bundling

7.7 Performance Analysis

Before we are going to analyze the performance of some complete decompositions of the up-down N counter, we give a brief summary of the possible decomposition steps. Below, $UDC(N)$ stands for $UDC(N, r? : ud, a! : efa)$ and
$CELL2(K)$ stands for $CELL2(K : int, r? : ud, a! : efa, sr! : ud, sa? : efa)$. A
similar correspondence holds for $PCELL2(K)$. (Recall that $CELL0$ is a special
case of $CELL2(K)$, viz., $CELL2(1) = CELL0$)

$$
\begin{aligned}
UDC(2N+1) &\rightarrow (\ CELL2(1)\ ,\ UDC(N)\ ) & (0) \\
UDC(2N+2) &\rightarrow (\ CELL2(2)\ ,\ UDC(N)\ ) & (1) \\
UDC(2N+3) &\rightarrow (\ CELL2(3)\ ,\ UDC(N)\ ) & (2) \\
UDC(2N+1) &\rightarrow (\ PCELL2(1)\ ,\ UDC(N)\ ) & (3) \\
UDC(2N+2) &\rightarrow (\ PCELL2(2)\ ,\ UDC(N)\ ) & (4) \\
UDC(2N+3) &\rightarrow (\ PCELL2(3)\ ,\ UDC(N)\ ) & (5)
\end{aligned}
$$

Let us consider the area complexity of some decompositions first. Our
first observation is that any of the cells above can be considered as a basic
component, since the number of states of each cell is bounded by some constant,
which is independent of N. A similar remark can be made for small counters,
like the 1, 2, and 3 counter. Accordingly, a first-order estimate for the area
complexity can be obtained by counting the total number of cells and end cells
of the decomposition. The area complexity of some decompositions can now be
calculated easily. A decomposition of the up-down N counter using any of the

cells $CELL2(K)$ or $PCELL2(K)$ for $K = 1, 2, 3$ obviously has area complexity $\Theta(\log N)$. In a similar manner as for the modulo-N counter, we can prove that this bound is optimal.

The analysis for the power consumption is a bit more difficult. Let us first consider the power consumption of a decomposition using only $CELL2(1)$. This decomposition is based on the usual binary number system. In the worst case a request propagates through all $\log N$ cells and comes back, and this can happen repeatedly. For example, when the count of the counter is $(N-1)/2$ (that is, the binary representation consists of all ones, except the most significant bit which is zero), and the sequence $r?up$; $a!ack$; $r?down$; $a!ack$ is performed repeatedly. Then, for each request, each bit has to flip once. Consequently, each external communication results in $\Theta(\log N)$ communications, and so the worst-case power consumption over all communication behaviors is $\Theta(\log N)$.

What is the power consumption of a decomposition using $CELL2(2)$ and $CELL2(3)$ with $UDC(1)$, $UDC(2)$, or $UDC(3)$ as end cell? This decomposition uses a redundant binary representation of the count. The important observation is that in both $CELL2(2)$ and $CELL2(3)$ in at most every other repetition step there will be a communication on the channels sr and sa. Accordingly, if in a communication behavior $2k$ external communication actions take place, at most $2k/2$ communication actions take place between the first and second cell, at most $2k/4$ communication actions take place between the second and third cell, and so on. Consequently, each communication behavior with $2k$ external communication actions has a power consumption of at most

$$(\sum_{i=0}^{\infty} 2k/2^i)/2k \leq 2$$

From this observation we may conclude that every behavior has a bounded power consumption, and therefore the decomposition using cells $CELL2(K)$ for $K = 2, 3$ has a bounded power consumption.

Would the power consumption change if we used a $PCELL2$ instead of a $CELL2$? From the programs for $CELL2$ and $PCELL2$ we derive that for each external behavior with $2k$ actions, the maximum number of internal communication actions that can take place in the parallel and the sequential version are the same. The only difference is in the order in which the actions can take place. For this reason, the results of the power consumption analyses for the sequential version are also valid for the parallel version.

Finally we analyze the response time of some decompositions. The 'sequential' decompositions are easy. A decomposition using any of the cells $CELL2$ has logarithmic response time, since in the worst case a request propagates through all $\Theta(\log N)$ cells to the end cell and then returns as an acknowledgement.

For the decomposition using the parallel versions of $CELL2$, we can use Theorems 2 and 3. We assume that the upper bounds and lower bounds for the response times of the cells (including the end cells) are given by Δ and δ respectively. Neither Theorem 2 nor 3 is directly applicable to a decomposition using only $PCELL2(1)$. However, if we assume that the environment gives input requests according to a worst-case scenario, then after an initial behavior each cell will communicate on channels r, a and sr, sa in each repetition step. According to Theorem 2, this decomposition then has a worst-case response time of $O(\log N)$ if $\Delta > \delta$. In order to attain this upper bound, however, we have to assume a very pessimistic delay distribution.

For $PCELL2(2)$ and $PCELL2(3)$ we observe that in at most every other repetition step there is a pair of communications on sr and sa. Accordingly, Theorem 3 is applicable, and we conclude that a decomposition using $PCELL2(2)$ and $PCELL2(3)$ has a bounded response time.

8 Concluding Remarks

We have illustrated some techniques in the design and performance analysis of asynchronous circuits. The techniques were illustrated by means of two examples: the modulo-N counter and the up-down N counter. For both examples we derived several designs that have an area complexity of $\Theta(\log N)$, a bounded power consumption, and a bounded response time. These bounds are optimal. Although the exercises have given us some insights in the alternatives in designing asynchronous circuits, there are still many problems that remain. We mention a few of them.

The final step of our derivations consisted of finding circuit implementations for the cells. Compared to many of the other steps in the design, this last step was rather large, often ad hoc, and without a proof. We did so, because our main emphasis was on the design of the parallel program rather than on the design of circuit implementations for the cells. In order to obtain a complete design method, however, it is important that this last step be investigated more closely, and that a systematic method be found to translate our specifications for the cells into circuit implementations. The choice of implementation for the data types will undoubtedly play a large role in this step.

For all designs, we first started with the design of a sequential program and then transformed this sequential program into a parallel program by reordering the communication actions. Can this be done for every design? That is, for every parallel program, does there exist a sequential program that can be transformed into the parallel program by reordering the statements and communication actions? If so, do there exist some general techniques for reordering communication actions so as to obtain a parallel program?

The performance analyses raised some interesting questions as well. For the area complexity we can compute a lower bound for any implementation of a component easily. For the power consumption and response time this is a different matter. How do we calculate the lower bounds for the power consumption and response time for any implementation of a particular component? For example, what is the lower bound of any implementation for an N-place stack or queue? Knowing what the lower bound is of the power consumption of a component, we may able to conclude whether we have found an optimal design. If not, we may want to search for a more efficient design.

Calculating the response time of an implementation was a nontrivial task. We gave some theorems that were very useful, but these theorems applied only to linear arrays of cells and only under certain conditions for the behavior of the cells. How do we calculate the response time for other networks of cells or when the cell's behavior does not satisfy these restricted conditions?

Finally, we mention the problem of formulating appropriate progress conditions. For any design it is important to know whether progress is guaranteed or not. For example, a design should be free from the danger of deadlock or livelock. Our correctness conditions are too weak to guarantee progress in general. (See Section 3.8.) Finding correctness conditions for progress that have a proper justification in the context of asynchronous circuit behaviors and, preferably, are easy to work with is still an open problem.

Acknowledgements

This work is supported by the Natural Sciences and Engineering Council of Canada under grant OGP0041920 and by a grant from the Information Technology Research Centre of Ontario.

Acknowledgements are due to Ad Peeters for his comments on the modulo-N counter decompositions, Kees van Berkel for directing our attention to the efficient decomposition of modulo-N counters and to analyzing the power consumption of decompositions, Marly Roncken and Joep Kessels who gave the idea for efficient modulo-N counter decompositions [19], Peter Mayo for giving the idea of using a redundant number system to decompose the up-down counter, and the International Council for Canadian Studies for providing financial support for the second author to conduct part of this research [30].

References

[1] C.H. van Berkel, *Handshake Circuits: An Intermediary between Communicating Processes and VLSI.* PhD Thesis, Dept. of Mathematics and Computing Science, Eindhoven University of Technology, 1992.

[2] K. van Berkel, VLSI Programming of a Modulo-N Counter with Constant Response Time and Constant Power, in : S. Furber and M. Edwards eds., *Asynchronous Design Methodologies*, IFIP Transactions A-28, North-Holland, 1993, 1–12.

[3] G.M. Brown, Towards Truly Delay-insensitive Circuit Realizations of Process Algebras, in: G. Jones and M. Sheeran eds., *Designing Correct Circuits*, Workshops in Computing, Springer Verlag, Berlin, 1990, 120–131.

[4] J.A. Brzozowski and C-J.H. Seger, A Unified Framework for Race Analysis of Asynchronous Networks, *J. ACM*, **36** (1), 1989, 20–45.

[5] S. Burns and A.J. Martin, Performance Analysis and Optimization of Asynchronous Circuits, in: Carlo H. Sequin ed., *Advanced Research in VLSI*, MIT Press, Cambridge, Mass., 1991, 71–86.

[6] W. Chen, J.T. Udding, and T. Verhoeff, Networks of Communicating Processes and Their (De-)Composition, in: J.L.A. van de Snepscheut ed., *Mathematics of Program Construction*, Lecture Notes in Computer Science 375, Springer-Verlag, Berlin, 1989, 174–196.

[7] W.A. Clark, Macromodular Computer Systems, in: *Proceedings of the Spring Joint Computer Conference*, AFIPS, Academic Press, London, 1967, 335–401.

[8] A. Davis, B. Coates, K. Stevens, Automatic Synthesis of Fast Compact Self-Timed Control Circuits, in : S. Furber and M. Edwards eds., *Asynchronous Design Methodologies*, IFIP Transactions A-28, North-Holland, Amsterdam, 1993, 193–208.

[9] A. Davis, B. Coates, K. Stevens, The Post Office Experience: Designing a Large Asynchronous Chip, in T.N. Mudge et al. eds., *Proceedings of the 26th Annual Hawaii International Conference on System Sciences*, Vol. 1, IEEE Computer Society Press, 1993, 409–418.

[10] E.W. Dijkstra, *A Discipline of Programming*, Prentice-Hall, 1976.

[11] D.L. Dill, *Trace Theory for Automatic Hierarchical Verification of Speed-Independent Circuits*, MIT Press, Cambridge, Mass., 1989.

[12] J.C. Ebergen, A Formal Approach to Designing Delay-Insensitive Circuits, *Distributed Computing*, **5** (3), 1991, 107–119.

[13] J.C. Ebergen and A.M.G. Peeters, Modulo-N Counters: Design and Analysis of Delay-Insensitive Circuits, in: J. Staunstrup and R. Sharp eds., *Designing Correct Circuits*, IFIP Transactions A-5, North-Holland, Amsterdam, 1992, 27–46.

[14] L.J. Guibas and F.M. Liang, Systolic Stacks, Queues, and Counters, in: P. Penfield Jr. ed., *Advanced Research in VLSI*, Artech House, 1982, 155–164.

[15] C.A.R. Hoare, *Communicating Sequential Processes*, Prentice-Hall, London, 1985.

[16] H. Hulgaard, S. Burns, T. Amon, and Gaetano Borriello, *An Algorithm for Exact Bounds on the Time Separation of Events in Concurrent Systems*, Technical Report 94-02-02, Department of Computer Science and Engineering, University of Washington, Seattle, 1994.

[17] E.V. Jones and G. Bi, Fast Up/Down Counters Using Identical Cascaded Modules, *IEEE Journal of Solid-State Circuits*, 23 (1), 1988, 283–285.

[18] M.B. Josephs and J.T. Udding, Delay-Insensitive Circuits: An Algebraic Approach to Their Design, in: J.C.M. Baeten and J.W. Klop eds., *CON-CUR 1990*, Lecture Notes in Computer Science 458, Springer-Verlag, Berlin, 1990, 342–366.

[19] J.L.W. Kessels, Designing Counters with Bounded Response Time, in: W.H.J. Feijen and A.J.M. van Gasteren eds., *C.S. Scholten Dedicata: van oude machines en nieuwe rekenwijzen*, Academic Service, Schoonhoven, 1990, 127–140.

[20] L. Lavagno and A. Sangiovanni-Vincentelli, *Algorithms for Synthesis and Testing of Asynchronous Circuits*, Kluwer Academic Press, 1993.

[21] X.D. Lu and P.C. Treleaven, A Special-Purpose VLSI Chip: A Dynamic Pipeline Up/Down Counter. *Microprocessing and Microprogramming*, 10 (1), 1982, 1–10.

[22] A.J. Martin, Programming in VLSI: From Communicating Processes to Delay-Insensitive Circuits, in: C.A.R. Hoare ed., *Developments in Concurrency and Communication*, Addison-Wesley, Reading, Mass., 1990, 1–64.

[23] C.E. Molnar, T.P. Fang and F.U. Rosenberger, Synthesis of Delay-Insensitive Modules, in: H. Fuchs ed., *1985 Chapel Hill Conference on VLSI*, Computer Science Press, Rockville, Maryland, 1985, 67–86.

[24] R.E. Miller, *Switching Theory*, Vol. 2, Chapter 10, Wiley, New York, 1965, 199–244.

[25] S.M. Nowick and D.L. Dill, Synthesis of Asynchronous State Machines Using a Local Clock, *Proceedings of the 1991 IEEE International Conference on Computer Design: VLSI in Computers and Processors*, IEEE Computer Society Press, 1991, 192–197.

[26] R.M.M. Oberman, *Counting and Counters*, Macmillan Press, 1981.

[27] B. Parhami, Systolic Up/Down Counters with Zero and Sign Detection, in: M.J. Irwin and R. Stefanelli eds., *IEEE Symposium on Computer Arithmetic*, IEEE Computer Society Press, 1987, 174–178.

[28] M. Rem, Trace Theory and Systolic Computations, in: J.W. de Bakker, A.J. Nijman and P.C. Treleaven eds., *PARLE, Parallel Architectures and Languages Europe*, Vol. 1, Springer-Verlag, Berlin, 1987, 14–34.

[29] F.U. Rosenberger, C.E. Molnar, T.J. Chaney, T.-P. Fang, Q-Modules: Internally Clocked Delay-Insensitive Modules, *IEEE Trans. on Comp.*, C-37, No. 9, 1988, pp. 1005–1018.

[30] J.P.L. Segers, *The Design and Analysis of Asynchronous Up-Down Counters*, Research Report CS-93-29, Computer Science Department, University of Waterloo, 1993.

[31] J. Sparsø and J. Staunstrup, Design and Performance Analysis of Delay-Insensitive Multi-Ring Structures, in: T.N. Mudge et al. eds., *26th Annual Hawaii International Conference on System Sciences*, Vol. 1, IEEE Computer Society Press, 1993, 349–358.

[32] Jan L.A. van de Snepscheut, *Trace Theory and VLSI Design*, Lecture Notes in Computer Science 200, Springer-Verlag, 1985.

[33] I.E. Sutherland, Micropipelines, *Communications of the ACM*, **32** (6), 1989, 720–738.

[34] Alan M. Turing, Lecture to the London Mathematical Society on 20 February 1947. In: Carpentar BE, Doran RW eds., *Charles Babbage Institute Reprint Series for the History of Computing*, vol. 10, MIT Press, Cambridge, Massachusetts, 1986.

[35] J.T. Udding, A Formal Model for Defining and Classifying Delay-Insensitive Circuits and Systems, *Distributed Computing*, **1** (4), 1986, 197–204.

[36] N.H.E. Weste and K. Eshragian, *Principles of CMOS VLSI Design*, Addison-Wesley, 1985.

[37] T.E. Williams, Analyzing and Improving Latency and Throughput in Self-Timed Rings and Pipelines, *Proceedings of 1992 ACM/SIGDA Workshop on Timing Issues in the Specification and Synthesis of Digital Systems*, Princeton University, March 1992.

Synthesizing Asynchronous Circuits: Practice and Experience

Al Davis

Department of Computer Science, University of Utah
Salt Lake City, UT 84112, USA

Abstract

Asynchronous circuit design has been a topic of study for over 40 years. Only recently has this study turned to the practical issues of building useful CAD tool support that approaches the level of functionality available in synchronous VLSI CAD systems. The existence of these asynchronous CAD systems enables today's designers to construct asynchronous VLSI devices of significant size and scope. This chapter provides a reasonably detailed description of the process of developing a set of VLSI CAD tools which permit correct controller circuits to be synthesized from finite state machine style specifications. The tools were developed in two stages. The first stage resulted in a set of tools called **MEAT**, which were used to implement a 300,000 transistor multiprocessor routing chip called the Post Office. This *Post Office experience* provided the motivation for the second stage which resulted in the **STETSON** tool set which is integrated with the Mentor VLSI tool suite and was used to synthesize a low-power infrared transceiver chip called ABCS. The chapter presents a somewhat chronological view of the development of the CAD methods and provides examples of how well these methods worked in a practical design setting.

1 Historical Motivation and Introduction

Today's digital designers are faced with a large array of stringent constraints. Design cycle times are shortening at an alarming rate. Clearly this is a side-effect of an intensely competitive marketplace but the reality is that today's designers spend their time on significantly different activities than they did a decade ago. It is often the case that the *time budget* to go from the initial design concept to working silicon can be as short as six months even for chips containing hundreds of thousands of transistors. Even multi-million transistor processor chips must be operational in slightly over a year from the time the architecture is frozen. Additional design requirements such as aggressive performance, area, and power goals exacerbate the difficulty.

The ability of the digital design field to keep up with these growing pressures has relied heavily on the capability of automated CAD tools which remove much of the low-level drudgery of the design processes in existence even a few years ago. With very few exceptions, there is simply not enough time to worry about every small detail of the design and perform hand optimized custom layouts for the implementation. The result is that designers must rely heavily on synthesis tools to automatically generate layouts from some high level hardware description such as a Verilog or VHDL program. Designs are often specified initially at the *behavioral* level to permit validation of the design's functional behavior. Once the functional behavior is considered correct then the design is further refined into a structural specification which describes the implementation details. At this point, the structural code can then be compiled into the actual implementation such as the geometry or layout in the case of a single VLSI device.

Today's synchronous designers have a large number of CAD options to choose from. Companies such as Mentor, Cadence, Synopsis, etc. vend commercial CAD tools which integrate a fairly large number of tool options, many produced by third parties. On the other hand, the asynchronous circuit design community has had very little in the way of acceptable CAD tools to support their design activities. This is in part a side effect of a primarily academic and research oriented discipline. There are also numerous asynchronous design styles. Many of these were mentioned in Chapter 1, and the other chapters in this book provide a more in-depth look at several of them. Certain asynchronous design styles do permit use of *some* of the existing synchronous CAD tools.

In particular the data path components of certain asynchronous designs and their synchronous counterparts are not significantly different. The biggest difference is that the timing of synchronous circuits is based on a periodic global clock whereas asynchronous circuits rely on a particular set of *signaling protocols* to provide appropriate control for a sequence of events. While this difference is significant for controller circuits, the difference for datapath circuits in some design styles is relatively minor. The use of *model* delays to provide asynchronous completion signals is similar to the synchronous circuit controller waiting for the specific number of clock cycles that corresponds to the completion delay of the enabled datapath component. The completion time of a synchronous datapath component is determined by performing a timing analysis on the final implementation and then picking the smallest number of clock cycles which is still greater than or equal to the modeled delay. In the asynchronous case, the request is passed through a delay element (often just a string of inverters) and then passed back to the calling module as the acknowledge.

This view that asynchronous controllers are fundamentally different but that datapath circuits are similar to synchronous circuits led to a design style has been used since 1970 when the author was first exposed to asynchronous circuit design through a series of courses taught by Chuck Seitz at the University of Utah. The initial methods were based on those described by Unger [45]. Since that time, they have been used in the design of a number of large asynchronous computer systems such as DDM1 [9], several asynchronous VLSI devices such as the ISM [7], as well as a number of synchronous, or mixed synchronous and asynchronous designs. Given this background, it is clear why the design style which subsequently evolved and the tools which support this style is similar to the synchronous design domain.

Many of today's asynchronous practitioners will argue vehemently that asynchronous circuit design is fundamentally different from synchronous design and hence it is somewhat silly to try and exploit any commonality between the two regimes. The work presented here argues the opposite point of view. Namely, if asynchronous designs are to make a practical impact, they must be capable of co-existing with synchronous designs. Furthermore if asynchronous methods are to ever be considered as a mainstream option, then synchronous designers must be educated to the point where they can exercise this option. The claim is often made in asynchronous circuit workshops and symposia that the best way to move asynchronous designs into the mainstream is to educate enough untainted fresh minds that in time they will take over. The perspective here views this as extremely unrealistic. This approach assumes that by exploiting the methodological and CAD commonalities, it is possible to expand the designer's options to the point where asynchronous components become a reality in future commercial designs rather than being either a rare commercial anomaly or merely a research topic.

Prior to 1989, the design at hand was the primary focus and no effort was spent on creating appropriate CAD support. In 1987, a group of people at HP Laboratories started the design of a scalable parallel processing system called Mayfly [10]. The Mayfly communication fabric was based on an interconnected set of fully-adaptive router chips. The performance of this router chip, called the *Post Office* was clearly going to be critical to the performance of the overall system. Furthermore it was decided that in order to avoid a number of severe global clock management costs, the Post Office would be asynchronous. Ken Stevens was the main designer of the Post Office, but some of the Post Office subsystems were designed by Bill Coates. The other Mayfly project members (other than the author) were Robin Hodgson and Ian Robinson.

At some point, probably in the summer of 1989, it became clear that our complete lack of CAD tools to support the Post Office design was becoming a nightmare for two reasons:

1. Some controller designs required so many states and had so many inputs that it was improbable that even the most conscientious hand design would be correct.

2. The rate at which the Post Office design was progressing was so slow that it was clear that we would not meet the project schedule.

The initial call for help came from Ken Stevens. In what turned out to be a rash claim, I responded with a comment to the effect that in just three weeks I could do a *quick and dirty* hack for an asynchronous logic minimizer that would get Ken out of the current bind. The Common Lisp code for the asynchronous minimizer, called *amin*, did indeed work in three weeks. The first practical test case was for a controller circuit that had been crafted by hand and was operational as a test chip. Amin ran for an entire weekend. After millions of garbage collections it was still only half way to producing a solution. We had just learned what all real CAD developers already knew, that algorithmic complexity is always important.

This somewhat humorous beginning led to a much more serious effort to develop a working tool, called MEAT, that was used extensively in the design of the Post Office chip. MEAT worked well, but not well enough. This led to a broader and more serious effort that involved numerous people from HP and from Stanford University and culminated in a second more highly evolved tool for asynchronous controller design called STETSON. MEAT was a stand alone tool which still required that the designer do the layout of the physical implementation manually. STETSON used many of the MEAT ideas, in particular it was based on a restricted multiple input change, finite state machine specification which we named *burst-mode*. A significant difference is that STETSON is not a stand alone tool. STETSON is integrated into the commercial CAD suite provided by Mentor Graphics and generates its implementation based on a library of standard cells.

The remainder of this document describes the development of the MEAT and STETSON CAD methods and discusses how well the tools worked in practical system design.

2 The Post Office

The Post Office was designed to support inter-node communication for the Mayfly parallel processing system [8]. Mayfly processing elements (PEs) communicate with each other via messages. Messages are inherently variable in length, but the physical transport of those messages is organized as a sequence of fixed length packets. The role of the Post Office is to handle all of the physical delivery aspects of packet communication. The Mayfly topology was designed to be extensible and permits an unbounded number of PEs to be interconnected. This implies that the *physical extent* of a Mayfly system is not fixed. The unknown physical extent of the system poses serious problems when considering the implied constraints on an implementation strategy which uses a common global clock. Clock skew is a possible headache for any synchronous design style and the problem is magnified as technology progresses [1]. In the case of extensible systems such as Mayfly, where the total number of boards is unbounded, the synchronous choice becomes totally intractable. We therefore chose an asynchronous design style for the Post Office implementation.

Another critical design constraint was the need for a high performance implementation. Message passing performance would clearly be critical to the success of the Mayfly system. Many proponents argue that asynchronous circuits are inherently faster since they are controlled by locally adaptive timing rather than the usual global worst-case clock frequency constraints. While we believe that this claim has merit, we feel that in general it is misleading. Asynchronous circuits require more components to implement the same function. This may result in longer wires, increased area, and reduced performance. When compared to a very well tuned synchronous design, a functionally equivalent asynchronous implementation may actually run slightly slower. The need for speed heavily influenced our particular asynchronous design style.

There are a large number of rather different design styles used by the asynchronous design community. One partition of design styles can be based on the type of asynchronous circuit target: locally clocked [34, 17, 9], delay-insensitive [25, 3, 46, 29], or various forms of *single-* and *multiple-* input change circuits [45]. Yet another distinction could be made on the nature of the control specification: graph based [35, 28, 48, 5], programming language based [25, 3, 46, 2], or finite state machine based [34, 17]. For the finite state machine based styles, there is a further distinction that can be made based on the method by which state variables are assigned [19, 43]. The design style space is large and each design style has its own set of merits and demerits. It is worthwhile to note that virtually all of the design styles focus on the design of the control path of the circuit.

All asynchronous design styles are fundamentally concerned with the synthesis of hazard free circuits. The methods which produce delay-insensitive circuits, while not perfect [26], are the most tolerant of variations in device and wire delays. This tolerance improves the probability that a properly designed circuit will continue to function under variations in supply voltage, temperature, and process parameters. We chose to slightly expand the domain of timing assumptions which must remain valid to retain hazard free implementation since this permits higher performance implementations at the expense of reduced operational tolerance. We also chose to pursue a finite state machine based style for two reasons: 1) the finite state machine concept is a familiar one for hardware designers like ourselves, and 2) the graph and programming language based models that we knew about were too slow for our purposes.

Compiled implementations based on programming language like specifications [46, 2], while elegant and robust, suffer in performance because they are presently compiled into intermediate library modules rather than into optimized transistor networks. The module of greatest concern is the C-element. C-elements are common circuit modules in asynchronous circuits and eliminating them completely is unlikely. However it has been our experience over the past decade that C-elements are similar to the proverbial GOTO statements in programming languages, i.e. too many of them are indications of serious performance problems. C-elements are latches and as such are synchronization points. Too much synchronization reduces parallelism and hence performance degrades. The finite state machine based design style does not use C-elements, although C-elements are used sparingly in interface circuits such as arbiters.

The methods developed by Martin [25, 3] are an exception in that while the initial specification of the circuit is a CSP [18] based program, the compilation process is partly interactive and results in optimized pull-up and pull-down circuits for CMOS implementation. In retrospect, we should have paid closer attention to these methods but our finite state machine bias made us somewhat blind to the capabilities of these methods since the initial specification would have required a significant stylistic change. While it is only conjecture at this point, it might make sense to create a finite-state machine front end for Martin's tools and see how well the resulting circuits fare against those generated by MEAT and or STETSON. However as designers, we were familiar with finite state machine methods and decided to maintain this flavor in our initial specifications. Part of this bias was purely stylistic, but in terms of transferring the results of our work to other commercial designers and CAD tools, it was a justifiable choice.

In order to achieve the necessary hazard free asynchronous finite state machine (**AFSM**) implementation, it is necessary to place constraints on how their inputs are allowed to change. The most common is the *single input change* or SIC constraint [45]. SIC circuits inherently require state transitions after each input variable transition. In cases where the next interesting behavior is in response to multiple input changes, the circuit response will be artificially slow, either due to too many state transitions or due to the external arbiters required to sequence the multiple inputs. *Multiple input change* or MIC circuit design methods have been developed [45, 6] but either required input restrictions or involved implementation techniques that were unsuitable for our purposes. As a result we developed a design style that we call **burst-mode** which permits a certain style of multiple input change. Our burst-mode implementation method does not require performance inhibiting local clock generation or flip-flops.

Prior to the Post Office design, anytime the output of an AFSM controller was in response to multiple input variable changes, we used an AND gate with selectively inverted inputs to account for the desired high-to-low or low-to-high input trajectories of the individual members of the input burst. The result was a new set of *conjunctively convolved* input variables to the real AFSM. The result was that the AFSM only needed to operate under SIC constraints and hence traditional methods could be used. As this practice evolved, we noticed an increasing number of instances where the resulting circuit was suboptimal in terms of logic complexity. Since traditional methods allowed us to minimize the logic of the SIC AFSM controller but the conjunctive *burst-mode* rendezvous logic (the AND gates) was not part of this minimization process. This provided the initial motivation to investigate a direct burst-mode AFSM design style that permitted optimization of the entire AFSM design. Another benefit of this method is that it works for **both** two and four cycle signaling protocols.

As our Post Office state machines became too complex for hand synthesis, we decided to create a tool kit that was capable of automatically synthesizing the transistor level circuits from burst-mode specifications. We called this tool kit **MEAT** (described in section 5). During the development of MEAT, we were fortunate to have Steve Nowick spend two summers with us. He incorporated David Dill's verifier [12] into the tool kit, and modified the verifier to accommodate our burst-mode timing model. Steve had considerable influence on our ideas. His locally clocked design style [34] is another outcome of these earlier interactions.

3 Post Office Behavioral Description

The Post Office [39, 40] is an autonomous packet delivery subsystem designed to be a co-processor to a controlling CPU. The interconnect topology is a wrapped hexagonal mesh used to create what we call a **surface**. Each sur-

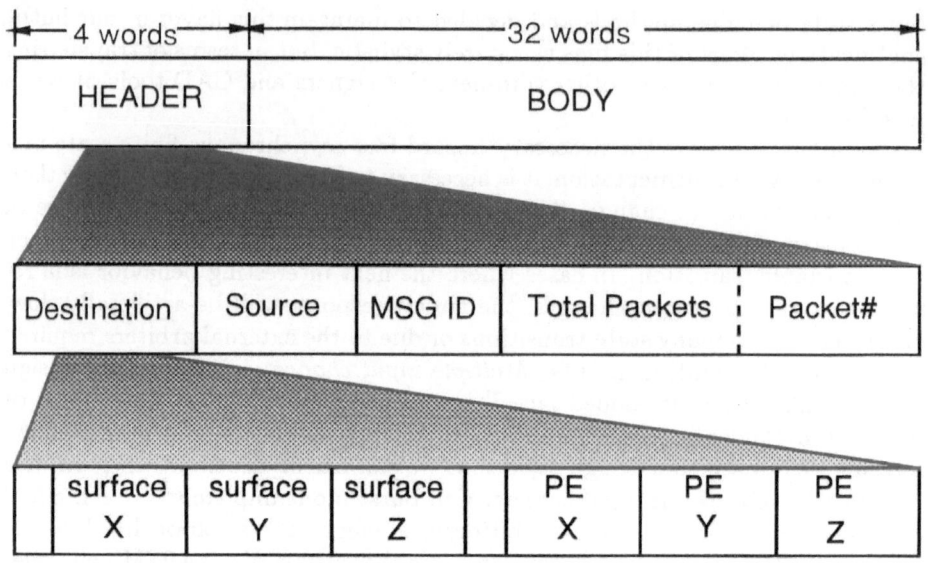

Figure 1: Packet Format

face has a hexagonal boundary and multiple surfaces can be interconnected by abutment. The Post Office is therefore a seven-ported device: six byte-wide external ports connect the Post Office to adjacent processing elements in the hexagonal mesh, and a 32-bit word-wide internal port provides a connection to the local CPU. The Post Office has been designed to permit all seven ports to be active concurrently.

The Post Office receives a packet from the CPU and delivers it over the communication network to a receiving Post Office, which in turn hands the packet to its CPU. Messages vary in length, but packet lengths are fixed at 36 words and have the format shown in Figure 1. The Post Office only uses the destination address field of the packet. The usage of the remaining 3 words of the header is dictated by Mayfly software convention. The Post Office contains an adaptive routing mechanism that can dynamically calculate the correct ports through which to route a packet. General communications efficiency can be significantly enhanced if the routing mechanism is capable of detecting congestion and routes packets around these congested areas. This mechanism also permits messages to be routed around nonfunctional nodes and ports.

Thus, the capability to dynamically route packets around congested or failed components implies that the order of packet arrival may vary from the order in which the packets were sent. The capability of reordering packets at their final destination is essential to ensure both deterministic behavior and to permit proper reassembly of multiple-packet messages by the destination CPU. This also implies that each packet contains a unique message identifier, the packet number, and the number of total packets in the message as shown in Figure 1. Although this slightly increases the amount of redundant information that must be transferred, it considerably reduces congestion related message delays. Congestion avoidance also increases the amount of packet traffic that can be in transit within the *postal* system before saturation effects are noticeable in terms of sharply increased packet latency times.

Messages are assumed to be delivered error-free in our current implementation and therefore no checksum or parity checking is performed. In order to avoid deadlock, a mechanism must be employed that either guarantees deadlock will not occur, or ensures that the probability of deadlock is much less than the probability of an unrecoverable component failure. The Post Office avoids physical deadlock by preventing unrestricted incremental resource claiming, thereby insuring that each Post Office will not be congested indefinitely. Note that livelock is still a possibility and there is no way that the Post Office can prevent a software deadlock situation where multiple user processes are all stalled waiting on each other. From the Post Office perspective, such a software deadlock situation looks exactly the same as the case when all PEs are busy but no messages are being sent.

For notational convenience, packets at a particular Post Office can be classified into 3 categories: transit, inbound, and outbound. Transit packets pass through a Post Office, inbound packets are destined for the local CPU, and outbound packets are created at the local node. The Post Office dynamically employs 4 types of routing algorithms: **virtual cut-through**, **best-path**, **no-farther**, and **random**. The algorithm choice depends on the specific congestion situation encountered by a particular packet. When a transit packet arrives at some external port, and the ideal external destination port of this Post Office is free, the packet will be forwarded immediately to the appropriate destination port. This method is called **virtual cut-through** [15]. The **best-path** method always takes the packet one step closer to its final destination. The **no-farther** routing neither increases nor decreases the distance of the packet from its final destination but allows it to orbit intervening congestion. The **random** method sends the packet out the first available port. In this case congestion may have the current Post Office surrounded, and it may be necessary to go farther away from the destination in order to make progress.

Virtual cut-through is only available for transit packets. If the ideal forwarding port is not available then the packet is stored in a central packet buffer pool in the Post Office. Inbound and outbound packets are always stored in the buffer pool prior to being forwarded to the local CPU or out onto the postal network. When an attempt to route a buffered packet is made, the best-path method is used. If it fails a **stagnation counter** is incremented. When the stagnation count for a particular packet exceeds its threshold then no-farther routing is also an option. Failure again increments the stagnation count. If a second threshold is exceeded then random routing is also an option. In any case where multiple available ports are found under any routing option situation, best-path will be taken first, no-farther second, and random last.

The Post Office routing algorithm is based on a labeling of PEs in the interconnected surface topology shown in Figure 2. Each of the PEs is assigned a unique location number which is a 3-tuple that corresponds to an <x,y,z>

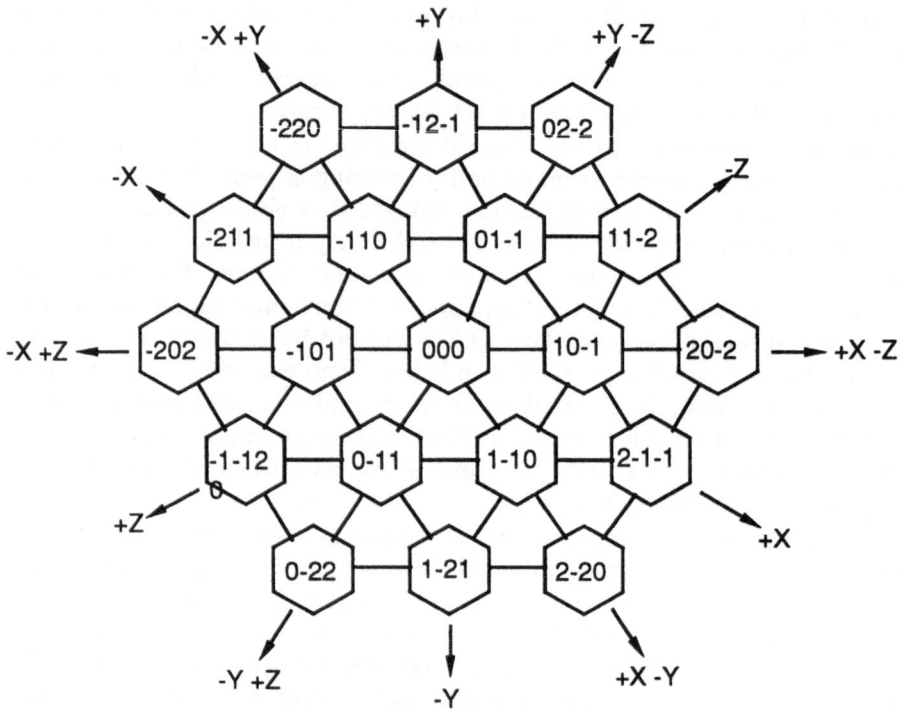

Figure 2: PE Labeling Convention

location in the surface. Note in Figure 2 that the X, Y, and Z axes correspond to the edges of the surface rather than the edges of the PE which are connected by the communication links. The reason for this choice is that it simplifies the routing algorithm. Each link traversed by a packet in either direction crosses 2 of the axes, one in the positive direction and another in the negative direction. Also note that the sum of the 3 location components is always equal to 0. This feature is used to detect illegal addresses. If a message is to be sent from the center PE to any other PE in the surface it is easy to understand how to make the decision as to which Post Office port should be used. For example, suppose a message is being sent from <0,0,0> to <1,-2,1>. The relative difference is <1,-2,1>. Since the y component is the largest and the value is negative, the **best-path** must be to send the message out either port that heads in the $-Y$ direction. The **no-farther** paths are the two paths which do not cross the Y axis.

The proper routing decision is much less obvious when the source PE is at the periphery of the surface, since a wrap link is a possibility. E.g. the $-X, +Y$ port of PE <-1,2,-1> is wrapped to the $+X, -Y$ port of node <1,-2,1>. The relative difference between these two paths is <-2,4,-2> which does not correspond to the possible single $-X, +Y$ move. One option would be to use a table based routing scheme; however the table size is **quadratically dependent** on the number of PEs and would be prohibitive. Therefore the Post Office computes the routing decision using a method described below. The result is a Post Office element which is fast, general, and scalable. The only configuration specific parameter that must be loaded into a Post Office register is a single integer n indicating a surface of size E–n, where n is the number of PEs on the surface edge. Figure 2 shows an E–3 surface.

The Post Office routing method removes the potential complexity imposed by the wrap lines by normalizing addresses to *center-based* coordinates. The method is:

1. Subtract the current packet location (the local PE label) whose value is stored at initialization time in a Post Office control register from the destination address. The result is an **unnormalized relative address**. For an E–n surface, a wrap line will be needed if one of the components of the resultant value is greater than n.

2. In order to convert the unnormalized relative address to a **normalized** form, which makes the current PE appear to be in position $<0,0,0>$, it is necessary to use a normalization vector. This vector has the value $<2n\text{-}1, n\text{-}1, n>$. For an E–3 surface, this value is $<5,2,3>$.

3. This normalization vector must then be aligned with the unnormalized relative address. The relative positions in the normalization vector must be preserved in the alignment, so a right end-around shift is used. There are two alignment conditions:

   (a) The 2n-1 component in the normalization vector is aligned with the component of largest magnitude in the relative address when no component of that address is zero.

   (b) Otherwise the n-1 component is aligned with the zero component in the unnormalized address.

4. The sign of the 2n-1 term of the normalization vector is set to match the sign of its aligned component. The signs of the n-1 and n components are the negation of the 2n-1 component's sign.

5. The normalization vector is subtracted from the unnormalized relative address to yield the normalized relative address.

6. At this point the routing decision is simple. The **best-paths** must reduce the largest magnitude component of the normalized relative address since this magnitude indicates the present distance of the packet to its final destination. In cases where there are no zeros in the normalized relative address and when there are two possible ports, one of which does not result in a zero relative address component at the next stage, the selection is made which does not result in a zero relative address component. A zero component in the normalized relative address indicates that there is only one **best-path** route. The goal is to keep multiple **best paths** available whenever possible.

114

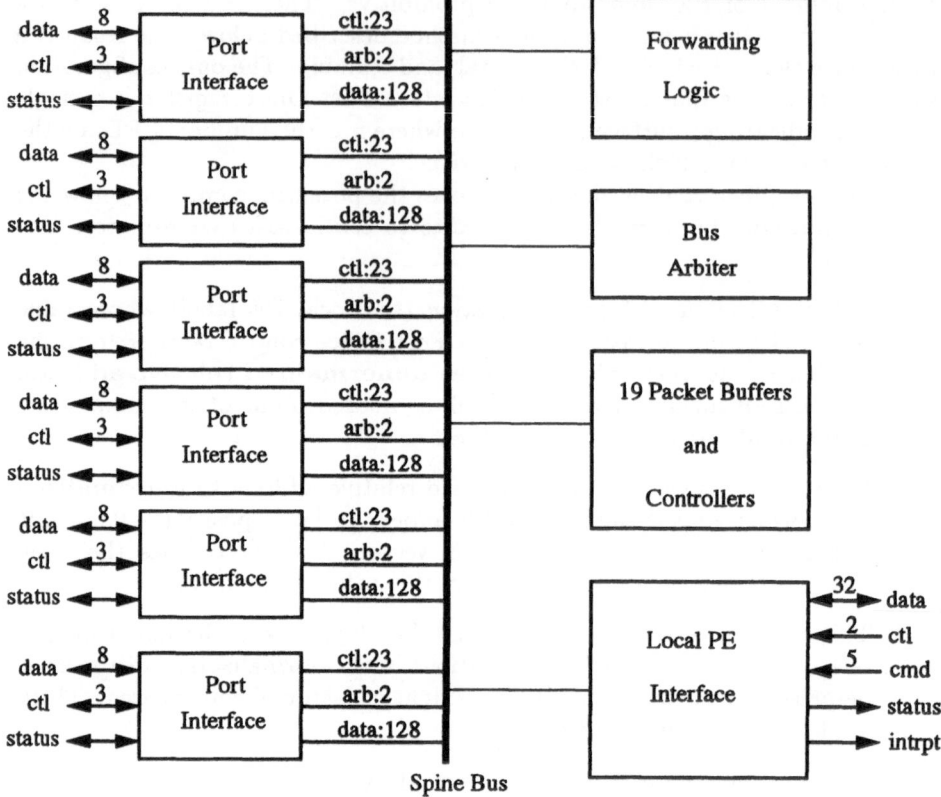

Figure 3: Post Office Block Diagram

4 High-Level Implementation

4.1 Functional Blocks

The Post Office consists of a set of asynchronous control logic, communication links, and datapath circuitry. Figure 3 shows the major logic blocks in the Post Office. There are five major components; the external port interface, adaptive routing and forwarding, bus arbitration block, central buffer controllers, and the internal PE interface. Each logic block is an independent agent, and can execute concurrently with the other components. The static buffer controller consists of 19 independent controllers, each with an associated packet buffer, seven of which can execute concurrently.

Submodules in the Post Office communicate via the *spine bus* using a four-cycle hand-shaking protocol. Each module must obtain mastership of the spine bus before its transaction can occur. There are 29 logic blocks which can obtain bus mastership. The Bus Arbiter block centrally services bus mastership requests. Whenever any module requires bus operations, it will assert its *Bus-Request* signal. When the requesting module receives the *Bus-Acknowledge* signal, it may drive the control and data signals on the bus. The arbitration logic is centralized as it results in faster servicing of bus requests when compared to distributed schemes. However, centralization requires a significantly higher wiring complexity.

Once a logic block has mastership of the bus, it may execute one or more transactions. The spine bus signals include three self-timed control lines ($\overline{Req}$, $\overline{Ack}$, and $\overline{Nak}$), a three-bit opcode, an eight-bit address, a three-bit sender address, a six-bit routing result addresses, and 128 bits of packet data. Typical transactions on the spine bus are to allocate a port for packet delivery, or transmit a 128 bit **line** of packet data between two blocks.

The CPU communicates with the Post Office via a 32-bit addressable register interface. The Post Office can run in either a polling or interrupt mode for either arriving and/or outgoing packets. A set of control registers can mask and disable the interrupts. There are also FIFO registers for loading and unloading packets, plus several control, status, and debugging registers.

Upon power-up or initialization, the current surface size and processor address must be loaded into control registers before packets can be correctly routed to their destination. The CPU must also write the normalization vector into the routing register. Data is normally transferred as an entire 144 byte packet. The only exception to this is on the PE interface which permits the CPU to quit reading or writing early by issuing a *done* command. This provides additional software protocol flexibility and increased performance.

All registers and memory in the PE interface are static. Although the Mayfly architecture is designed to service message traffic quickly [11], an inbound or outbound packet can remain unread or partially loaded in the buffers indefinitely since the CPU's software protocol will determine the timing of packet reads and writes.

The *packet buffer controllers* and *forwarding logic* blocks cooperate to centrally buffer and forward packets between the external ports. The forwarding logic is responsible for finding a free port for packet delivery using adaptive best-path, no-farther, and random routing. The buffer controllers snoop spine bus transactions between the forwarding logic and external ports to determine when to actively deliver and latch a packet to/from a port. Packet buffer storage requests are verified by the forwarding logic block. If there are insufficient buffers to guarantee deadlock free operation, the request will be rejected. Otherwise, the best-path address for the packet will be latched and a free packet buffer address will be returned to the requester.

Reliability requires that at least one Post Office node contain a valid copy of every packet being delivered. The PE interface loads each outbound packet into a central packet buffer. The source Post Office is responsible for keeping a valid copy of the message until it has been accepted and placed into the packet buffers of another Post Office (which could be several hops away due to cut-through). The first buffer location can then be freed for storing another packet. Centrally buffering outbound packets also increases bandwidth utilization of the external ports as packets are not forwarded until they are entirely received. The pipelined, cut-through transmissions can always proceed at full port-interface speed, independent of the PE interface logic. The speed of the PE interface may vary depending on the CPU speed, the amount of data to be transmitted, the drivers, or CPU state. The requirement that packets be stored in the buffers at the source and destination nodes increases the latency in lightly loaded networks. However, average latency will be much lower in cases where either local or global congestion is present.

The port interface controls transfers across an eight-bit bidirectional link connecting two Post Office chips. One interface becomes the sender, while the other becomes the receiver. Interface Hand-shaking uses three control wires and a two-cycle protocol. The port interface will receive a service request to forward a packet to the adjacent chip. If the interface is busy, either sending or receiving a packet, the delivery request will be rejected. When a delivery request is made to an idle interface, the interface must arbitrate for control of the external link. If this interface wins the arbitration, then the delivery is accepted and this interface becomes the sender and must reject the pending delivery request.

The first four bytes loaded by a receiving port interface contain the destination address of the packet. After latching the address the routing logic calculates the delivery port(s). The packet can then be immediately forwarded using virtual cut-through to the destination port, or to the packet buffers if this node is the final destination. When cut-through forwarding through the shortest path(s) cannot occur, the interface will attempt to store the packet in the packet buffers. If there is no room in the buffers due to deadlock prevention, the delivery to this node will be rejected.

4.2   Design Style and Circuit Elements

The Post Office consists of four types of components: finite state machines, RAM, arbiters, and some datapath logic such as subtracters, comparators, counters, registers, etc. All of these components were designed as self-timed modules, which were interconnected to form the larger elements in a hierarchical fashion. The largest self-timed element is the entire Post Office chip.

All state machines were designed from burst-mode AFSM descriptions. We assume arbitrary logic delay and insignificant wire delay. The environment is assumed to enforce burst-mode operation where all outputs will be driven before the next input burst arrives. The speed-independent model includes the isochronous fork assumption [47], and also applies to the verifier used to check our designs. Filtering or *triggering* of slowly rising input signals is used to ensure that this assumption doesn't create problems within state machines due to logic threshold variances.

Performance can be greatly enhanced if asynchronous *library* components such as C-elements, toggles, selectors, etc. are not used. The physical operation of sets of library components can usually be collapsed into asynchronous state machines, which are both smaller and faster than the library components. These AFSMs exhibit concurrency when designed using burst-mode. However, AFSMs cannot be designed where there are nondeterministic races between input signals. Hence there are only two library components which are required in our AFSM based design; the mutual exclusion or (ME) element and the C-element. Arbiters used in the Post Office consist of one or more ME and a finite state machine.

4.3 Datapath Logic

Data is transferred between elements using a two- and four-cycle self-timed bundled-data protocol [41]. This is a weaker model than the speed-independent model used for control signals inside state machines [27]. Also, some datapath circuits are controlled in a *clocked* domain, rather than in a self-timed fashion. These datapath components either contain a stoppable clock or are clocked by state machine outputs. The stoppable or raw clock signals are usually burst-mode generated in parallel with an associated set of asynchronous hand-shake signals. The minimum delay of these hand-shake signals must be greater than the maximum delay required by the clocked circuitry.

An example can be found in the port interfaces. The number of bytes which have been transmitted is recorded by a shift register. A state machine is signaled when the last byte has been transmitted, indicating the completion of the data delivery phase. The counter consists of an eight transistor dynamic shift register (plus one transistor for initialization). It is clocked by a state machine that latches and enables the packet data onto the external bus. The easily verifiable timing assumption is that the shift register *done* signal will arrive before the completion of the latch/enable cycle.

Making timing assumptions for datapath logic has two consequences. First, the logic is smaller and simpler because no completion signals are generated. Second, there is an increased potential for errors, as correct operation is now dependent upon physical circuit layout, process, and environmental parameters. Hence this is rarely used except for easily verifiable datapath components where completion signal generation is too costly in terms of speed or area.

Each RAM block is the size of a packet, 1152 bits. They are configured in 128 × 9 arrays, so it is fairly efficient to detect when a word line has been driven. Data validity and precharge is sensed on a single bit line pair, with the assumption that all bit lines will exhibit similar delay.

Many resources in the Post Office are accessible by other components. Accesses to these resources are made in a nondeterministic but mutually exclusive fashion. For example, two port interfaces may attempt to transfer data along the spine bus. Also, port interfaces A and B may both request forwarding a packet through port interface C. These two cases require different arbitration circuits. In the first case, the two accesses will be serialized on the spine bus; the loser of the arbitration will be blocked until the resource is freed. In the second case, the accesses should *not* be serialized! The winner will transfer its packet through port C, but the loser should not be blocked until the winner's transfer completes. Rather, it should abort the operation and be allowed to move on to other operations. Standard arbitration will serialize accesses for the first example, but a *naking* or *nonblocking arbiter* is required for the operation

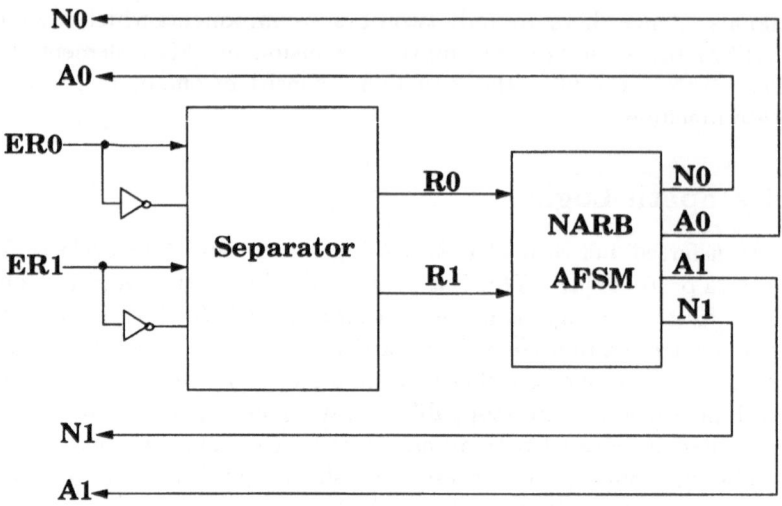

Figure 4: Nacking Arbiter Block Diagram

specified in the second example.

There are 13 naking arbiters in the Post Office. We found nonblocking arbiters to be a simple yet intriguing asynchronous circuit and posted it as a design and implementation exercise to our peers who are designing asynchronous tools and circuits. Our solution consists of a "separator" (which is built out of ME elements and a few NAND gates) and a state machine synthesized by MEAT. The block diagram of this circuit is shown in Figure 4. The circuit for the separator is shown in Figure 5 and the state machine is shown in Figure 6.

5 MEAT

The MEAT tools are sufficiently fast that alternative design options can be explored. The designer is freed from the task of understanding the underlying transformations required to produce hazard-free asynchronous circuits. Asynchronous circuits are specified for MEAT as burst-mode Mealy state machines. This style of specification provides a powerful way of encapsulating concurrency, communication, and synchronization in an accurate and easily understood form. The input specification is compiled into a set of CMOS complex gates. The result is an implementation which is efficient in both speed and area.

State flow diagrams are used to model the behavior of state machines implemented using MEAT. They provide an intuitive method for defining control functionality, and are similar to the flow charts and state diagrams that are commonly taught in multiple disciplines today [13]. A finite state machine is modeled as a directed graph, where the nodes represent states and arcs represent transitions between states. Each arc is labeled with the set of input firings

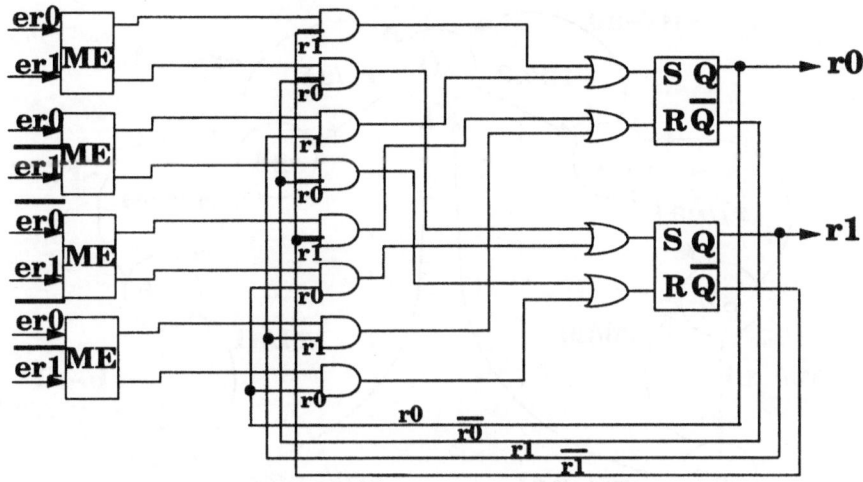

Figure 5: Nacking Arbiter Separator

which trigger the transition and an associated set of output firings. These state diagrams can easily represent parallelism and synchronization, and are reasonably compact when compared to other graphical specification methods.

MEAT state diagrams allow a constrained form of MIC operation, which we refer to as **burst-mode**. When a state change is triggered by a conjunction of input signal transitions (an input burst), these signals are allowed to change in any order and at any time. Allowing MIC operation simplifies the definition of synchronization operations and tends to more closely match the designer's mental model of the hardware. Presently MEAT does not contain a state graph editor so a textual specification format is used. The more natural graphical state machine description may be trivially mapped to the textual version: each arc in the state diagram is mapped to a single statement in the text file, which indicates the source and destination states along with the associated input and output bursts.

Burst-mode state diagrams are reasonably compact when compared to petri-nets, m-nets, STG's, and other graphical representations. These diagrams work well for transition (2 cycle) or level-mode (4 cycle) signaling protocols. Figure 7 shows an example of an STG (a), enhanced STG (b), and burst-mode state diagram (c) for an asynchronous D flip-flop. In this paper we assume positive logic, hence $a\uparrow$ corresponds to a high transition on signal a. In the textual version $a\uparrow$ is represented simply as a and $a\downarrow$ as $a\sim$. The corresponding textual entry version for MEAT is:

```
:fsm Asynch-Flip-Flop      ;name of FSM for documentation.
:in  (D Clk)               ;list of input variables.
:out (Q)                   ;list of output variables.
:init-in  ()               ;initial value of inputs, default zero.
:init-out ()               ;initial value of outputs, default zero.
:init-state 0              ;initial state, default is State 0.
```

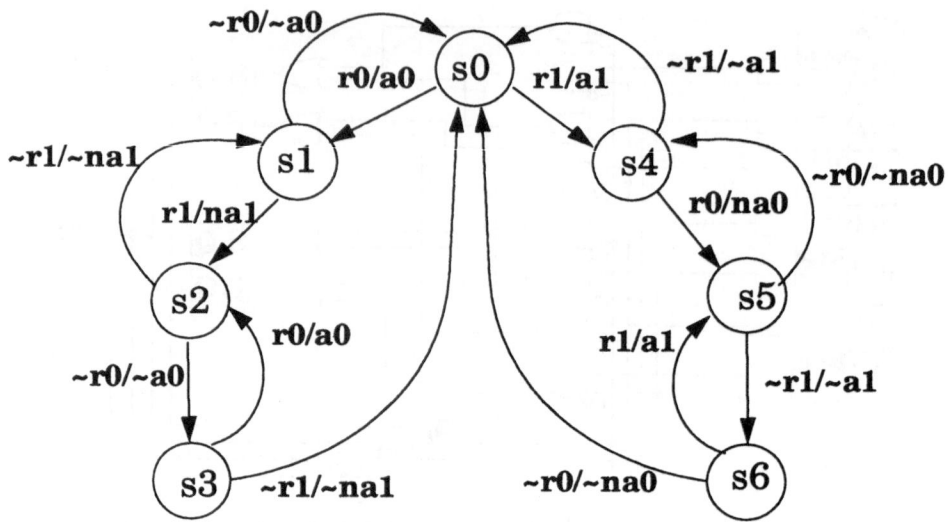

Figure 6: Nacking Arbiter State Machine

```
:state 0 (Clk~) 0 ()          ;specification of state transitions:
:state 0 (D * Clk) 1 (Q)      ;  format is <current state> <input burst>
:state 1 (Clk~) 1 ()          ;                <next state> <output burst>
:state 1 (D~ * Clk) 0 (Q~)
```

The first automated task performed by MEAT is to generate a primitive flow table [45] from the textual FSM specification. This is a two-dimensional array structure which captures the behavior represented by the state diagram in a more detailed form. Each row of this table represents a node in the state diagram; each column represents a unique combination of input signals. Each entry in the table thus represents a position in the possible state space of the FSM.

For each entry, the value of the output signals and the desired next state may be specified. If a next-state value is the same as that of the current row, the state machine is said to be in a *stable state*. If the next-state value specifies a different row, the table entry represents an *unstable state*. A simple way of understanding the flow table is to note that horizontal movement within a row represents changes in the values of input signals, while vertical movement within a column represents a *state transition*. All of our specifications are given in *normal form*, that is, each unstable entry in the table must lead directly to a stable state.

Each allowed input burst will result in a particular path through the FSM state-space, starting at the stable entry where the burst begins. Other entries in the same row may be visited during the course of the input burst. In order for MIC behavior to be correctly represented, it must be guaranteed that the circuit will remain stable in the initial row until the input burst is complete. This is an important point and is a cornerstone of the burst-mode methodology.

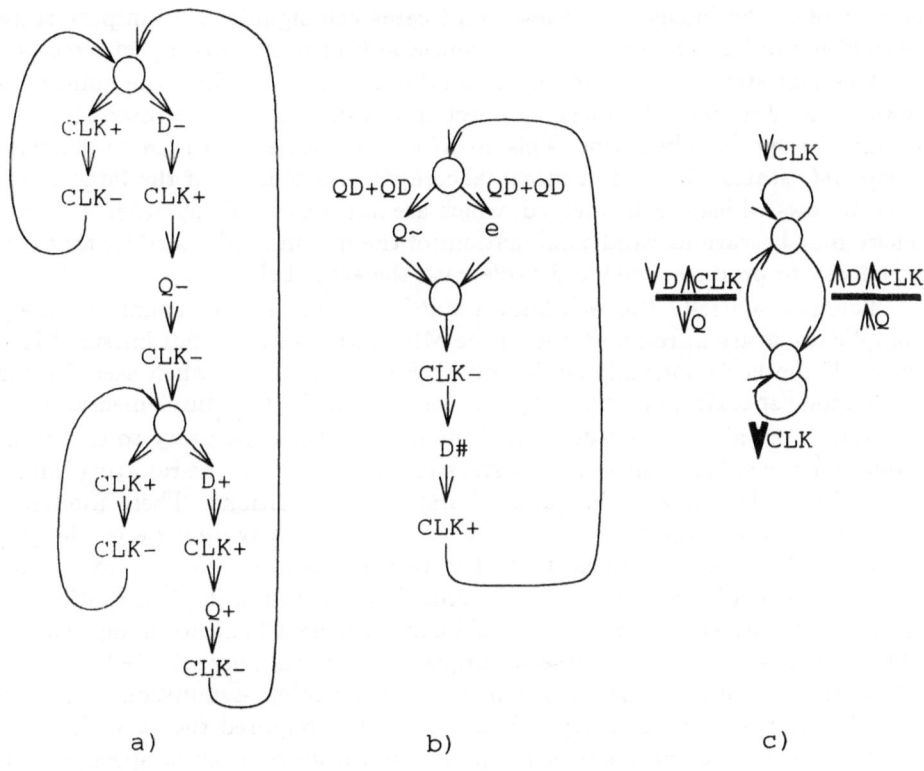

a) b) c)

Figure 7: Sample Flip-Flop Specifications

In essence, any minterm formed from input variables which can be reached during the course of an input burst must be *covered* by a stable entry in the flow table. The minterm defined by the completion of the burst will correspond to an unstable state which will cause a transition to the target row and fire the output burst.

The output burst, if any, may occur concurrently with the state change, or may be constrained to happen *after* the state change has occurred. To allow the flexibility for the later synthesis stages to choose either option, signals in the output burst are labeled as don't cares in the unstable exit state of the flow table. Since all state transitions are STT *Single Transition Time*, the monotonicity of output voltage changes is guaranteed, regardless of whether the value of a given output transition in an unstable entry is mapped to logic level zero or one.

Any entry in the flow table not reachable by any allowed sequence of input bursts is labeled as a *don't care* and can take on any value for the outputs or next-state values. As in the case of output bursts discussed above, it is not immediately evident which values will lead to the simplest circuit. Therefore, the assignment of specific values to the don't care entries is deferred for as long as possible. The inclusion of these don't cares can significantly simplify state reduction and boolean minimization, and also lead to more compact circuits.

The next step in the design process is to attempt to reduce the number of rows in the flow table by merging selected sets of two or more rows while retaining the specified behavior. This involves first calculating the set of *maximal compatible* states. The set of maximal compatibles consists of the largest sets of state rows which can be merged, which are not subsets of any other such set. There may be various valid combinations of the maximal compatibles that can be chosen to produce a reduced table with the same behavior.

This is essentially the well-known state-reduction problem; unfortunately complications are introduced due to the MIC nature of the input bursts. "Traditional" methods normally apply only to SIC circuits, and when used for our burst-mode specifications may produce hazards in the final implementation.

Nowick et. al. [31] have developed the modifications necessary to the state-reduction and subsequent synthesis steps to guarantee that the resulting implementation will be hazard-free under burst-mode conditions. These modifications are not presently incorporated into MEAT. Currently we use a verifier [12] on the synthesized implementation. The verifier has been modified to operate with explicit timing assumptions. Hazards detected in the implementation are then reviewed to see if the circuit would exhibit correct behavior under reasonable delay assumptions. If these assumptions fall within acceptable bounds of fabrication and operational constraints then the timing assumption is entered into the verifier. If an unacceptable assumption is required then the circuit is fixed either by manual repair or by modifying the state-machine specification. The manual repair usually involves the addition of an appropriate inverter chain to delay the critical path race.

The final choice of minimized states is an example of the *binate covering* problem. There are three constraints on this choice. First, and obviously, only compatible states may combined (*compatibility* constraint). Second, each state in the original design must be contained in at least one of the reduced states (*completeness* constraint). Third, selecting certain sets of states to be merged may imply that other states must also be merged (*closure* constraint). Grasselli and Luccio [16] have developed a tabular method for determining a closed cover of states, which is also in the process of being incorporated into MEAT. At present, MEAT requires the user to manually determine and enter a state covering. If any of the necessary constraints are not satisfied, MEAT will inform the user that the covering is invalid.

A new flow table representing the behavior of the minimized FSM is then generated by merging the specified rows of the original flow table. It should be noted that it is not always true that minimizing the number of states will simplify the hardware or increase performance. However, a reduced state machine can result in fewer state variables which in most cases does indeed result in a smaller and faster implementation.

A set of state variables must then be assigned to uniquely identify each row of the reduced flow table. These state variables are used as feedback signals in the final circuit. In contrast to synchronous control logic design, state codes may not be randomly assigned, but must be carefully chosen to prevent races. The MEAT state assignment algorithm is based on a method developed by Tracey [43]. The Tracey algorithm has the advantage that it produces STT state assignments which minimize delay in the implementation. In cases where two or more state variables must change value when transitioning to a new state, all variables involved are allowed to change concurrently, or *race*. However, it must be guaranteed that the outcome of the race is independent of the order in which the state variables actually transition in order to produce a *non-critical* race which exhibits the correct asynchronous operation. Several valid assignments may be produced, and each will be passed to the next stage for evaluation. Each state assignment will result in a unique implementation.

After state codes are assigned, the next synthesis stage computes a canonical sum of products boolean expression for each output and state variable. A modified Quine-McCluskey minimization algorithm is used. The resulting expression includes all essential prime implicants, and possibly other prime implicants and additional terms necessary to produce a covering free of logic hazards. It may be possible for each output or state variable to be specified using several other minimal equations. The large number of don't care entries typically present in the flow table causes the standard algorithm to be rather inefficient and increases the likelihood that more than one minimal expression will be found. The MEAT implementation contains optimizations for don't care dominant functions. Each possible solution is given a heuristic "weight" that indicates the expected speed and area cost of implementation using complex CMOS gates. When multiple state assignments have been produced in the previous step, the total weight of each unique SOP (sum of products) equation is then used to choose between the various instantiations.

The minimized equations produced in the previous step are then used to automatically produce transistor netlists, suitable for simulation, representing complex CMOS gates. An interface to the Electric [36] design system is used to generate a schematic diagram to help guide the layout process, which unfortunately has not yet been automated. The complementary nature of CMOS n-type and p-type devices is exploited to generate a single, complex, static gate through simple function preserving transformations. These transformations can increase performance while reducing the area and device count. As a SOP equation is *folded* into a complex gate, the number of logic levels required to generate the output can be reduced. If the function is too large to be implemented as a single module, it can easily be broken up into a tree of complex gates with 2 or more logic levels, but better overall performance [42]. Typical state machine implementations have response times between 3 and 5 2-input NAND gate delays.

Our complex gate design generates negative logic outputs (low voltage levels for asserted signals). A convention of positive logic levels is assumed for all signals external to the state machine, requiring that the outputs be inverted. This is a feature for performance reasons as the gain of the inverter can be used as a driver to increase signal strength and reduce rise and fall times. When outputs need to drive a large load, a buffer tree can be used.

All state machines also require a reset signal to place the storage logic into the correct initial state. Storage in these state machines is implemented via the state variables. If a single complex gate is used to generate the output, the state storage is reset by NOR-ing the output with the reset line. For complex gate trees, a resettable NAND gate is used. Although the performance of the NOR gate is not optimal, the load on the feedback lines is local to the state machine and typically small so a large gain is not required.

6 Design Issues and Examples

Figure 7 essentially shows how AFSM designs are specified using MEAT. Rather than presenting a series of more complex designs which will show roughly the same thing, we will present a number of design vignettes which illustrate interesting points in the design space, and an example of MEAT usage.

6.1 A Story about C-element Design

During a 1986 course on asynchronous circuits taught by Ivan Sutherland and Bob Sproull, the discussion turned to the design of the common C-element. At that time, the standard C-element consisted of a 2-high stack, followed by an inverter. This element had the problem that it was a *dynamic* gate. If the two inputs remained at different voltage levels for long enough the C-element's state would be lost and cause an invalid output transition. This was clearly unacceptable for general asynchronous applications. Static versions of the circuit were created by including a weak "trickle charge" inverter to maintain correct voltage on the internal node in the absence of it being directly driven by the 2-high stack.

The trickle charge inverter was a problem for several reasons. First, it reduced the performance of the circuit. When the internal node $\bar{c}$ (in Figure 8a) needed to be flipped to a different voltage, the trickle inverter would be actively driving the circuit one way, while the 2-high stack was actively driving it another way. The 2-high stack needed to charge the node, as well as dissipate the current supplied from the trickle inverter. This caused increased power consumption due to the existence of a DC path between the power rails during a state change. Secondly, the inherent gain of an inverter is greater than the gain of a 2-high stack. This design requires the 2-high stack to overpower the inverter to flip the state of the device. Unless the drive of the 2-high stack is significantly greater than the inverter, the node becomes susceptible to noise problems which could result in hazards. This gain difference can only be overcome by reducing the size of the inverter and increasing the size of the 2-high stack. Hence the sizing of the components becomes critical. Increasing the size of the 2-high stack slows the circuit by requiring additional input drive. Decreasing the width and increasing the length of the inverter reduces the reliability of the inverter and the portability to other processes.

After the day's discussion, we spent several hours attempting to come up with a better C-element design which eliminated the trickle inverter, yet did not

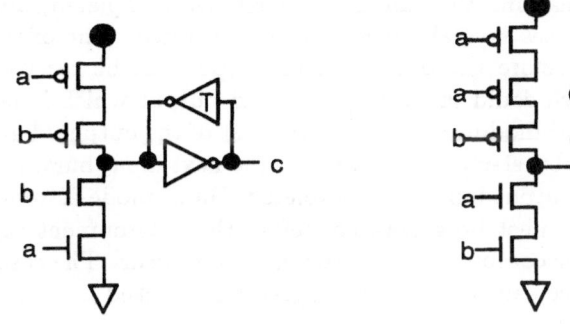

a) Trickle inverter C-element

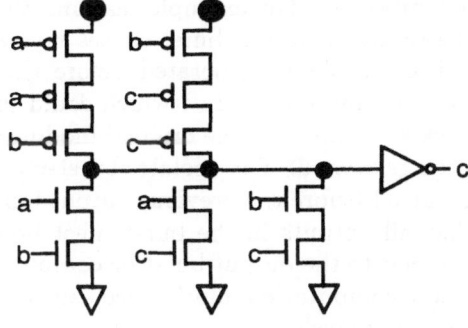

b) Complex gate for c = ab + ac + bc

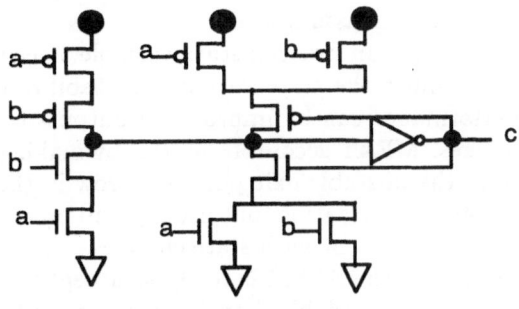

c) Optimized complex gate C-element

Figure 8: C-Elements, Hand Optimized Matched by MEAT

add significant complexity to the component. Ultimately a design was found which was compact and efficient. This design has been widely used in a number of sites. This design required 4 more transistors than the trickle charge design. However, the 2-high stack could be of optimally sized transistors and there was no fight to drive the internal node $\overline{c}$. Although this circuit was larger, and the inputs drive twice the number of devices, it was significantly faster than the original design and avoided the power consumption, noise, portability, and function problems of the old design.

Several years later, curiosity lead us to see what MEAT would produce for a C-element. The exact same circuit was produced from MEAT in an instant. MEAT generated equations for the circuit shown in Figure 8b and the back-end schematic generated the equivalent but optimized version shown in Figure 8c.

6.2 Using Burst-Mode to Increase Performance

Burst-mode assumes that inputs and outputs are generated as discrete sets, or bursts. In general, this violates delay-insensitive and speed-independent assumptions. For example, assume that an input burst has completed, and the resulting output burst causes several outputs to be generated. One of the outputs could be generated before the others. This output can be received by a destination module which could in turn generate an output which is fed back as an input to the original module even *before* the rest of the outputs have been generated. This violates burst-mode operation as the next input burst has occurred before the previous output burst has completed. Burst-mode assumes that all outputs in the burst must be generated before the environment can respond to the output burst or computation interference may occur. The cases where computation interference can occur can be flagged and checked by circuit timing analysis.

MEAT's burst-mode MIC model is similar to the fundamental mode assumptions for traditional SIC AFSM designs. Namely we assume that once an input burst has arrived the AFSM will settle in a stable state before the next burst can arrive. If this assumption cannot be met then external arbitration will be required to enforce the assumption.

If an input burst changes an internal state variable, speed-independent operation will generally require the state variable to stabilize before the output can be changed. Performance can be improved if outputs can change concurrently with state changes. MEAT accomplishes this by making the transitioning output a don't care in the unstable exit point of a row in the flow table. This places a priority on logic minimization, but usually will produce a circuit which can generate an output concurrent with state changes. The fundamental mode assumption guarantees that the AFSM is ready to accept the next input burst when it arrives, as the state variable transition has completed and the logic has stabilized. Unger has shown that it is possible to weaken this fundamental mode assumption [44], although his method is not presently incorporated into MEAT.

6.3 When Speed-Independent Circuits Fail: The Isochronous Fork

Ideally all asynchronous circuits should be designed as delay-insensitive modules. However, performance requirements may force one to make weakening assumptions about circuit behavior. Many of these assumptions are realistic, as physical devices and wires do not require unbounded delays to generate and propagate signals. However, care must be used to assure that the circuit complies to these assumptions under all operating conditions or the design will be unsafe and costly failures may occur.

Simplifying assumptions are best exploited when they are constrained to a fixed extent physical domain as is the case with AFSM modules. Hierarchical composition of these modules can then proceed conforming to delay-insensitive rules since all of the external interfaces can be designed to avoid timing assumptions. Inside an AFSM, the relative delay of wires and gates can be more easily controlled, analyzed, and modified as the constraints are all local. When these timing assumptions apply outside an individual module then the entire

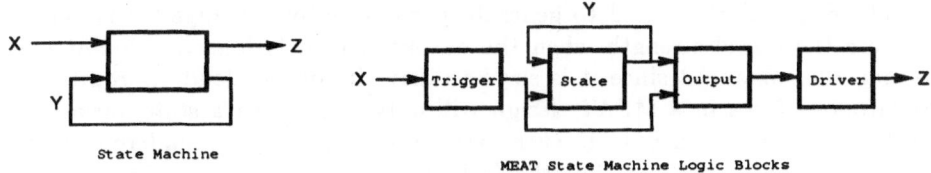

Figure 9: State Machine Generation

system must be analyzed to assure compliance with the timing assumption set. At this point there is little to distinguish the circuit from a synchronous one.

A common performance and synthesis assumption made by many asynchronous circuit designers is that of speed-independence. The assumption that wire delay is zero leads to the **isochronous fork assumption**. This implies that multiple devices driven by a single component react to the signal change at very nearly the same time. This model works well for situations where the transistors are slow and the paths are fast. Unfortunately this model becomes less valid as IC technology progresses and is suspect even today.

Furthermore, whenever the rise or fall time of an isochronous fork is greater than the switching delay of any physical device, failure may occur due to variances in switching thresholds. Noise, long wires, and high-capacitance paths exacerbate the problem. Within a particular AFSM module, this problem can be managed successfully but it is difficult between modules. Martin [24] and Van Berkel [47] have both described circuit failures due to paths which did not behave in an isochronous fashion. Both failures were the result of using C-elements in module interfaces. C-elements inherently contain an isochronous fork, since the output of the C-element will be an output of the module as well as being fed back locally to maintain the C-element's state.

The philosophy we have used in the MEAT tool and in the design of our circuits is to remove isochronous forks from external interfaces. MEAT state machines are broken into the partitions shown in Figure 9. Our philosophy is that we would gladly increase the cost and difficulty of designing modules if it simplifies the composition of systems. Timing assumptions are always easier to analyze and fix in a small, local cell rather than across a series of modules.

The trigger box has two functions. First, high capacitance inputs (inputs with a slow rise time) will be passed through an inverter or Schmitt trigger. This will reduce the load on the input line, which can increase circuit performance. It also results in crisp rise and fall times of signals internal to the AFSM. Secondly when an unasserted input signal is required by the state or output boxes, the trigger box will invert that signal. Each input will have its inverted and uninverted signal shared by all function blocks in the state machine to eliminate hazards and create a smaller implementation. The isochronous forks created by sharing the inverters are easily controlled within the local domain of the AFSM since components within a particular AFSM are physically close. Hence wire delays of the internal signals and the trigger box delay are normally insignificant.

The driver block is used to generate positive output voltage levels and to increase the signal strength when the output is heavily loaded. Circuit performance is enhanced since it is sized to drive its output load appropriately. Isochronous forks in a MEAT design will only exist when a state variable is used directly as an output. In such cases, the output can be buffered by one or two inverters to assure that the fork is isolated within the AFSM. While this decreases the performance of the circuit, the module can function in a delay-insensitive manner and can be used safely without analyzing its load in a broader context.

This design style has been tested continuously over the last five years. We have designed several large asynchronous circuits which have usually worked the first time, using simulators to verify correct composition of the modules. The result of this experience has led to a high confidence factor in the practical value of the method.

6.4 An AFSM example

In order to illustrate exactly what MEAT does, we will transcribe an actual synthesis run using MEAT to create a Post Office state machine called the SBUF-SEND-CTL. The behavior is initially specified as a burst-mode AFSM as shown in Figure 10. This example is taken from the suite of Post Office state machines publicly available for use by other researchers [21, 34].

The specification of sbuf-send-ctl from Figure 10 is textually entered for MEAT as follows:

```
:fsm sbuf-send-ctl
:in  (Deliver Begin-Send Ack-Send) ;list of input variables
:out (Latch-Addr IdleBAR Send-Pkt) ;list of output variables
:state  0 (Deliver)
        1 (IdleBAR * Latch-Addr)
:state  1 (Deliver~)
        2 ()
:state  2 (Begin-Send)
        3 (Latch-Addr~)
:state  3 (Begin-Send~)
        4 (Send-Pkt)
:state  4 (Ack-Send)
        5 (Send-Pkt~)
:state  5 (Ack-Send~)
        0 (IdleBAR~)
:state  4 (Deliver)
        6 ()
:state  6 (Deliver~ * Ack-Send)
        7 (Send-Pkt~ * Latch-Addr)
:state  7 (Ack-Send~)
        2 ()
```

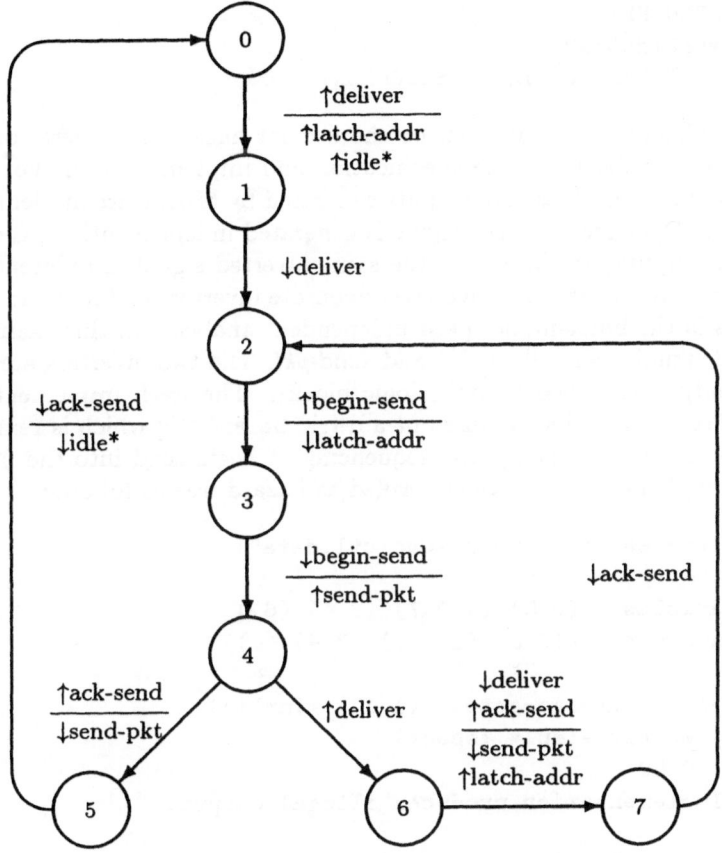

Figure 10: Sbuf-send-ctl State Machine

The following is a transcript from a MEAT session. The specification resulted in a single implementation with two state variables.

```
> (meat "sbuf-send-ctl.data")

Max Compatibles: ((0 5) (1 2 7) (3 4) (6))
Enter State set: '((0 5) (1 2 7) (3 4) (6))

SOP for "Y1":
  18: DELIVER + Y1*BEGIN-SEND~
SOP for "Y0":
  28: BEGIN-SEND + Y0*ACK-SEND~ + Y0*DELIVER
SOP for LATCH-ADDR:
  12: Y1*Y0~
SOP for IDLEBAR:
  30: ACK-SEND + BEGIN-SEND + Y0 + Y1
```

```
SOP for SEND-PKT:
  12: YO*BEGIN-SEND¯
HEURISTIC TOTAL FOR THIS ASSIGNMENT: 100
```

The implementation can then be checked for hazard-free operation by the verifier. The verifier reads the specification and implementation. For this example, the state variables and outputs generated by MEAT are implemented as two-level AND/OR logic. Each signal is generated independently of the others. Only direct inputs are shared, so the same inverted signal in different output logic blocks will use separate inverters. Separate inverters will result in verification errors in the burst-mode speed-independent analysis. In this example, the *begin-send* signal is shared by *Y1* and *send-pkt*. The two inverters are merged and the output is forked to both logic blocks. The fresh implementation is then checked. The verifier points out a d-trio hazard [45] which is removed by adding an inverter to change the sequencing of begin-send into the *Y0* logic. The final implementation is then verified as hazard free as follows:

```
> (verifier-read-fsm "sbuf-send-ctl.data")

Max Compatibles: ((0 5) (1 2 7) (3 4) (6))
Enter State set: '((0 5) (1 2 7) (3 4) (6))

> (setq *impl* (merge-gates '(1 11) *impl*))
> (verify-module *impl* *spec*)
10 20 30 40 50
Error:  Implementation produces illegal output.

> (setq *impl* (connect-inverter 10 6 *impl*))
> (verify-module *impl* *spec*)
10 20 30 40 50 60 70 79 states.
T
```

The canonical SOP equations generated by MEAT are then transformed into complex gates for implementation. The CMOS circuit for *Y0* is shown in Figure 11. The complex gates are then manually implemented using the Electric [36] layout editor. The physical layout is then simulated with COSMOS [4] to check for layout errors. Cooperating sets of state machine cells are interconnected to form larger modules, integrating clocked datapath logic when necessary.

6.5   D-Trio Hazards, Assumptions, and Possible Elimination

Figure 12 shows a static **d-trio** or nonessential function hazard which is found in some of the state machines produced by MEAT. D-trio hazards are fundamental and cannot be removed in every case, but they will be detected by the verifier In this cases the hazard occurs because the input burst resulted in an

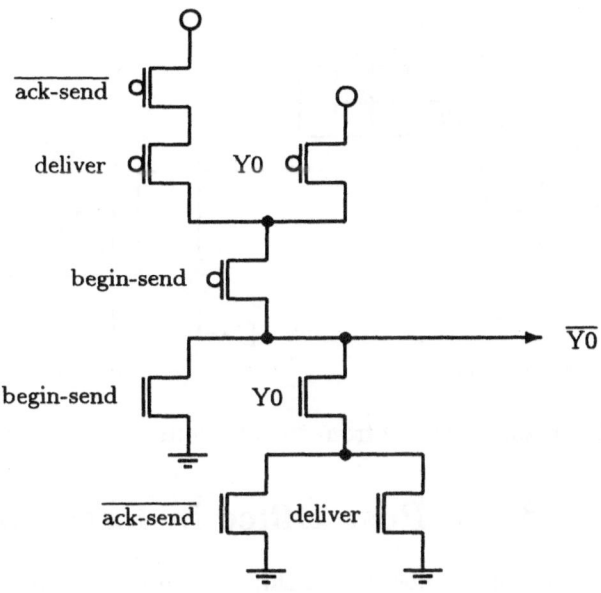

Figure 11: Complex CMOS Gate for sbuf-send-ctl Y0

internal state change while the output burst contained no transition for the *Done* signal. The d-trio hazard in this example can produce a static 1-hazard on the *Done* signal. The input burst is perceived by the *Done* output logic *after* the state change burst thereby creating the hazard.

The $W8$ signal of the logic with the d-trio also contains an isochronous fork. If we ignore the potential threshold deviations then timing analysis shows that the physical behavior will not exhibit the hazard. However, this circuit cannot be included in a system without analyzing the driver, load, and stray capacitance on the $W8$ input or errors will result.

By modifying the trigger logic in the Sendr-Done state machine shown in Figure 12, we can both eliminate the d-trio hazard and the external isochronous fork. This incurs *no* performance penalty. The $\overline{W8}$ signal to the *Done* logic remains delayed by a single inverter, while the $W8$ signal to the state logic becomes doubly inverted rather than fed directly into the logic from the input.

The double inversion has the effect enforcing correct *sequencing* of the order of arrival of the $\overline{W8}$ signal to the *Done* logic. Transitions on $\overline{W8}$ will always be perceived by the *Done* logic before changes in the state variable, resulting in hazard-free circuit operation. Transitions are ordered such that the assertion of the state variable is not critical to the performance of the circuit, so the double inversion of $W8$ into the state logic does not negatively impact performance.

132

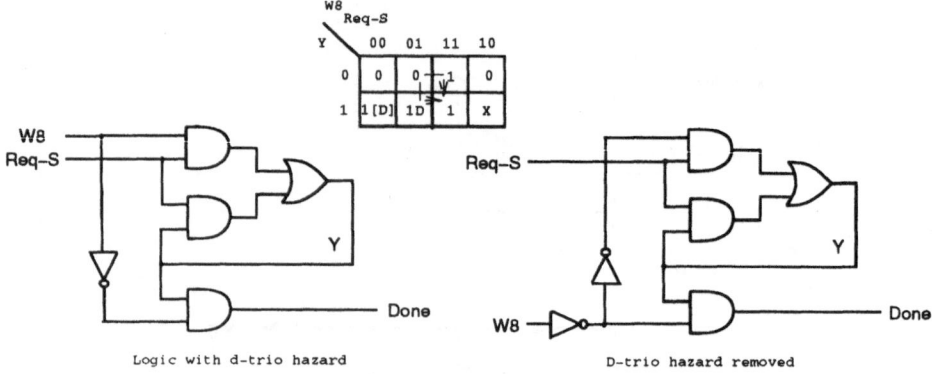

Figure 12: Hazard removal from "Sendr-Done" state machine

7 MEAT and the Post Office in Retrospect

The first silicon version of the Post Office chip was only partly operational. It was also approximately only half as fast as expected. As the causes for the performance and functional problems were uncovered several methodological errors became painfully clear. In particular our focus and CAD development (e.g. MEAT) had centered on controller circuits since the Post Office is a control intensive chip. There are 94 different AFSM's, many of which are replicated six times, once for each of the six external ports. To some extent the modular composability of asynchronous circuit modules had led us astray. Our focus should have included the overall design. The complex communicating AFSM control path of the Post Office required proper coordination of multiple AFSM controller modules. We discovered some functional errors in the first silicon that were the result of design errors involving more than 2 AFSM controllers. There was no problem with pairwise AFSM combinations. The difficulty came with the need to sustain a high level of internal concurrency involving multiple AFSM's.

The other problem was the somewhat cavalier attitude taken towards datapath circuits. Since MEAT could only generate optimized transistor schematics for controller circuits, all the datapath design and all the controller layout were done manually. This was a time consuming exercise and the resulting floor plan was the major culprit in terms of the factor of two *performance surprise.* In particular, the spine bus was so long that it became the major performance bottleneck in the chip. Since modification of the design to improve the floor plan would require significant manual layout time, and the project deadline was fast approaching, we had no option but to fix the functional problems and live with less performance.

While these problems were certainly real, there were several encouraging aspects of the design and the methodology. Since the Post Office design instigated the development of the MEAT tools, approximately a fifth of the Post Office control path design was done manually prior to the development of MEAT. The remaining four-fifths was implemented using the MEAT tools. The automated part of the design took one quarter of the total design time and was virtually error free. The nature of these errors are the basis for the subsequent sections of this chapter. Since it is based on traditional finite state machine control, the design style has proven to be a very natural transition for existing hardware designers. The MEAT synthesis techniques have generated compact high-performance circuits that work, and the complexity of the synthesis algorithms have proven to be tractable for large designs.

The present status of the Post Office is that second silicon is *adequately* operational. In testing the second silicon we were able to generate enough concurrency in the chip to create power starvation in the core which can cause failure. The cause of this problem was inadequate power distribution wire sizes. Power analysis had been performed for a 3 micron process and we had naively assumed that a fresh analysis was unnecessary when the process scaled to 1.2 microns. Fortunately a simple software modification prevents this extremely rare situation from occuring in the Mayfly systems. Packets can still flow concurrently on all seven ports of the Post Office. Each of the external ports permits a byte of information to be transmitted every 50 nanoseconds. Each packet is 144 bytes and hence takes 7.2 microseconds to transmit. The internal port is a 32-bit word-wide data path which is capable of sending or receiving a word every 50 nanoseconds. The current CPU that we use is only a 16 MHz device using a multiplexed address and data bus. The CPU cannot keep up with the Post Office and therefore actual packet loads and unloads take approximately 10 microseconds.

The Post Office was fabricated using MOSIS revision 6 design rules in a 1.2 micron CMOS process. The circuit contains 300,000 transistors and has an area of 11 × 8.3 mm. Although this is a research circuit, it is of sufficient complexity to demonstrate the validity of the design style for realistic systems component design.

The goal in the development of the MEAT tool was to generate fast, compact, efficient circuits. Showing the excellent performance that can be achieved with asynchronous designs is an important part of forwarding this technology to the general circuit design community. While experienced asynchronous designers understand that there are more benefits in the asynchronous approach than speed, it is clear that the dominant metric in evaluating circuit design styles in the commercial arena is performance. Our Post Office design was no exception; as long as the circuit was fast nobody cared how we did it except us. We view this as a sad reality, since it restricts the impact of the conceptual elegance of asynchronous circuits to the academic community.

The problems that we encountered in the Post Office design proved that MEAT, like any CAD tool, is incomplete. The back-end only produces schematics. Manual layout is prohibitively time consuming. Some form of automatic layout is necessary unless we abandon the complex gate approach in order take advantage of standard cell and technology mapping approaches. However, automatic layout is a difficult task and should also include automatically sized transistors for the performance needs of the design. Using standard cells will result in some lost performance but makes the synthesis task easier. It was also painfully clear that we needed to make a connection to existing CAD tool frameworks to permit synthesis of datapath as well as control path circuits, and to allow behavioral simulation to be performed on the entire design.

These realizations plus the fact that the verifier occasionally found hazards in the design were the motivation for a subsequent asynchronous synthesis CAD effort called **STETSON**. Steve Nowick's efforts to modify David Dill's asynchronous verifier [12] created a realization that our flow table based state and logic minimization methods in MEAT were not quite correct for burst-mode operation. Changing from SIC to even the restricted MIC behavior of burst-mode required further thought. In order to pursue this work, a research project at the HP Stanford Science Center was set up to provide a collaborative effort between the asynchronous research efforts at Stanford and those in the Mayfly group at HP laboratories. HP had recently made a corporate decision to move to the Mentor GDT IC tool platform and since HP was sponsoring the project, the obvious choice was to improve on the MEAT ideas in a way that was both correct and could be integrated into the Mentor GDT framework which is called Falcon.

The STETSON work was primarily done by five people. Bill Coates, one of the Mayfly project members, did the work on integrating the tools into the Mentor GDT framework and developed an improved version of the state machine minimization routines. Steve Nowick developed a set of criteria for minimizing burst-mode AFSM's which permitted correct hazard free logic min-imization for unclocked burst-mode AFSM's as well as an alternative method which uses state holding latches for AFSM implementation. The latter formed the basis for his Ph.D. dissertation [33] called the *locally-clocked* method. Ken Yun's dissertation [49] developed a third synthesis path called the *3D* method. Since the problem of generating a quality custom layout automatically was extremely difficult and since our corporate sponsors were increasingly moving towards standard cell and gate array design methods, we decided to develop the STETSON methods to generate standard cell implementations. A key problem in doing this was the ability to do hazard non-increasing technology mapping onto the standard cell library. The solution to this problem became the Ph.D. dissertation of Polly Siegel [38].

The project which drove the development of the STETSON tool kit was a chip which we called **ABCS**. ABCS is a low-power infrared communications receiver intended for hand held digital assistant type devices. Alan Marshall, an HP laboratories member from Bristol England, came to Stanford for two years to participate in the effort. He was the main designer of the ABCS chip and his efforts catalyzed much of the STETSON development. The author and Prof. David Dill of Stanford were certainly involved, but the day to day efforts and the bulk of the credit for STETSON belongs to Bill Coates, Steve Nowick, Alan Marshall, Polly Siegel, and Ken Yun.

The remainder of this chapter is devoted to the description of their efforts.

8 STETSON AFSM Synthesis

The information in this section is based on a more detailed paper by Bill Coates and Steve Nowick [30]. The goal of this effort was to produce hazard free gate level implementations for burst-mode AFSM specifications that did not require the post synthesis verification step required by designs synthesized by the MEAT tools.

While the STETSON tool set also permits the 3D [49] and locally clocked [33] options, the unclocked method described here has several advantages. Unlike the locally clocked method, there is no need for local clock generation logic or transparent latches. Unlike the 3D method, a state transition requires no multi-stepping through intermediate states. Additionally this method does not rely on the addition of delays to the critical path as in other MIC methods [22, 14]. The result is improved performance.

The AFSM is specified by the same burst-mode AFSM description used in MEAT, although the syntax is slightly different for STETSON. There are two restrictions on this specification:

- **Maximal Set Property:** no exit burst from a state can be a proper subset of an exit burst from the same state.

- **Unique Entry Point Property:** A given state must always be entered with the same set of input values.

These two properties were also true of MEAT AFSM's and are necessary to prevent ambiguity in the AFSM's behavior. If the maximal set property were not present then the AFSM would be presented with the need to react to part of a burst rather than waiting for a larger burst. Such a distinction is silly in an asynchronous environment. If the unique entry point were not observed then sometimes a state would be waiting for input variable A to go high, while another time it would wait for it to go low.

The first step of the synthesis is to transform the burst-mode specification into an unminimized flow table. For example the specification shown in Figure 13 is transformed into the flow table shown in Figure 14.

An output of nil in the specification indicates no change to the current outputs. For inputs a + after a variable name indicates a low to high transition for that input. Similarly a − indicates a high to low transition. Since the specification tracks every transition of every input variable, the choice of four or two cycle signaling protocols does not influence the methodology and becomes simply a designer choice. Hence mixing two and four cycle protocols is a simple task, and the method of choice can be suited to the constraints of the design. It is sometimes the case that two-cycle protocols are faster, and it is often the case that four-cycle protocols require less logic to implement. The result, as was true for the Post Office, is that it is often appropriate to mix the two styles.

In the flow table, horizontal movement indicates a change in the input state and a vertical movement indicates a change in the state of the AFSM. The combination of the two is the total state of the machine. Note that one aspect

136

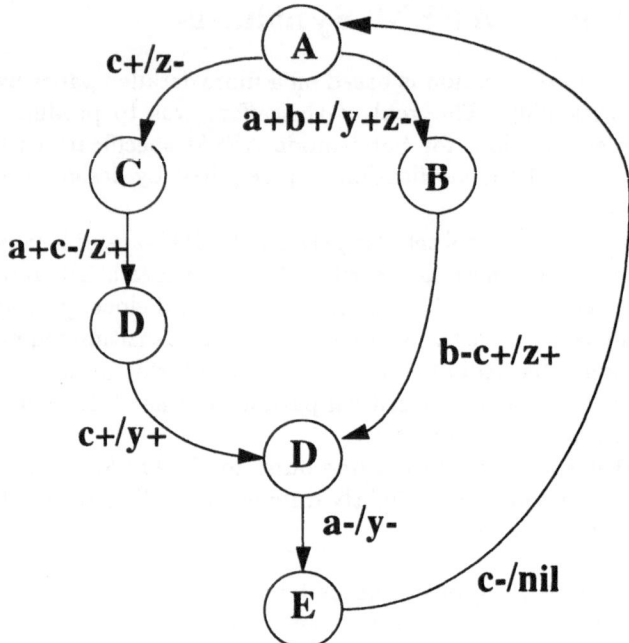

Figure 13: Sample Burst-Mode Specification

of the burst-mode flow table is the covering of all possible *trajectories* from an initial state to a final state target of some burst change. This is illustrated from state A where two exit possibilities exist. The c+ exit is a simple SIC transition, but the a+b+ exit represents a true burst where input concurrency can occur. This means that either a can go high and then b, or vice versa, or they can go high with so little separation as to appear to be changing concurrently from the perspective of the circuit. Until the burst is complete, the flow table indicates that no state change will occur.

In classical methods such as those used by MEAT, states are then merged based on compatibility relations in the columns of the flow table. States are compatible if they are output compatible, and if they are output compatible then their next state sets must be compatible as well. These methods are well described in Unger [45]. The problem is that certain legal state minimizations can result in a reduced flow table which cannot be implemented as a hazard free logic circuit. Nowick and Coates [30] have shown that for the unminimized flow table there is a hazard free logic implementation, and that using the classical methods certain state combinations may create a reduced flow table for which no hazard free cover exists. These situations caused the need for post minimization verification in the MEAT tools.

As an example, consider rows A and D in Figure 14. All columns are output compatible, and the only implied next state compatibility is in column labeled **100** which indicates that A and D are compatible if A and D are compatible.

Next State, Outputs=yz

inputs = abc

		000	001	011	010	110	111	101	100
	A	A, 01	C, 00	——	A, 01	B, 10	——	——	A, 01
	B	——	——	——	——	B, 10	B, 10	E, 11	B, 10
Current State	C	C, 00	C, 00	——	——	——	——	C, 00	D, 01
	D	——	——	——	——	——	——	E, 11	D, 01
	E	——	F, 01	——	——	——	——	E, 11	——
	F	A, 01	F, 01	——	——	——	——	——	——

Figure 14: Sample Unminimized Flow Table

This is always true and hence at first glance states A and D could be merged to reduce the AFSM and result in a column shown in Figure 15.

The Karnaugh map covers for the z output variable are shown in the resulting merged state row for AD. Output z has a 1 to 0 transition after input burst a+b+ and a 1 to 1 transition for input burst c+. The result is that any product term which intersects the c+ transition also intersects the a+b+ product term. Hence there is a logic hazard for the z output. The problem becomes even more complex for the next state logic. This is a curious result in that although a hazard free solution exists prior to minimization no solution may exist after classical minimization.

Nowick studied the effects of burst-mode trajectories and came up with a set of state merging constraints that must be added to the classical output and next state compatibility constraints. The result is a new state algorithm (called dhf-compatibility) which guarantees that a hazard free two-level logic sum of products implementation exists. The key to this algorithm is to realize that burst-mode trajectories from initial sub-cube to the final destination fill a binary n-cube area of the boolean space. This area is easy to visualize using the traditional Karnaugh map models. In short if two spaces (call them S1 and S2) intersect and are dhf-compatible then they must satisfy the following constraints:

- If S1 and S2 are both 1 to 0 transitions, then they must share a common initial point, OR they must intersect at their destination points, OR they must not intersect at any point.

- If S1 is a 1 to 0 transition, and S2 is a 1 to 1 transitions, then S2's initial

Next State, Outputs=yz inputs = abc

		000	001	011	010	110	111	101	100
Current State	A	A, 01	C, 00	——	A, 01	B, 10	——	——	A, 01
	D	——	——	——	——	——	——	E, 11	D, 01

Next State, Outputs=yz inputs = abc

		000	001	011	010	110	111	101	100
Merged Current State	AD	A, 01	C, 00	——	A, 01	B, 10	——	E, 11	AD, 01

Figure 15: Sample Unminimized Flow Table

point may lie in the S1 space but the S1 destination point may not lie in the S2 space.

Detailed arguments of why these constraints are necessary can be found in [30]. Experimental results indicate that the output latency and the logic complexity of the unclocked method are comparable to that of the 3D method, but the next-state latency of the unclocked method is an improvement over the inherent multi-stepping to the next state model inherent to the 3D method. For applications where cycle time is critical the unclocked methods have an advantage, whereas there is little difference in the unclocked and 3D methods for applications whose performance is dominated by output latency.

9 Hazard Non-Increasing Technology Mapping

Once an AFSM has been specified and transformed into a set of equations which implement the specification, we essentially have a netlist for the implementation. The next step, given the standard cell bias for the STETSON tools, is to map this implementation onto the standard cell library. A detailed account of this approach can be found in [38]. Automatic technology mapping techniques have been employed for synchronous design styles [20, 37, 23]. These algorithms allow translation of a technology-independent logic description into a technology-dependent implementation. However, these techniques by them-selves are not suitable for asynchronous design styles since they do not take hazards into account.

In particular we were interested in a *generalized fundamental-mode* asyn-chronous design style [32], since it separates the combinational portions of the design from the sequential portions, similar to many synchronous design styles. Each step of algorithmic technology mapping is examined for its influence on the hazard behavior of the modified network. Our approach was to modify an existing synchronous technology mapper, CERES, to adapt it to work for generalized fundamental-mode designs.

Our burst-mode design style extends the fundamental mode assumption imposed by Mealy and Moore machines to allow multiple-input change bursts in a particular state. The synthesis methods that incorporate this burst-mode or *generalized* fundamental-mode design style produce logic under the assumption that a burst of input changes can occur, in any order, and the outputs and feedback variables of the combinational portion will settle before the next set of input changes are applied. As in the case of single-input change fundamental-mode, no hazards can be tolerated during the input bursts. However, both single-input change hazards and multi-input change hazards must be considered for the technology mapping step.

There are two basic classes of hazards: *function* and *logic* hazards. Function hazards are a property of the logic function and can only be eliminated through appropriate placement of delay elements, whereas logic hazards are purely a property of the implementation. If a network has a function hazard for a given transition, then it cannot have a corresponding logic hazard for that same transition.

Within the two classes of combinational hazards, we can have *static* or *dynamic* hazards. Given that a transition is being made between two points I_1 and I_2 in the input space $\{0, 1\}^n$, then static hazards apply to transitions where $f(I_1) = f(I_2)$, and dynamic hazards apply to cases where $f(I_1) \neq f(I_2)$. Within the class of logic hazards, we will further distinguish between *single-input change* (s.i.c.) hazards and *multi-input change* (m.i.c.) hazards.

Static logic hazards will occur whenever the transition is not properly cov-ered by a single gate; that is, whenever the implementation does not contain a single gate that maintains the output value throughout the input transition. Dynamic hazards are applicable to both single-input change and multi-input change situations. A dynamic hazard occurs when, during an expected $0 \rightarrow 1$ $(1 \rightarrow 0)$ transition of the output, a $0 \rightarrow 1 \rightarrow 0 \rightarrow 1$ $(1 \rightarrow 0 \rightarrow 1 \rightarrow 0)$ transi-tion occurs. For single-input change conditions, this corresponds to a situation where a literal and its complement fan-out to several paths. For multi-input change conditions, a dynamic hazard can occur when a single gate is turned on momentarily during the transition. For the generalized fundamental-mode asynchronous design style, all logic hazards are of interest, and we must make sure that new logic hazards are not introduced during the technology mapping operation.

The technology mapping approach taken by algorithmic technology mappers typically divide the problem into three major steps: decomposition, partitioning, and matching and covering. First, the initial network, which is represented as a directed acyclic graph (DAG), is decomposed into a multi-level network composed of canonical two-input, single-output gates. Next, the circuit is partitioned into sets of single-output *cones* of logic, where a cone of logic represents the gate-level network as traced from a given output to the inputs. Each cone is then treated independently by the mapper. All possible matches to library elements are then found for subnetworks within each logic cone. Finally, an optimal set of matching library elements is selected from the set of matches to realize the network.

The decomposition step partitions the network into an equivalent network composed of 2-input, 1-output base functions. This process can be performed by recursively applying DeMorgan's rule and the associative law to the network. Both operations have been shown to be hazard-preserving for all logic hazards [45]. Thus, the modified network composed of 2-input, 1-output gates has identical hazard behavior to that of the original network.

Partitioning breaks the decomposed network at points of multiple fanout into single-output cones of logic. This is a heuristic simplification that is required to convert a multi-output logic network into a collection of single-output cones of logic, so that simpler algorithms can be employed to find the best cover for the network [20, 23]. Each subnetwork is then mapped separately by the matching and covering algorithm described in the next subsection. Given that we start with a hazard free network and preserve this behavior (within the partitions) in the covering step, the partitioning step cannot alter the hazard behavior of the network.

The matching and covering steps involve identifying equivalence between a subnetwork and a library element, and replacing that subnetwork with the equivalent library element. This step must not introduce any new hazards into the design.

Different algorithms are used for the matching and covering steps. Keutzer and Rudell [20, 37] use tree pattern matching techniques for matching elements within the library to portions of the circuit. In MIS, a complete set of patterns composed of different decompositions of two-input, one-output gates is generated for each library element. Standard pattern matching techniques are used to compare the library element with a portion of the network to be mapped. As long as the patterns themselves do not have hazards, these techniques will work to preserve the hazard behavior of the circuit, provided that the initial pattern decomposition is hazard-preserving and that the library elements do not have hazards. However, if a hazardous library element is selected, then the hazard behavior of the subcircuit must be examined before the match is accepted.

Some mappers, such as CERES, use Boolean techniques to detect equivalent networks. These techniques decouple the structure of the subnetwork from the matching process, which means we cannot reason about the transformations of one subnetwork into the other. However, we can build on theorems presented by Unger in [45] to show that if we replace a portion of the circuit with an equivalent circuit with similar hazard behavior, the resulting circuit is still hazard free for the transitions of interest.

Siegel [38] has shown that if we start with a set of hazard free library elements, the covering step will not introduce new hazards and the resulting cover will introduce no new hazards. Additionally, if we have library elements that contain hazards, we need to make sure that the hazards they contain are a subset of those in the portion of the network that is being matched.

Since the STETSON synthesizers (unclocked, locally clocked, and 3D) all generate hazard free implementations for burst-mode inputs it is then the mappers job not to introduce any new hazards. In order to do this, the library must be augmented with hazard information. The matching step must then check to see if the hazards contained in a candidate library cell are subset of the hazards contained in the logic equations. If this is true then a match succeeds and the final implementation has does not contain added hazards and hence is hazard free for burst-mode transitions of interest.

The functionality of each library element is expressed in Boolean factored form (BFF). We use the BFF expression as an accurate and convenient representation for both the functionality and structure of the particular library element. This BFF expression is then analyzed for logic hazards when the library is read into the mapper, and the logic hazard behavior of each library element is added as an annotation to the library element for later use during the matching phase.

The matching algorithm must now be modified to take the logic hazard information into account. If a hazardous library element is selected as a match for a particular subcircuit, then the subcircuit of interest must be examined to see if the same logic hazards exist. Note that we are not interested in the function hazard behavior of either the library element or the subnetwork, since function hazards are purely a property of the function itself and thus are independent of implementation.

The procedure for doing the comparison is much easier than the initial computation of the hazards for each library element, since we already know the transitions of interest. For each logic hazard in the library element, we must look at the subcircuit to see if the same logic hazard exists. As soon as we find a hazardous transition in the library element that is not in the subcircuit, we can stop, since that library element cannot be used safely. At this point the library element is eliminated from consideration for this match.

We examined a typical commercial standard cell library to see how many elements contained logic hazards. Only the multiplexers, which represented a small fraction of the library, contained hazards. So most matching elements will be logic hazard-free and the normal synchronous algorithms will be used with negligible overhead.

Table 1 shows the (logic) hazardous elements that are present in several libraries. The LSI and CMOS3 libraries are commercial ASIC libraries. The GDT library is a standard-cell library produced specifically for a particular chip, and includes many complex and-or-invert gates. (Note that many of these elements had several instances with different drive capability but only

142

one is shown for brevity.) For these libraries, the only library elements with
logic hazards were multiplexors.

Lib	Hazardous Elements	#	Total Elements	% Hazardous Elements
LSI9K	Muxes	12	86	14%
CMOS3	Muxes	1	30	3%
GDT	none	0	72	0%

Table 1: Libraries and their hazardous elements

The remaining problems we must solve then are how to efficiently char-
acterize the logic hazard behavior of the library elements, and how to easily
determine whether the subcircuit has the same logic hazards. We address these
problems in the next section.

Table 2 shows mapping results and mapping runtimes for some asynchronous
benchmark circuits. The area costs are relative to the particular library.

The SCSI design was mapped from logic equations produced by the locally
clocked method [31]. The ABCS design is a part of the control logic for the infrared communications chip which is and was mapped from equations generated
using the 3D method.

Design	Library					
	CMOS3			LSI		
	CPU	Delay	Area	CPU	Delay	Area
chu-ad-opt	.6s	24ns	152	3.7s	2.1ns	9
dme-fast-opt	.8s	11.8ns	232	3.8s	1.7ns	13
dme-fast	.6s	11ns	136	3.7s	1.7ns	9
dme-opt	.7s	22.4ns	184	3.9s	2ns	10
dme	.6s	17.8ns	168	3.7s	1.9ns	10
oscsi-ctrl	10.6s	96.5ns	3552	16.1s	7.3ns	172
pe-send-ifc	2.3s	47.2ns	864	5.3s	3.5ns	40
vanbek-opt	.6s	19.5ns	144	3.7s	2.6ns	9
dean-ctrl	33.6s	126ns	11320	43.5s	10.3ns	565
scsi	20.7s	95ns	6888	27.9s	7.2ns	330
abcs	9s	74.7ns	3288	13.3s	6.8ns	168

Table 2: Mapping results for synchronous vs asynchronous mapper (Depth of
5). Benchmarks were run on a DEC 5000.

10   STETSON and ABCS

10.1   Control Path Synthesis

Figure 16 shows STETSON's tool flow options for synthesizing controller circuits. The oval shape at the beginning of the process is implemented by a program written by Bill Coates called **bm-check**. This program takes a textual burst-mode controller specification and checks it to insure that the specification meets the requirements for a legal burst-mode specification, namely adheres to the maximal set and unique entry point properties. There are a number of switches that can be set to control which output files should be generated. Bm-check can generate behavioral VHDL or Verilog code, an extensive set of burst-mode test vectors to exercise all of the paths in the AFSM, or input files for any of the three synthesis methods.

The next step is to create minimized state machines from any or all of the AFSM synthesizers, namely the unclocked, locally clocked or 3D methods. The result of this step is a set of hazard free but unminimized logic equations for the implementation. The asynchronous logic minimizer is then used to produce hazard free minimal equations. The asynchronous logic minimizer a constrained variant of the familiar Quine-McCluskey algorithm which has also been optimized for the don't care dominant nature of burst-mode specified AFSM's. The logic produced by the asynchronous minimizer is typically larger than that produced by an equivalent synchronous minimizer due to the need to provide redundant covers in the logic in order to avoid hazards. However in the large number of cases that we have examined, the overhead with respect to the number of product terms has been no more than 6%. For a large system chip, the area spent on control circuitry is typically a small fraction of the total chip area. We view this 6% area overhead for control circuits as being a negligible cost given the modular composition, circuit reuse, and incremental improvement benefits of asynchronous systems.

The technology mapper then creates a technology dependent netlist using the cells in the cell library. At this point the Mentor AutoCells package performs the place and routing chores to produce the final implementation. Switch level simulation and VHDL simulation results can then be compared using the same set of test vectors to validate that the implementation and the specified behavior are consistent. This comparison step was useful in early designs when we did not trust our toolset implementation to be correct. However, for the last year, we have not encountered a single instance where the comparison did not indicate a match. However, considering the cost in terms of time and dollars spent per processing run, we continue to run the match path.

10.2   Datapath Synthesis

Our design style views datapath circuits as they are viewed in synchronous designs. We typically use bundled data and either two or four cycle signaling protocols and use model delays associated with the datapath elements to provide completion signals. Hence datapath synthesis is performed by writing

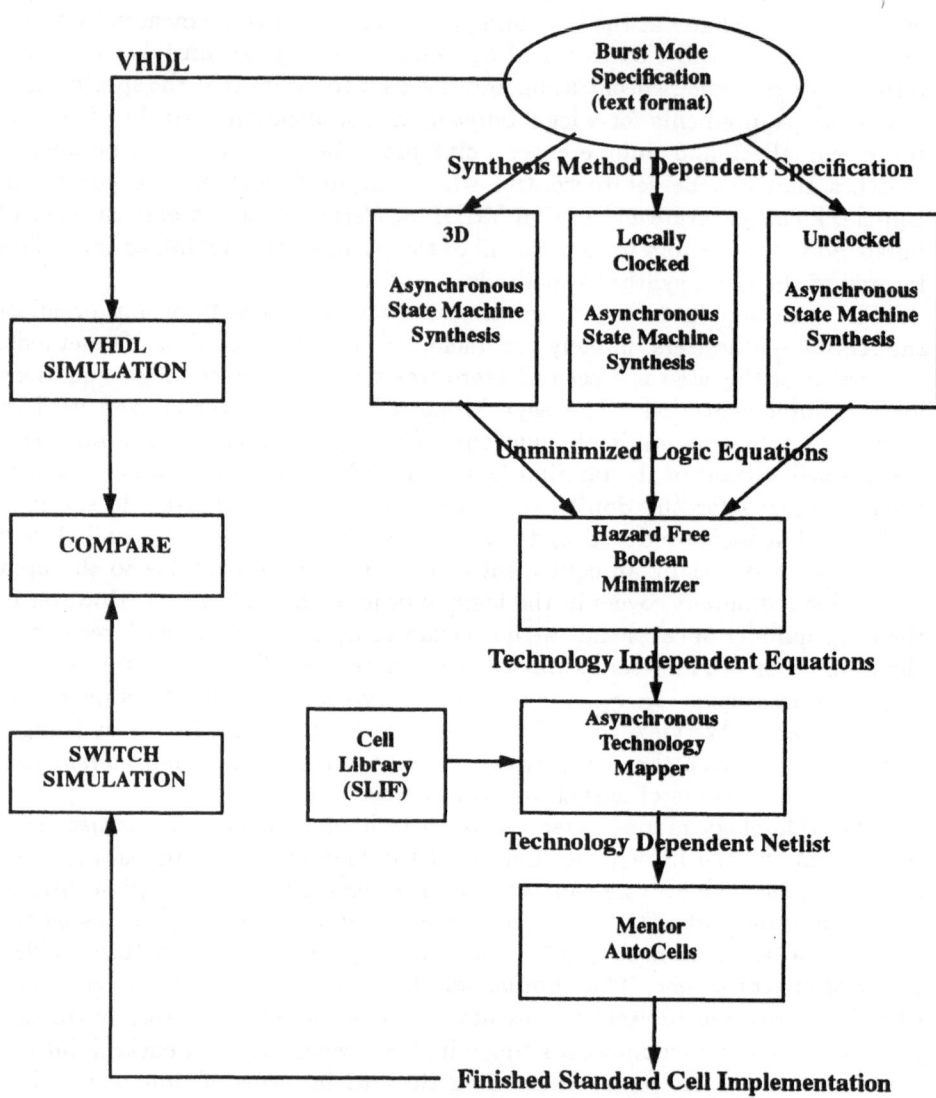

Figure 16: STETSON Tool Flow

VHDL models, which are then synthesized into standard cell circuit blocks by the Mentor VHDL compiler. Behavioral VHDL models can be used in concert with the synthesized AFSM models generated by bm-check to run simulations on system wide behavior of multiple control and data path modules.

Model delays are determined using standard timing analysis tools. The margins that we use are typically smaller than those employed by synchronous designers since issues like clock skew do not have to be accounted for in our asynchronous design style which uses point to point control signals from controlling AFSM modules to their respective controlled datapath modules. Most companies build their own set of datapath generators which are used to generate datapath blocks like memories, register files, ALU's, etc. and these tools can be used inside the STETSON framework. All that is needed is to run the timing analyzer on the result in order to determine the model delay and then generate an inverter chain that corresponds to this delay into the synthesized module.

10.3 ABCS

The STETSON toolset was used to synthesize a low-power infrared receiver chip suitable for hand held digital assistant style applications. Prior to submission of the ABCS design, several small MOSIS *tinychip* test fragments were designed and submitted to check the validity of our synthesis tools. They all worked as expected on first silicon. Given our Post Office experience this came as a pleasant surprise.

The biggest problems in the development of the ABCS design came from two basic sources:

1. The design and the tool developments were concurrent activities. Hence much of the design testing and modeling had to wait or even drive some of the tool development activities which significantly extended the design cycle time.

2. The desire to produce the ABCS design in a mainstream fashion meant that we needed to produce a VHDL specification of the data and control path elements. The Mentor VHDL compiler has some serious performance flaws. The result was an extremely slow and painful development and modeling cycle.

Nonetheless, the ABCS chip was submitted and ran on first silicon.

11 Conclusions

The work reported in this chapter was done over a five or six year period. We started this effort somewhat naively thinking that automating what we had been doing by hand for decades would be reasonably straight-forward task. We were totally wrong. The evolution of the burst-mode idea, initially important from the standpoint of performance, eventually required that we rethink the entire methodology. MEAT was the first step, and while it drastically cut the design cycle time, the need for post verification of the synthesis and the need for manual layout were critical flaws. Another critical flaw was that while control path circuits are often the most complex and error prone aspects of a design, the datapath logic consumes the most area on the final chip, and MEAT had no support for datapath synthesis. The final problem with MEAT was that we still had no way to perform behavioral analysis on the total design.

The STETSON effort was the logical next step. It was fortunate that we were able to conduct this effort in an academic setting. The mere fact that multiple students were involved provided options that would not have been pursued otherwise. The result is a toolset that can be used for large scale designs, uses rather than ignores commercial CAD capabilities, and permits synchronous designers to maintain most of their familiar design methods.

As with any CAD effort, the STETSON design has many aspects which can be improved. There is considerable designer art required to decompose a large control path specification into a set of communicating AFSM's which achieve a reasonable balance of cost and performance. Similarly, our method of always using model delays for the datapath elements rather than using more directly asynchronous methods is not appropriate for designs which require more robust behavior. We view these areas as opportunities for future research.

12   Acknowledgments

It should be clear from the text of this chapter that the author is merely acting as a scribe for the efforts of a number of people who made the MEAT and STETSON tools a reality. The same is true for the Post Office and ABCS chips. The author started the MEAT effort and produced the logic minimization methods that were used in MEAT but these were subsequently modified by Steve Nowick. Bill Coates did all of the work on the state machine minimization part of MEAT and continued this effort with help from Steve Nowick to produce the unclocked state machine synthesis method in the STETSON tools. Ken Yun provided the 3D option in STETSON and Steve Nowick provided the locally clocked option. Polly Siegel developed the methods and implementation for doing the hazard non-increasing technology mapper for STETSON. Ken Stevens was the primary designer of the Post Office with help from Bill Coates. Alan Marshall was the designer of the ABCS chip. The entire STETSON effort is indebted to the guidance and insights of David Dill.

The activities reported here were truly a group effort. The author played only a small part, however as scribe for these efforts he assumes all responsibility for errors in the transcription.

Finally, we owe a great deal of gratitude to Hewlett Packard Laboratories, and in particular Dick Lampman and Stephen Rosenberg, whose support made all of this possible.

References

[1] H. B. Bakoglu. *Circuits, Interconnections, and Packaging for VLSI.* Addison-Wesley, 1990.

[2] Erik Brunvand and Robert Sproull. Translating Concurrent Programs into Delay-Insensitive Circuits. In *IEEE International Conference on Computer Aided Design: Digest of Technical Papers*, pages 262–265. IEEE Computer Society Press, 1989.

[3] Steven M. Burns and Alain J. Martin. *The Fusion of Hardware Design and Verification*, chapter Synthesis of Self-Timed Circuits by Program Transformation, pages 99–116. Elsevier Science Publishers, 1988.

[4] Carnegie-Mellon University. *User's Guide to COSMOS.*

[5] Tam-Anh Chu. On the models for designing VLSI asynchronous digital systems. Technical Report MIT-LCS-TR-393, MIT, 1987.

[6] Henry Y. H. Chuang and Santanu Das. Synthesis of multiple-input change asynchronous machines using controlled excitation and flip-flops. *IEEE Transactions on Computers*, C-22(12):1103–1109, December 1973.

[7] William S. Coates. "The Design of an Instruction Stream Memory Subsystem". Master's thesis, University of Calgary, December 1985.

[8] A. Davis. "The Mayfly Parallel Processing System". Technical Report HPL-SAL-89-22, Hewlett-Packard Company, March 1989.

[9] A. L. Davis. The Architecture of DDM1: A Recursively Structured Data-Driven Machine. Technical Report UUCS-77-113, University of Utah, Computer Science Dept, 1977.

[10] A. L. Davis. Mayfly: A General-Purpose, Scalable, Parallel Processing Architecture. *Lisp and Symbolic Computation*, 5(1/2):7–47, May 1992.

[11] A. L. Davis, B. Coates, R. Hodgson, R. Schediwy, and K. Stevens. Mayfly System Hardware. Technical Report HPL-SAL-89-23, Hewlett-Packard Laboratories, April 1989.

[12] David Dill. *Trace Theory for Automatic Hierarchical Verification of Speed-Independent Circuits. An ACM Distinguished Dissertation.* MIT Press, 1989.

[13] William I. Fletcher. *An Engineering Approach to Digital Design.* Prentice-Hall, 1980.

[14] A. D. Friedman and P. R. Menon. Synthesis of asynchronous sequential circuits with multiple-input changes. *IEEE Transactions on Computers*, C-17(6):559–566, June 1968.

[15] R. M. Fujimoto. *VLSI Communication Components for Multicomputer Networks*. PhD thesis, Univ. of California at Berkeley, August 1983.

[16] A Grasselli and F. Luccio. A Method for Minimizing the Number of Internal States of Incompletely Specified Sequential Networks. *IEEE TEC*, June 1965.

[17] A. B. Hayes. Stored State Asynchronous Sequential Circuits. *IEEE Transactions on Computers*, C-30(8), August 1981.

[18] C. A. R. Hoare. Communicating sequential processes. *CACM*, 21(8):666–677, Aug 1978.

[19] Lee A. Hollaar. Direct implementation of asynchronous control units. *IEEE Transactions on Computers*, C-31(12):1133–1141, December 1982.

[20] K. Keutzer. DAGON: Technology binding and local optimization by DAG matching. In *24th Design Automation Conference*, pages 341–347. IEEE/ACM, 1987.

[21] L. Lavagno, K. Keutzer, and A. Sangiovanni-Vincentelli. Synthesis of Verifiably Hazard-Free Asynchronous Control Circuits. In Carlo H. Sequin, editor, *Proceedings of the 1991 UC Santa Cruz Conference on Advanced Research in VLSI*. MIT Press, 1991.

[22] G. Mago. Realization methods for asynchronous sequential circuits. *IEEE Transactions on Computers*, C-20(3):290–297, March 1971.

[23] F. Mailhot. *Technology Mapping for VLSI Circuits Exploiting Boolean Properties and Optimizations*. PhD thesis, Stanford University, 1991.

[24] A. J. Martin, S. M. Burns, T. K. Lee, D. Borkovic, and P. J. Hazewindus. The Design of an Asynchronous Microprocessor. In C.L. Seitz, editor, *Advanced Reserach in VLSI: Proceeedings of the Decennial Caltech Conference on VLSI*, pages 351–373. MIT Press, 1989.

[25] Alain Martin. Compiling Communicating Processes into Delay-Insensitive VLSI Circuits. *Distributed Computing*, 1(1):226–234, 1986.

[26] Alain Martin. The Limitations to Delay-Insensitivity in Asynchronous Circuits. In William J. Dally, editor, *Sixth MIT Conference on Advanced Research in VLSI*, pages 263–278. MIT Press, 1990.

[27] C. Mead and L. Conway. *Introduction to VLSI Systems*. McGraw-Hill, 1979. Chapter 7.

[28] Teresa Meng. *Synchronization Design for Digital Systems*. Kluwer Academic, 1990.

[29] Charles E. Molnar, Ting-Pien Fang, and Frederick U. Rosenberger. Synthesis of Delay-Insensitive Modules. In Henry Fuchs, editor, *Chapel Hill Conference on Very Large Scale Integration*, pages 67–86. Computer Science Press, 1985.

[30] S. M. Nowick and B. Coates. UCLOCK: automated design of high-performance unclocked state machines. In *Proceedings of the 1994 IEEE International Conference on Computer Design: VLSI in Computers and Processors*, pages 434–441, October 1994.

[31] S. M. Nowick and D. L. Dill. Synthesis of asynchronous state machines using a local clock. In *1991 IEEE International Conference on Computer Design: VLSI in Computers and Processors*. IEEE Computer Society, 1991.

[32] S. M. Nowick, K. Y. Yun, and D. L. Dill. Practical asynchronous controller design. In *1992 IEEE International Conference on Computer Design: VLSI in Computers and Processors*. IEEE Computer Society, 1992.

[33] Steve Nowick. *Automatic Synthesis of Burst-Mode Asynchronous Controllers*. PhD thesis, Stanford University, 1993.

[34] Steven M. Nowick and David L. Dill. Automatic synthesis of locally-clocked asynchronous state machines. In *1991 IEEE International Conference on Computer-Aided Design*. IEEE Computer Society, 1991.

[35] S. S. Patil. Coordination of asynchronous events. Technical Report TR-72, MIT Project MAC, June 1970.

[36] Steven M. Rubin. *Computer Aids for VLSI Design*. VLSI Systems. Addison-Wesley, 1987.

[37] R. Rudell. *Logic Synthesis for VLSI Design*. PhD thesis, U. C. Berkeley, April 1989. Memorandum UCB/ERL M89/49.

[38] Polly Siegel. *Technology Mapping for Asynchronous Designs*. PhD thesis, Stanford, January 1995.

[39] Kenneth S. Stevens. The Communications Framework for a Distributed Ensemble Architecture. AI 47, Schlumberger Palo Alto Research, February 1986.

[40] Kenneth S. Stevens, Shane V Robison, and A.L. Davis. "The Post Office – Communication Support for Distributed Ensemble Architectures". In *Proceedings of 6th International Conference on Distributed Computing Systems*, pages 160 – 166, May 1986.

[41] I. E. Sutherland, R. F. Sproull, C. E. Molnar, and E. H. Frank. Asynchronous Systems, Volume I. Technical report, Sutherland Sproull and Associates, Palo Alto, CA, January 1985.

[42] Ivan E. Sutherland and Robert F. Sproull. Logical effort: Designing for speed on the back of an envelope. In Carlo H. Sequin, editor, *Proceedings of the 13th Conference on Advanced Research in VLSI*, pages 1–16. UC Santa Cruz, March 1991.

[43] J. H. Tracey. Internal state assignments for asynchronous sequential machines. *IEEE Transactions on Electronic Computers*, EC-15:551–560, August 1966.

[44] S. H. Unger. A Building Block Approach to Unclocked Systems. In *Proceedings of the 26th HICSS Conference*, January 1993. To appear.

[45] S.H. Unger. *Asynchronous sequential switching circuits*. Wiley-Interscience, 1969.

[46] C. H. (Kees) van Berkel. *Handshake circuits: an intermediary between communicating processes and VLSI*. PhD thesis, Technical University of Eindhoven, May 1992.

[47] Kees van Berkel. Beware the Isochronic Fork. *Integration, the VLSI journal*, 13(2):103–128, 1990.

[48] Peter Vanbekbergen, Francky Catthoor, Gert Goossens, and Hugo De Man. Optimized synthesis of asynchronous control circuits from graph-theoretic specifications. In *International Conference on Computer-Aided Design*. IEEE Computer Society Press, 1990.

[49] K. Yun and D. L. Dill. Automatic synthesis of 3D asynchronous finite-state machines. In *Proceedings of the International Workshop on Computer-Aided Design*, pages 576–580, November 1992.

VLSI Programming of Asynchronous Circuits for Low Power

Kees van Berkel

Philips Research Laboratories, WAY4,
Prof. Holstlaan 4, 5656 AA Eindhoven, The Netherlands

Martin Rem

Eindhoven University of Technology,
P.O. Box 513, 5600 MB Eindhoven, The Netherlands

Abstract

In this chapter we analyze the potential of asynchronous circuits for low power consumption. We set out by reviewing the mechanisms of energy dissipation of digital CMOS ICs in general and clocked circuits in particular. For many applications the generation and distribution of the clock signal account for more than half the power dissipation, directly or indirectly. Much of this wasted clock power — and often much more — can be saved by applying asynchronous circuit techniques.

This will be illustrated by a variety of simple, but practical examples: N-fold repeaters, modulo-N counters, ripple and wagging implementations of shift registers, $\pm 2^M$-incrementers, parallel-to-serial converters, and systolic computations.

We apply a programming approach to the design of these circuits using the CSP-based VLSI-programming language Tangram. So-called handshake circuits form an intermediate architecture between Tangram and asynchronous circuits. The transparent compilation of Tangram enables a simple analysis of power, performance, and costs of generated VLSI circuits. This method has been applied to several industrial applications, including a 2-IC realization of a DCC error corrector comprising 155 k transistors.

1   Introduction

Research on asynchronous circuits is booming, see Figure 1. But, what are asynchronous circuits good for? Which qualities of asynchronous circuits make them so appealing that asynchronous circuits can be considered an attractive alternative to clocked circuits for practical applications? Of course, there are the intrinsically asynchronous "niche" applications, including clock dividers, wake-up controllers, RAM controllers, routers, and some interface circuits.

For main-stream VLSI, asynchronous circuits *potentially* offer two advantages:

- higher throughput (on average) or lower latency, by avoiding synchronization with ("waiting for") clock edges;

- lower power consumption, by avoiding redundant voltage transitions on wires.

This article focuses on the potential of asynchronous circuits for low power consumption.

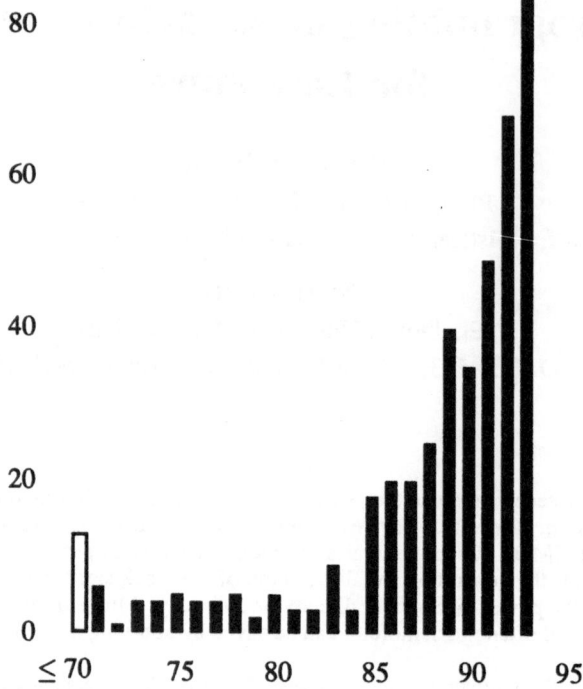

Figure 1: Yearly publication count on asynchronous circuits. (Source: async.bib of EUT [26].)

Integrated circuits in operation burn power. The IC of a digital wrist watch generally consumes less than 1 microwatt [44], a high-performance monolithic microprocessor more than 10 watts [7]. For both applications, and many others, the costs of power may be a real concern for a variety of reasons, related to

- costs of power generation, storage, and distribution;

- costs of associated heat collection and disposal;

- problems caused by the associated electro-magnetic radiation.

Although the subject of low power consumption of ICs is drawing considerable attention ("cool chips are hot"), this interest is only of recent date. The drive for low power is fueled by the explosion in portable products, including Compact Disc players, Digital Compact Cassette players, cordless telephones, pagers, games, notebooks, Personal Digital Assistants, and measuring equipment. Playing times of only a few hours for personal audio, notebooks, and cordless telephones are clearly not very consumer friendly. Also, the required batteries or battery packs are voluminous and heavy, often leading to bulky and unappealing products. In many of these portable products the fraction of the power consumed by the digital ICs is significant, often between one and two thirds. The remaining part is typically

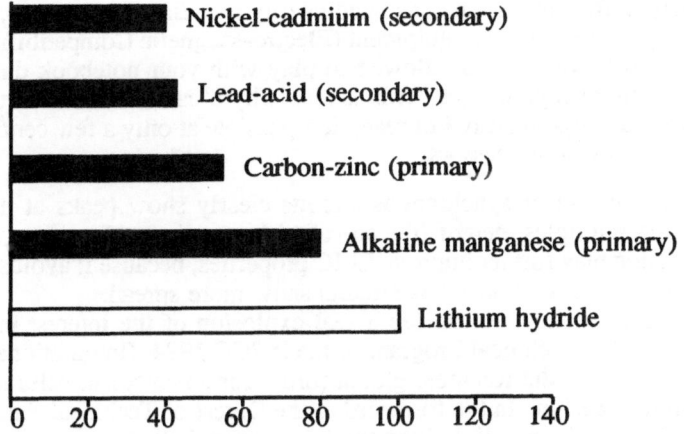

Figure 2: Energy density of several battery types [watthours/kilogram]. A primary battery converts chemical energy into electrical energy by irreversible chemical reactions. Secondary cells involve reversible reactions and are hence rechargeable. (Source: Battery Reference Book [5].)

consumed by LCD backlights, servomotors, lasers, radio transmitters, and analogue electronics.

One way to reach longer playing times of portable products is to increase battery capacities. Figure 2 shows the energy density of several battery types. A density of 40 watthour per kilogram means that a light bulb of 40 watts shines for one hour if powered by one kilogram of the battery type. Note that the range of energy densities is rather small, and that the popular rechargeable NiCd batteries perform relatively poorly. Industry is investing heavily in research on novel battery types, including Lithium-based batteries. At best they seem to offer twice or three times the density, but they will be relatively expensive. Improvements in battery technology alone provide a small factor, whereas an order of magnitude is needed.

A more promising approach to longer playing times of battery-powered systems is to reduce the energy consumption of ICs. Reduced energy consumption yields additional benefits.

- Less power consumption allows cheaper, smaller, and lighter power supply units. (Consider transformers, switches, fuses, rectifiers, capacitors, and relays in many non-portable appliances connected to the mains.)

- Less power consumption at the same supply voltage results in smaller supply currents, allowing narrower metal tracks for power distribution, and resulting in smaller voltage drops along these tracks. Smaller supply currents also reduce the magnitude of ground bounces and substrate crosstalk.

- Less power consumption implies less heat generation, possibly allowing cheaper IC packages or the riddance of a (noisy) fan. Furthermore, the silicon die will be cooler, offering better performance.

- Finally, a fraction of the consumed energy is radiated into space, possibly affecting other electronic equipment (Electro-Magnetic Compatibility, or EMC). This is why you are not allowed to play with your notebook during take-off or landing of a jumbo jet. This is also why a car stereo set requires a lot of tin to make high-quality FM reception possible at only a few centimeters from the DCC signal processors.

The radiation spectra of synchronous circuits clearly show peaks at the clock frequency and at multiples thereof (the so-called harmonics). Note that asynchronous circuit operation may further improve EMC properties, because it avoids these peaks. Radiation spectra of asynchronous circuits show more spread.

In the early nineties we can see a real explosion of the interest in low power. Almost 25% of the Technical Program of the ISSCC 1994, (International Solid-State Circuits Conference, the foremost global forum for advances in solid-state circuits) is on low-power design. In the foreword to the Digest of Technical Papers, Program Chairman Dennis Monticelli defined the low-power challenge as

"...to maintain functional integration and performance while simultaneously reducing power consumption by a large factor."

Also:

"Nothing less than the creation of a low-power culture in the design environment is necessary to make big strides."

Contributions to the reduction of power consumption can be made at all stages of the product creation process including algorithms, architectures, (embedded) software, circuit techniques, IC technology, and IC packaging. The purpose of this report is to show that asynchronous circuit techniques can make a significant contribution in many cases by avoiding the power waste of their synchronous counterparts.

In Section 2 we examine the energy consumption in CMOS circuits in general and in synchronous CMOS circuits in particular. The purpose of this section is to provide a simple model for the analysis of power consumption and to investigate how and how much power can be saved.

Asynchronous control circuits form the subject of Section 3. We shall encounter N-fold repeaters and modulo-N counters as examples that illustrate an asymptotic power advantage over their synchronous counter parts. Much of the advantage also applies more generally to the distribution of control in asynchronous circuits.

Section 4 deals with storing and moving data. A variety of shift-registers shows the type of trade-offs between power, costs, and throughput typical for such data structures.

By applying a more intricate interleaving of control and data manipulation, interesting solutions can be obtained for $\pm 2^M$-incrementers and parallel-to-serial converters. This is shown in Section 5.

In Section 6 the trade-off between power and area is investigated for systolic computations. Using the problem of window maxima as vehicle, three types of implementations are presented, with attention to some of the methodical aspects of their design.

In a concluding section we discuss the low-power potential of asynchronous circuits in practice by reviewing several asynchronous ICs. An additional low-power benefit of asynchronous circuits arises by dynamically adapting the power supply voltage. This results in a form of "just-in-time" processing that exploits the absence of clocks. Finally, a first attempt is made to compare the low-power potential of asynchronous circuits with that of clocked circuits.

Most examples are described in the VLSI programming language Tangram. Tangram is based on Hoare's Communicating Sequential Processes (CSP [12]). Viewed differently, Tangram is very similar to traditional programming languages such as C or Pascal. The most visible extensions to such a language are parallel execution of statements and procedures, and synchronized communications among these parallel procedures [41, 35, 30]. Reasoning about chip area, speed, and power at the Tangram level is possible, because of the transparent translation into asynchronous circuits using handshake circuits as an intermediate architecture [34, 41, 35]. Our views on asynchronous VLSI and the role of CSP-based formalisms in their design are in part inspired by the work of Martin at Caltech [19].

2 Energy consumption in digital CMOS ICs

"In real systems, the cost of power, cooling, and electrical bypassing often exceeds the cost of the chips themselves. Hence any discussion of the cost of computation must include the energy cost of the individual steps of the computation process." (Carver Mead in [22], page 322).

Most CMOS circuits are designed in such a way that they consume no power when quiescent. (Here we ignore the leakage current; see below.) A CMOS gate dissipates energy, that is, converts electrical energy into heat, only when its output makes a transition. The amount is more or less fixed. In this section we develop a simple model for estimating the energy consumption of a CMOS circuit. The model, "counting gate-output transitions" in essence, is used to analyze the power consumption of clocked circuits, and that of the asynchronous circuits. The limitations of this simple model will be discussed as well.

2.1 Digital CMOS

Figure 3 shows a CMOS gate schematically. The gate consists of a path of PMOS transistors by which wire w can be pulled up and a path of NMOS transistors by which that wire can be pulled down. In practical circuits the state that both the p-path and the n-path are conducting is avoided. Let C_L denote the load capacitance of the gate, that is, the capacitance of wire w, including the capacitance of the transistor gates connected to w. Let V_{DD} denote the supply voltage.

When wire w makes a transition from the ground level (0 volt by convention) to the supply level (here V_{DD} volts), the quantity of $Q = C_L V_{DD}$ coulomb charge is transferred from the + side of the battery to w. During the down transition, charge Q is transferred from w to the − side of the battery. In the process, charge Q has lost V_{DD} volts in potential, costing $Q V_{DD} = C_L V_{DD}^2$ in energy. Hence, the *switching energy* for a single transition amounts to $\frac{1}{2} C_L V_{DD}^2$. For a supply voltage of 5 volts and a (typical) load capacitance of 0.1 picofarads, the switching charge Q is 0.5 picocoulombs, or about 8 million (!) electrons. The switching energy per transition is then $1\frac{1}{4}$ picojoules.

However, evewn more energy dissipation is involved. During a transition there is generally a brief period in which both the n-path and the p-path conduct. The *short-circuit energy* dissipated by an unloaded inverter caused by the short-circuit

156

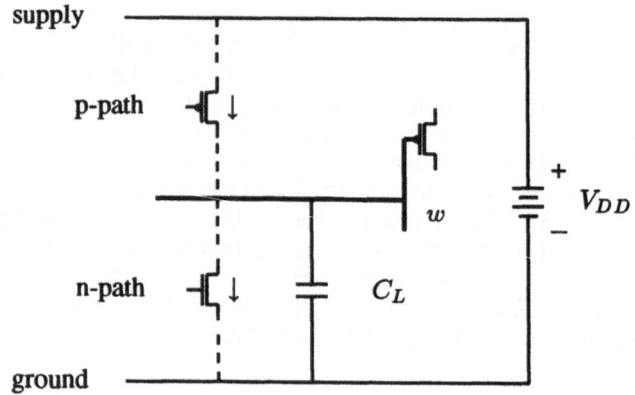

Figure 3: CMOS gate to illustrate the energy per transition.

current $I(t)$ during a time interval τ in which the input makes a linear transition from 0 to V_{DD} or vice versa is given by

$$V_{DD} \int_0^\tau I(t)dt = \frac{\beta\tau}{24}(V_{DD} - 2V_T)^3$$

for $V_{DD} \geq 2V_T$ [42]. Here we assume that the gains of the NMOS and PMOS transistors are both equal to β and that the respective threshold voltages are both equal to V_T. With $V_T = 1$ volt, $\beta = 0.2$ milliamperes/volt^2, and $\tau = 1$ nanosecond, a typical value for the short-circuit energy of an inverter is 0.2 picojoules per transition. Loading the inverter tends to reduce the short circuit current, because the transistor that is being switched off has a smaller drain-to-source voltage compared to the unloaded case. When $V_{DD} < 2V_T$, the short-circuit energy is negligible. Extensive simulations show that for carefully designed circuits, where transition times are kept minimal, the short-circuit energy is approximately 20% of the switching energy [42]. With a 5 volts supply voltage and a 1.0 micron transistor length, this figure may even be rather optimistic[1].

The total energy per (gate-output) transition[2] is the sum of switching energy and the short-circuit energy. For our typical wire of 0.1 picofarads this is about 1.5 picojoules.

There is even a third contribution in the energy consumption of a CMOS circuit, namely leakage. A typical gate leaks 0.1 to 0.5 nanoamperes, or about 1 to 2 nanowatts power at 5 volts. This is equivalent to about 1000 transitions/second. The leakage power of a gate can therefore only be ignored if that gate switches with a frequency of at least 10 kilohertz.

---

[1]Velocity saturation of electrons and holes tends to have more effect on the transition time of heavily loaded gates (the pulling transistor being in saturation) than on the short-circuit current (with drain-source voltages close to $\frac{1}{2}V_{DD}$).

[2]Most texts on dissipation of ICs (for example [45]) focus on the *power* consumption of a *gate*, rather than on the *energy* consumption of a *gate transition*. This implies a bias for synchronous circuits, since it assumes a clock frequency and a measure for gate activity. It also leads to the notion of power-delay product, which *is* the energy per transition.

So far we considered the energy consumption of a single gate, loaded with capacitance C_L. How can we generalize yhis to the energy consumption of circuits composed from gates? The proposal is to simply count transitions, assuming an average value for the load capacitances. That is, we assume that for a given computation

$$\sum_i N_i \frac{1}{2} C_i V_{DD}^2 \;=\; \frac{1}{2} C_L V_{DD}^2 \sum_i N_i$$

where the summation ranges over all gates, N_i is the number of output transitions on gate i, C_i the load capacitance of gate i, and C_L the average load. Histograms of extracted capacitances of several (standard cell) layouts show sharp peaks around C_L and little spread (see Figure 4). The peak at 60 femtofarads corresponds to gate outputs inside standard cells; the peak at 80 femtofarads to cell-external gate outputs.

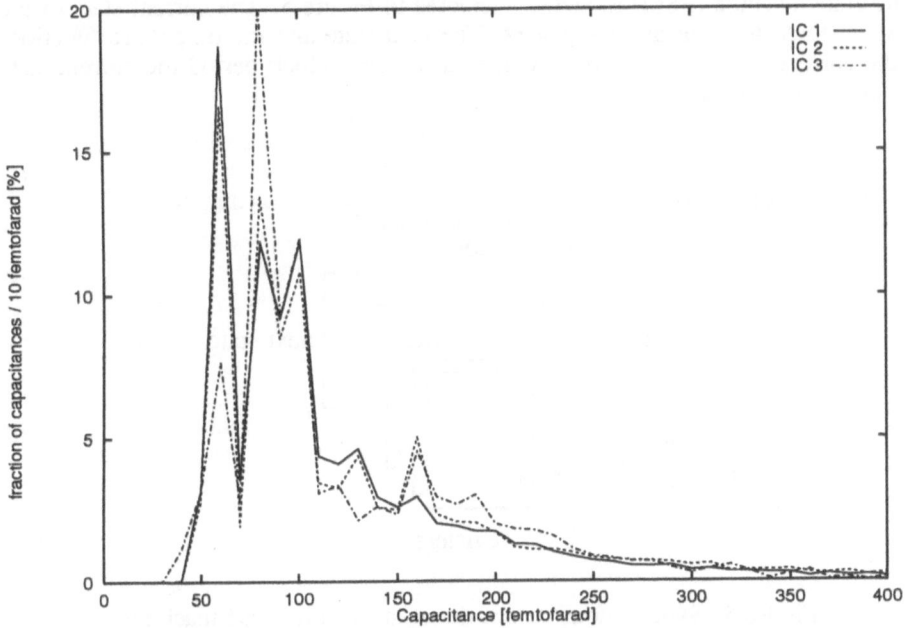

Figure 4: Histogram of extracted capacitances of three standard-cell layouts. The circuits are discussed in Section 7. The bin size is 10 femtofarads.

In summary, when we compare the energy consumption of computations on (different) circuits we shall simply count gate-output transitions. However, there are two pitfalls:

1. high-fanout gates and very long wires, leading to exceptionally large capacitive loads (if these wires switch frequently);

2. gates that switch infrequently so that leakage may dominate.

To reduce the energy consumption of an IC, there are two basic strategies: to reduce the energy per transition or to reduce the number of transitions.

158

Reduction of the energy per transition can simply be achieved by lowering V_{DD}. This has a super-quadratic effect, because the switching energy reduces quadratically, and the short-circuit energy reduces cubically. However, the circuit also becomes slower. To compensate for the loss in speed one can either introduce more parallelism (the "Berkeley approach" [4]) or lower threshold voltages (the "Stanford approach" [3]). In Section 7.2 we investigate an asynchronous side of tuning V_{DD}.

Shrinking circuits to technologies with smaller feature sizes also helps to reduce power dissipation, because transistor-gate and wire capacitances decrease. However, this decrease is less than linear, and hence rather marginal. Reduction of the number of transitions is less straightforward, as we shall see.

2.2 Synchronous circuits

We base our analysis of the energy consumption of synchronous circuits on the finite-state machine (FSM) model, as depicted in Figure 5. The current state of the FSM is stored in an array of registers. The next state and the output are functions of the current state and the input. Once during each clock period the current state is replaced by the next state.

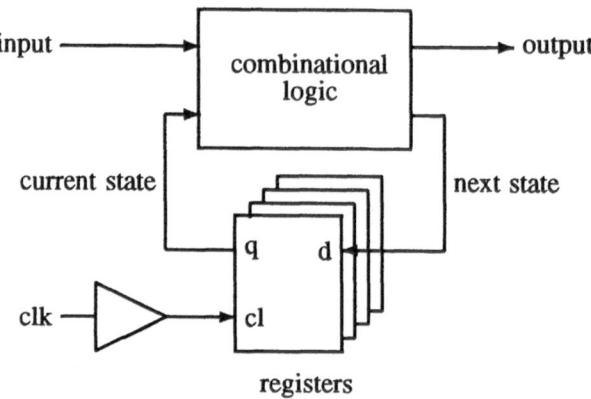

Figure 5: Synchronous circuit drawn as a finite-state machine.

A popular expression for power consumption of a synchronous circuit is (e.g. [45] page 237):

$$\frac{1}{2} f_c \alpha C_{total} V^2 \qquad (1)$$

where f_c denotes the clock frequency, C_{total} is the sum of the load capacitances of all gate outputs in the circuit, and α the average *activity* of the circuit. The activity of a gate is the average number of output transitions per clock cycle. For instance, if a gate output changes once during each cycle we have $\alpha=1$. If glitches occur, α may exceed one, but typically α is well below one. Unfortunately, expression (1) does not make the power consumed by the clock visible. In most synchronous circuits the clock drives the largest capacitive load at an activity of two!

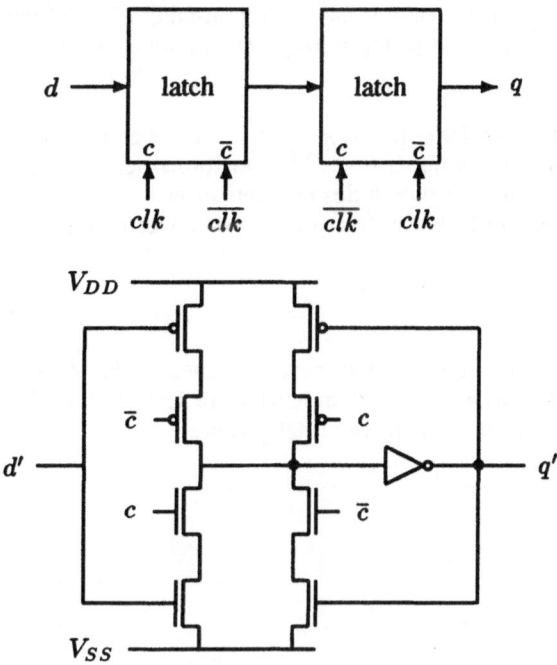

Figure 6: A typical edge-triggered D-register (top), comprising two identical CMOS latches (bottom).

In order to refine Expression (1) we look inside the typical clocked register of Figure 6. It consists of two identical latches, enabled on complementary clock phases. The master latch (left) is enabled when *clk* is high, and locked when *clk* is low. The slave latch (right) is enabled when *clk* is low. Note that if input *d* remains unchanged over a clock period, then output *q* won't change either and the transitions of the clock are redundant. This is potentially a major source of energy waste, because a clock fans out to hundreds (thousands) of registers and each register contributes to the clock load.

The energy wasted in these redundant clock transitions depends on the particular register circuit. The register of Figure 6 contributes substantially to the overall clock load with eight transistors. Dynamic register circuits using pass transistors contribute considerably less with only two transistors. (See [45] for a nice collection of different register circuits.) Here we take a modest C_L (the average capacitance) as the contribution of a register circuit to the clock load, that is, we count the clock power per register effectively as only two transitions per clock cycle.

Now we are ready to refine Expression (1). Let N_f be the number of flip-flops and N_g the number of gate equivalents in the combinational logic. Then the power consumption measured in transitions equals

$$f_c(2N_f + \alpha(4N_f + N_g))$$

Here we count 2 transitions for enabling a register and 4 transitions for changing the state of a register. Let k be the ratio of the number of combinational gates over the number of flip-flops, that is $N_g = kN_f$. Then above expression simplifies to

$$f_c N_f (2 + \alpha(k + 4)) \tag{2}$$

For example, a circuit of 1000 flip-flops and $k = 10$ operating at 10 megahertz with $\alpha = 0.1$ would consume about 3.4×10^{10} transitions per second, or approximately 50 milliwatts (assuming 1.5 picojoules per transition).

An interesting quantity is the fraction of the power consumption caused by the clock:

$$\frac{2}{2 + (k + 4)\alpha}$$

This ratio is plotted in Figure 7 for various values of k. Audio applications, for example, typically have $k \approx 10$ and an activity of less than 10%. Hence, the clock power is then as much as 60% of the total power.

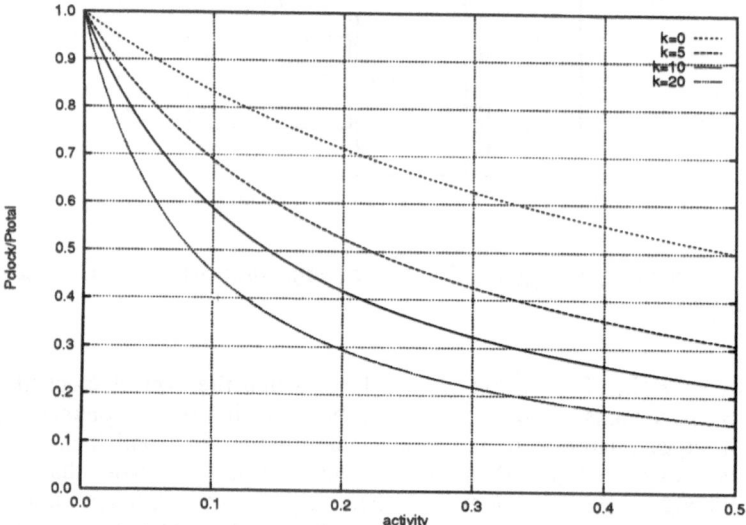

Figure 7: Clock power as a fraction of the total power in a synchronous circuit as a function of the circuit activity for various k (see text).

The circuit activity is often low (below 10%), for one or more of the following reasons.

- The clock frequency often exceeds the sample frequency by several orders of magnitude, in order to allow for time sharing of resources, such as busses, i/o pads, embedded memories, and multipliers. Typical audio ICs have sample rates of 100 kilohertz and are nevertheless clocked at several megahertz for this reason.

- Large ICs consist of a number of more-or-less independent modules. These modules generally have different *optimal* clock frequencies. Nevertheless, the number of different clock frequencies on an IC is kept small (usually one), to avoid problems with interfacing, clock distribution, and testing.

- The clock frequency must be such that a worst-case workload (in terms of number of clock cycles) can be handled in the allocated time. This generally implies an excessively high clock frequency for the average case, which becomes particularly expensive for a sub-module when the worst-case is an exception, or when stand-by is the rule.

Of course, not all clock power can be avoided. Registers (or latches) must be enabled to store new values, albeit at a rate lower than f_c. The problem is then: *when* must each individual register be enabled? For certain circuits, such as modulo-N counters and N-fold repeaters of Section 3 this can be predicted in a relatively straightforward and cheap fashion. For arithmetic circuits this is not so easy: when successive samples are not correlated, about half of the input bits will change; but which ones? Let the *enable efficiency* ε of a circuit be the fraction of effective register enables over the total number of register enables. The power required for the latch enabling can then be reduced from $2f_c N_f$ (see Expression (2)) to:

$$2f_c \frac{\alpha}{\varepsilon} N_f$$

For a synchronous circuit we have $\varepsilon = \alpha$. If each latch is enabled only when its state requires modification, we have $\varepsilon = 1$. If a group of latches is enabled collectively and half the bits change in state, we have $\varepsilon = 0.5$.

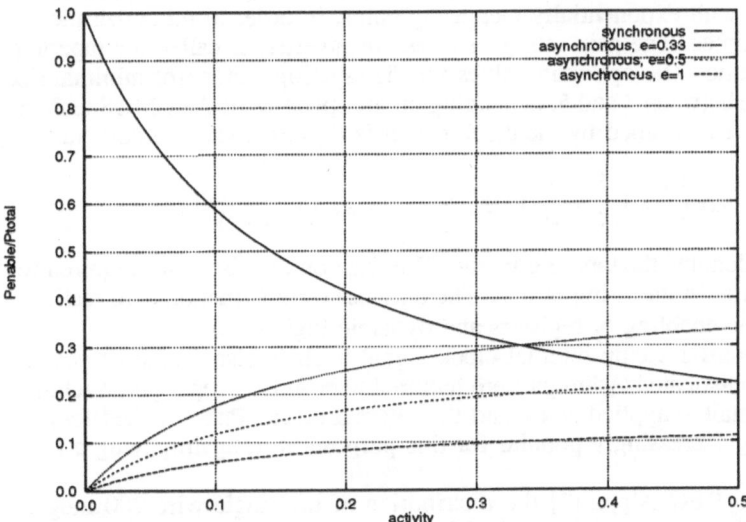

Figure 8: Enable power as a fraction of the total power in a *synchronous* circuit as a function of the circuit activity for various values of the enable efficiency ε.

In Figure 8, the enable power is plotted versus activity as fraction of the total power of a *synchronous* circuit, for several values of the enable efficiency ε. The power difference between the (normalized) clock power and enable power can be saved.

The combinational logic of Figure 5 is often realized using a clock as well. The role of a clock in these so-called dynamic-logic circuits generally is to precharge (pull-up) the output of a gate during one clock phase and to discharge (pull-down) that wire *conditionally* during the next phase. The earlier mentioned leakage currents have little effect on the outputs that remain floating during that next phase: the ratio of C_L over a usual leakage current is in the range of tens of microseconds, several order of magnitude in excess over a typical clock period. Dynamic techniques are standard in CMOS ROMs and PLAs. A revival of dynamic logic structures such as domino logic and DCVSL (see e.g. Chapter 5 in [45]) can be expected, because for combinational logic they tend to be more power efficient: each gate output makes either zero or two transitions, about one in average (depending on the specific form). For deeper combinational circuits, standard CMOS circuits exhibit glitches (redundant pairs of up/down transitions) which may, in extreme cases, dominate all other forms of power consumption [10]. DCVSL has been applied to asynchronous circuits in [13].

Dynamic forms of combinational logic may increase the clock capacitance dramatically. Dynamic ROMs, for instance used as instruction memory or table of filter constants, may easily exceed the clock power of the registers. When these dynamic structures are not used productively during each clock cycle, we have a similar situation as with our registers: their activity becomes less than one, and asynchronous techniques can be used to improve their enable efficiencies.

The *total clock load C_{clock}* is formed by the combined clock capacitances of all registers and dynamic logic gates, and the clock wire(s) connecting these. The clock driver that drives this capacitance (the triangle in Figure 5) is generally a string of inverters with exponentially increasing gains, in order to minimize the clock's rise and fall times. The gain ratio of successive inverters is called the tapering factor or step-up ratio. The optimum values for the tapering factor (for minimal rise and fall times) lie between 3 and 5, depending on the specifics of the CMOS technology [11]. The power consumed by the clock driver is (ignoring short-circuit currents)

$$f_c C_{clock} V^2 (1 + \frac{1}{t} + \frac{1}{t^2} + ..) \quad \approx \quad f_c C_{clock} V^2 \frac{t}{t-1}$$

where t denotes the tapering factor. The fraction of this power required to drive the last inverter in the string is $\frac{1}{t-1}$. The *only* use of this fraction is to realize *synchrony* among the enabling of registers and dynamic logic gates.

Generating a chip-internal clock signal itself is also a sink of power. Increasingly sophisticated techniques are required to ensure a proper phase relation between the externally-supplied clock and the internal clock. Phase-locked loop circuits are becoming increasingly popular for this purpose [45], again paying a price for synchrony.

In the DEC Alpha [7] the distribution of the single-wire 200 megahertz clock requires 7.5 watts out of a total of 30 watts. The clock driver and predriver dissipate another 4.5 watts! The fraction of power required for distributing a high-frequent clock (with limited skew and short transition times) tends to increase with circuit complexity.

2.3 Asynchronous circuits

We compare the power consumption of asynchronous circuits with clocked circuits, with reference to Figure 5 and the analysis of synchronous CMOS circuits of Section 2.2.

Obviously, asynchronous circuits can do without a clock driver. As we have seen above, this saves power even if we cannot save on the number of latch enables. It becomes more interesting if we can enable latches less frequently and thereby exploit the low activity in many digital applications. For control circuits this is investigated in Section 3. Attractive examples to demonstrate this are N-fold repeaters and modulo-N counters. Both examples exhibit an $O(\log N)$ advantage over their clocked counter parts. A generalization of some of the arguments explains why distribution of control offers a power advantage for asynchronous controllers, and generally not for clocked control circuits.

Moving data around from latches to latches (without computing new values) is investigated in Section 4. This is a class of circuits where the enable efficiency is essentially 0.5, assuming that successive input samples are not correlated. Using N-place shift registers as a vehicle, a wide range of different implementations is studied and compared in terms of circuit cost, performance, and power. Although the enable-efficiency is the same for all implementations, their power consumption varies considerably. Many of these alternatives have no clocked counter part, because the traditional bi-partition of phase-1 (enabled by clk) and phase-2 (enabled by $\overline{clk}$) latches (as required for synchronous operation) is not applicable.

If combinational logic is realized using static CMOS, there is no intrinsic power advantage in using asynchronous circuits. There may be an indirect advantage, because sometimes the clock dictates additional circuitry to match the timing requirements of that clock. In Section 5 this is illustrated by an N-bit wide incrementer/decrementer and parallel-to-serial converters. If dynamic circuits are used asynchronous circuits become advantageous if their activity is less than one.

The examples so far deal with the number of transitions required to realize a given computation. An alternative is to reduce the energy per transition by lowering the supply voltage. In Section 7 we describe one technique for doing so that takes advantage of the asynchronous operation of a circuit. In this concluding section we also review the measured power consumption of a number of complex asynchronous ICs designed at Philips Research Laboratories.

With the examples in the next sections we hope to learn when, why, and how to save power by applying asynchronous circuit techniques. We shall see that the power advantage will only rarely arise by simply copying the synchronous 'finite-state machine' architecture. Usually, "low power" requires design effort, or in our case a VLSI-programming effort. Also, often there is a price to be paid in circuit area and/or performance in order to reduce power. The amount of power saving and the trade offs with area and speed depend on the particular style of asynchronous circuits. In particular a number of design choices, such as 2-phase versus 4-phase handshaking and single-rail (bundled data) versus double-rail data encoding make a difference. Interestingly, in analyzing and illustrating the above factors we can abstract from these choices, by using handshake circuits. In the next five sections we discuss the above themes in some detail and illustrate them by carefully selected, practical examples. We analyze these examples by counting transitions, where a transition represents an energy of about one picojoule. But ...

Quantity	Energy	Remark
uv photon	-18	
neural transition	-13	varies with size
CMOS transition	-12	gate with 100 fF load
α-particle	-12	from space or IC package
sequential action	-11	cf. Section 3.2
8-bit assignment	-10	cf. Section 4.2
PCB transition	-10	10 pF load
ARM instruction	-8	
8bit-access 16Mb SRAM	-8	
DEC Alpha instruction	-7	
flight of a mosquito	-7	[3]
correcting DCC word	-6	cf. [37]
NiCd penlight battery	3	
can of beer	6	600 kJ
lead-acid car battery	6	5kg × 40 WH/kg
kg coal	7	
daily human consumption	7	2500 kilocalories
daily human energy use	9	
man-made nuclear explosion	14	Trinity (July 16, 1944)
1906 San Francisco earthquake	17	8.3 on the Richter scale
yearly world consumption	21	few 10^{10} tons coal equivalent
nova	37	[3]
big bang	73	[3]

Table 1: The $\log_{10}$ of the energy for various quantities [joules].

2.4 How much is a picojoule?

In Table 1 we compare the energy of a transition of a 1994 CMOS gate (about 1 picojoule) with a variety of energy quantities. Note that neural transitions are an order of magnitude more efficient (in average). However, with a reduction in supply voltages and finer geometries this relation may soon reverse.[4] Interestingly, the energy released by an α-particle is of the same order of magnitude as a CMOS transition. No wonder that the α-particle emission from minute traces of radioactive elements in the IC-package material may cause soft errors in dynamic circuits.

A typical action in a Tangram circuit is an assignment, which consumes about 100 picojoules for a word width of eight bits. Note that the execution of a single instruction of the most efficient commercially available 32-bit microprocessor around, the ARM, dissipates two orders of magnitude more. The instruction of a DEC Alpha consumes even a thousand times more energy! This clearly suggests that microprocessors, due to their general-purpose nature, are not particularly power efficient.

Alternatively, we can view a single (rechargeable) nickel-cadmium penlight battery as a source for CMOS transitions:

---

[3]Kindly provided by Jim Garside (Manchester University).

[4]A neural transition may have much more effect, however, since a neuron tends to fan out to hundreds of other neurons.

Energy of a nickel-cadmium penlight
≈   15 grams × 40 watthours/kilogram (cf. Figure 2)
≈   600 milliamperehours × 1.2 volts
≈   2.5 kilojoules
≈ $2,000,000,000,000,000$ 5-volts transitions

This calculation shows that the popular "nicad" is good for two quadrillion (!) 5-volts transitions. One of the major VLSI-design challenges of the next decade is: *make each transition work!*

3   Control

This section focuses on the control part of digital circuits, where handshaking reduces to simple request-acknowledge signaling, and where transport, storage, and manipulation of data can be ignored. After reviewing the basics of handshake signaling, we introduce three handshake components: the repeater, the sequencer, and the mixer. The latter two suffice to implement N-fold repeaters with remarkable cost, performance, and power properties. A fundamental drawback of this particular implementation of N-fold repeaters is the peculiar dependence of their energy dissipation on N, compounded by an $O(\log N)$ jitter. Both issues are addressed in the implementation of modulo-N counters. Their power and count frequencies are $O(1)$. The implementation of modulo-N counters requires one more handshake component: the selector. We conclude this section with a number of more general observations on control circuits, in particular related to the low-power consumption of asynchronous distributed controllers.

3.1   Handshake signaling

A clock signal in synchronous circuits is used for a variety of purposes, including the enabling of latches and the precharging of nodes in dynamic circuits. This clock signal is global in nature. When used in combination with positive edge-triggered flip-flops, for instance, the states of all flip-flops are updated shortly after the rising edge of the clock. This new state then ripples through the combinational circuits to compute the next state. It is generally required that the circuit is quiescent well before the next rising edge of the clock, by a timing margin called the set-up time. This simple requirement, together with the global nature of the clock, permits a variety of structured design methods. At the same time, the global nature of the clock is also a limitation. It requires increasingly more design effort to maintain this global synchrony, that is, to bound clock skew.

Structured design of asynchronous circuits requires a timing discipline to replace that of a global clock. In this section we shall argue that a consistent application of handshake signaling, at all design levels, is such a discipline.

Handshake signaling between two partners is commonly realized by using two wires, one for signaling a request, the other for signaling an acknowledge to that request [31]. Assume both wires to be low in their initial state. The first *2-phase* handshake amounts to the sequence "request up ; acknowledge up". The next 2-phase handshake is then "request down ; acknowledge down". The partner that has control over the request wire plays the active role (i.e., takes the initiative and starts

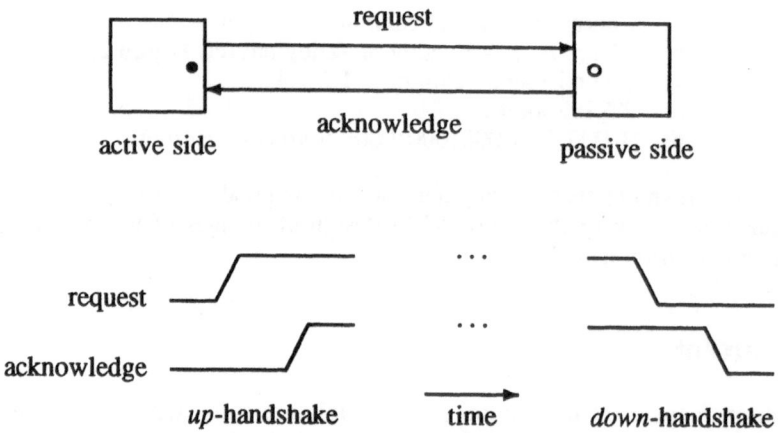

Figure 9: A handshake configuration (top; symbols ● and ○ denote the active and passive sides respectively) and two 2-phase handshakes (bottom).

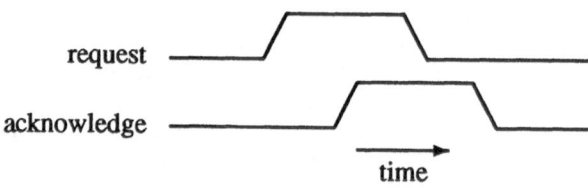

Figure 10: A four-phase handshake.

a handshake), the other partner plays the passive role (i.e., waits and acknowledges), see Figure 9.

The request can be interpreted as a local clock signal. The acknowledge is then an explicit indication of the completion of the action triggered by the request. In synchronous circuits this completion is left implicit. A disadvantage of the 2-phase handshake protocol is that the state of the wires after a handshake differs from the state before that handshake (both high after both low, or vice versa). A *4-phase* handshake consists of two successive 2-phase handshakes, and hence leaves the wires in their initial state (here both low; see also Figure 10). In comparison with 2-phase handshaking, 4-phase handshaking may take more time and energy (especially with long wires), but generally results in simpler and cheaper circuits. Furthermore, 4-phase signaling allows self-initialization, a form of circuit initialization ("reset") that does not require additional hardware [37, 38]. The 2-phase protocol has been applied by [33, 1, 8] and others; 4-phase handshake signaling for instance by [19]. There has been considerable debate about 2-phase versus 4-phase signaling, resembling the debate on 2-stroke versus 4-stroke combustion engines. Because 2-phase signaling is conceptually simpler and 4-phase signaling appears faster and cheaper, there seems

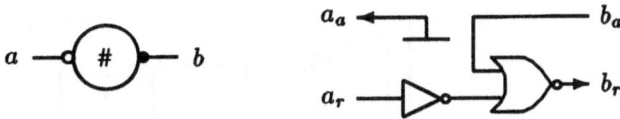

Figure 11: Symbol and circuit for a handshake repeater.

to arise a consensus on "think 2-phase; build 4-phase". In the sequel we apply 4-phase handshake signaling.

3.2 Control handshake components

This section introduces three basic handshake components, namely the repeater, the mixer, and the sequencer. The repeater is used to control the unbounded repetition of a (compound) action. The sequencer controls the sequential execution of two (compound) actions, for instance the writing in and reading from a memory. The mixer provides access to a shared resource to two parties. These are the three most frequently used handshake components in the control parts of many practical applications.

A repeater has passive port a and active port b. After a request through a the repeater responds with an unbounded series of handshakes through b. The symbol for a repeater and its implementation are depicted in Figure 11. In the initial state the two input and two output wires are low. After wire a_r is pulled up, the repeater responds with $b_r \uparrow$; a subsequent acknowledge $b_a \uparrow$ yields $b_r \downarrow$, etcetera. The request through a is never acknowledged. Hence a_a is grounded. The energy required for one complete (4-phase) cycle of this repeater, denoted by E_{rep}, is two transitions.

A more interesting handshake component is the sequencer of Figure 12. It has passive port a and active ports b and c. A request through a triggers a complete handshake through b, followed by one through c, followed by the completion of the handshake through a. Its behavior in terms of transitions on individual wires is captured by the regular expression given below, in which internal state variable x (initially high) has been introduced to distinguish between the states directly before and directly after the b handshake.

$$(a_r \uparrow \quad b_r \uparrow \quad b_a \uparrow \quad x \downarrow \quad b_r \downarrow \quad b_a \downarrow \quad c_r \uparrow \quad c_a \uparrow \quad a_a \uparrow \quad a_r \downarrow \quad x \uparrow \quad c_r \downarrow \quad c_a \downarrow \quad a_a \downarrow)^*$$

The handshakes through b and c are mutually exclusive. Given this state encoding, the circuit of Figure 12 follows straightforwardly. It is built around a $\overline{set}$-$reset$ latch (the circuit enclosed by the dashed box). A high on b_a (with a_r high) resets the latch (pulls x down) and a low on a_r (with b_a low) sets the latch (pulls x up). Notice that only transistors $P1$ and $N1$ are involved in (re)setting x. The other four transistors and the inverter are applied to make x static. They can be of minimum size, or can even be omitted if the sequencer is invoked frequently enough. The energy required for one complete cycle of the sequencer E_{seq} is 10 transitions, that is 2 transitions on wires x, y, b_r, $\overline{b_r}$ (inside the AND gate), and c_r.

168

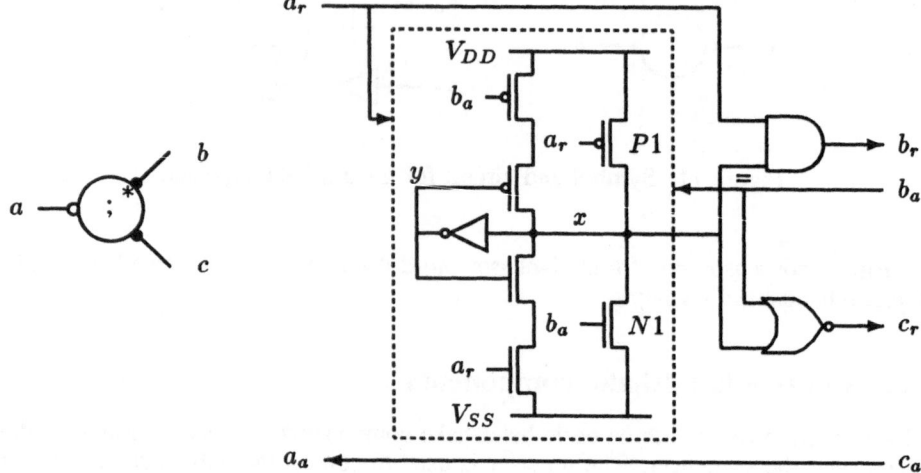

Figure 12: Symbol and CMOS circuit for a handshake sequencer. The asterisk labels the active port that is served first.

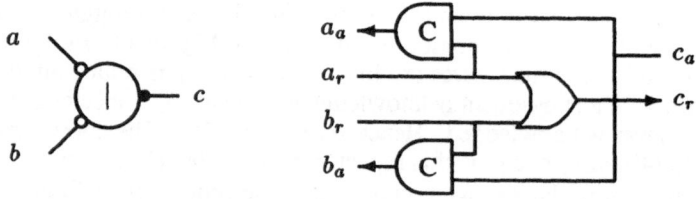

Figure 13: Symbol and circuit for a handshake mixer.

The mixer of Figure 13 has passive ports a and b, and active port c. Each handshake through a passive port encloses a handshake through c. Its correct operation requires mutual exclusion of handshakes through a and b. This requirement must be fulfilled by the mixer's environment. The OR gate simply passes the incoming request by making c_r high. An acknowledge arriving at c_a either results in $a_a \uparrow$ or $b_a \uparrow$, depending on the source of the incoming request. The acknowledge is generated by a so-called Muller-C element. This is a kind of latch, of which the output is set when *both* inputs are high, and reset when *both* inputs are low. The latter is essential for the mixer to prevent a premature $a_a \downarrow$ after $a_r \downarrow$. The energy required for one complete cycle of the mixer E_{mix} is 8 transitions (2×2 in both the C-element and the OR gate), serving either a or b.

This small selection of handshake components has prepared us for a first design of a handshake circuit: a 2^M-fold repeater.

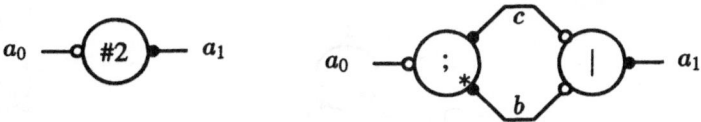

Figure 14: Duplicator: symbol and handshake circuit.

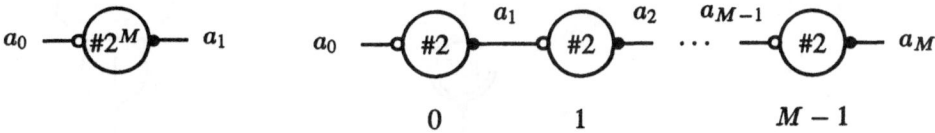

Figure 15: 2^M-fold repeater composed of M duplicators.

3.3 2^M-fold repeaters

By connecting the two active ports of a sequencer to the passive ports of a mixer we obtain the *duplicator* handshake circuit of Figure 14: each handshake through a_0 encloses two handshakes through a_1. That is, a request through a_0 triggers a handshake along b, followed by one along c. The latter two are merged by the mixer through handshake port a_1. The energy required for one duplication E_{dup} equals $E_{seq} + 2E_{mix}$, which amounts to 26 transitions.

M duplicators, $M > 0$, can be chained to form a 2^M-fold repeater, with a_0 as passive port and a_M as active port (Figure 15). Let a_i, $0 < i < M$, be the handshake channel connecting duplicators $i - 1$ and i. Observe that for each handshake through a_0 we have 2^i handshakes along a_i. Viewed differently, duplicator i does twice as much work as duplicator $i - 1$, four times as much as duplicator $i - 2$, etcetera. Consequently, duplicator $M - 1$ dissipates more than all other duplicators together!

For a given cycle frequency f at port a_M, the power dissipated by duplicator $M - 1$ is $f E_{dup}$. The power of the 2^M-fold repeater is then bounded by $2f E_{dup}$ and hence independent of M.[5] A 2^M repeater running at 1 megahertz consumes 2×10^6 E_{dup} per second, or about 50 million transitions/second, or about 70 microwatts in a 5-volts CMOS realization.

3.4 N-fold repeaters

Figure 16 shows how an N-fold repeater can be constructed for arbitrary N, $N \geq 1$. The scheme is simple. For $N = 1$ the repeater reduces to a pair of wires connecting the respective request and acknowledge wires. For even values of N, the repeater consists of a $N/2$-fold repeater and a duplicator. For odd N, $N \geq 3$, an $(N-1)$-fold

---

[5]Strictly speaking, this is not always true. An N-fold repeater operating at a frequency well below 1 kilohertz dissipates most of its of energy in leakage (pitfall 2 of Section 2.1). Its power dissipation is then, of course, proportional to M.

170

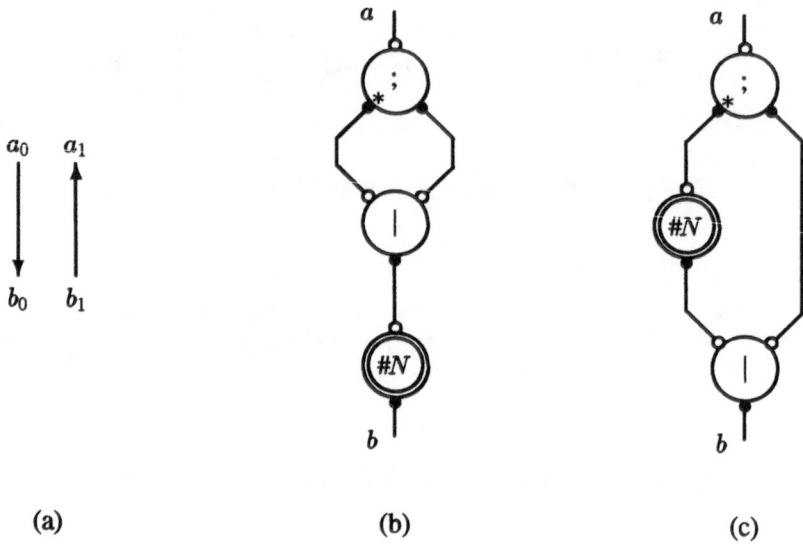

(a)   (b)   (c)

Figure 16: A 1-fold repeater (a), a $2N$-fold repeater (b), and an $N + 1$-fold repeater (c). The latter two are based on N-fold repeaters.

repeater is used, and a sequencer and mixer are added to extend the repetition cycle with one b handshake.[6]

The area for an N-fold repeater, denoted by $A_{Nrep}(N)$, is given by $A_{Nrep}(1) = 0$ and for $N > 1$ by

$$A_{Nrep}(N) = A_{Nrep}(N \textbf{ div } 2) + (1 + N \textbf{ mod } 2)A_{dup}$$

See also Figure 17. Note that

$$
\begin{aligned}
A_{Nrep}(2^M) &= M A_{dup} \\
A_{Nrep}(2^M - 1) &= 2(M - 1)A_{dup}
\end{aligned}
$$

Hence, $A_{Nrep}(N)$ is bounded by $\log_2(N)A_{dup}$ from below and $2\log_2(N)A_{dup}$ from above.

The energy for an N-fold repetition, denoted by $E_{Nrep}(N)$, is given by $E_{Nrep}(1) = 0$ and for $N \geq 1$:

$$E_{Nrep}(N) = 2E_{Nrep}(N \textbf{ div } 2) + E_{dup} + (N \textbf{ mod } 2)(E_{seq} + N E_{mix}) \qquad (3)$$

The last term only contributes for odd N (see Figure 16.c). The average energy per tick, $\frac{1}{N}E_{Nrep}(N)$, is shown in Figure 18. Note that

$$E_{Nrep}(2^M) = (2^M - 1)E_{dup}$$

---

[6]This scheme is *not* optimal. Using sequencers and mixers only, cheaper realizations may be found for some N. See also the appendix on the Cheapest Unary Expression.

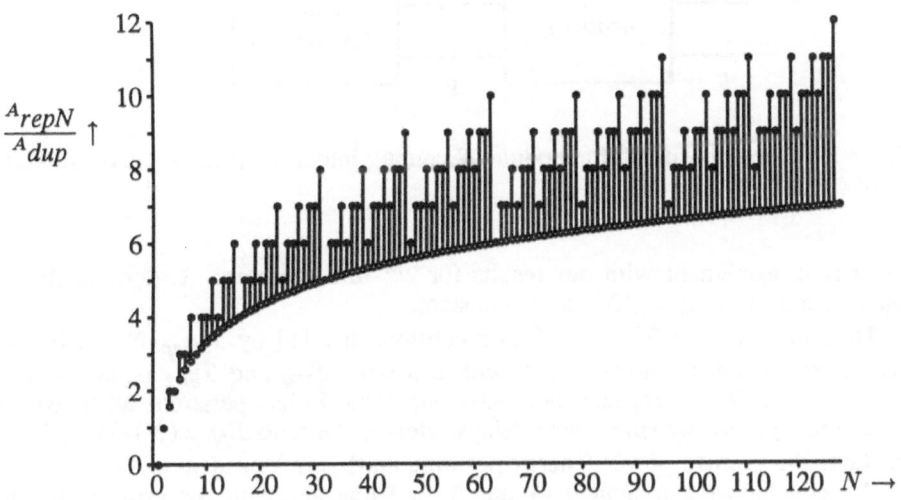

Figure 17: Bullets show $A_{Nrep}(N)$, the area of N-fold repeaters. Circles show $\log_2 N$. Both are normalized on A_{dup}, the area of a duplicator.

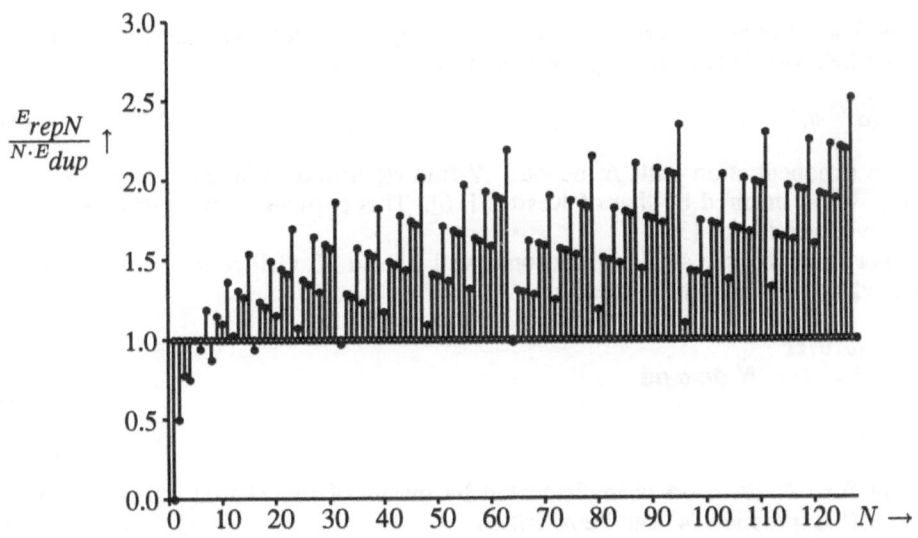

Figure 18: Bullets show the energy per tick of N-fold repetition $\frac{1}{N} E_{Nrep}(N)$, with $E_{Nrep}(2) = E_{dup}$ as reference and unit.

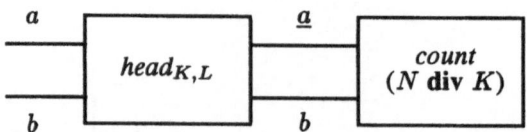

Figure 19: Decomposition of a modulo-N counter into a cell $head_{K,L}$ and a modulo-(N **div** K) counter.

which is in agreement with our results for 2^M-fold repeaters. Unfortunately, the upper bound for $\frac{1}{N}E_{Nrep}(N)$ is *not* constant.

The time required for an N-fold repetition, denoted by $T_{Nrep}(N)$, satisfies a recurrence relation identical to (3), with constants T_{seq} and T_{mix} replacing E_{seq} and E_{mix}. An N-fold repeater is strictly sequential in its operation: all transitions are ordered, spaced by single gate delays. Hence, the ratio $E_{Nrep}(N)/T_{Nrep}(N)$ is constant: the average power (energy per time unit) is independent of N.

We conclude the treatment of the N-fold repeater with the observation that its performance is highly dependent on N, both in *average* response times and in energy. For N a power of 2, both the *average* energy per tick and the response time are nevertheless attractive. Still, the worst case response times are $O(\log N)$. This excessive "jitter" may be unacceptable in performance critical situations. Modulo-N counters, implemented with internal parallelism, offer a solution to both problems.

3.5 Modulo-N counters

A modulo-N counter $count_N$, $N \geq 1$ is a component with external ports a and b, whose behavior is specified by the regular expression

$$(a^N\ b)^* \tag{4}$$

where exponentiation with N denotes N-fold repetition. We apply a systolic decomposition inspired by that of Kessels [16]. This particular implementation is an improved version of [36].

For small N, say smaller than some well chosen K, Cell $count_N$ can be implemented by the Tangram expression

$$
\begin{aligned}
&\textbf{forever} \\
&\textbf{do}\quad \textbf{for}\ N\ \textbf{do}\ a\ \textbf{od} \\
&\ ;\quad b \\
&\textbf{od}
\end{aligned}
\tag{5}
$$

where **for** N **do** a **od** is implemented by means of the N-fold repeater of Section 3.4. Expression (4) can be rewritten as

$$(a^{N\ \textbf{mod}\ K}\ (a^K)^{N\ \textbf{div}\ K}\ b)^*$$

For $N \geq K$, this suggests a decomposition of the counter into a modulo-(N **div** K) counter and cell $head_{K,L}$ as depicted in Figure 19. The channels connecting the head cell with the sub-counter are named $\underline{a}$ and $\underline{b}$ respectively. The behavior of $head_{K,L}$ is given by the regular expression

$$(a^L \ (\underline{a} \ a^K)^* \ \underline{b} \ b)^*$$

where $L = N \ \textbf{mod} \ K$. The task of the $(N \ \textbf{div} \ K)$-fold repeating has been delegated to the subcounter. For $head_{K,L}$ we derive[7]

$$(a^L \ (\underline{a} \ a^K)^* \ \underline{b} \ b)^*$$

$$= \ ((a^L \ \underline{a} \ a^{K-L})^* \ a^L \ \underline{b} \ b)^*$$

$$= \ (a^L \ \underline{a} \ a^{K-L} + a^L \ \underline{b} \ b)^*$$

$$= \ (a^L \ (\underline{a} \ a^{K-L} + \underline{b} \ b))^*$$

Casting the above regular expression in Tangram yields

> **forever**
> **do**   **for** L **do** a **od**
> ;   **sel** $\underline{a}$; **for** $K - L$ **do** a **od**
> | $\underline{b}$; b
>   **les**
> **od**

where the bracket pair **sel les** encloses Tangram's selection construct. Here it offers the environment the choice to communicate either along channel $\underline{a}$ or along $\underline{b}$. Repeated application of the decomposition of Figure 19 results in an array of $N \ \textbf{div} \ K$ cells of type $head_{K,L}$, and a tail cell based on Tangram Expression (5).

Before we address the handshake circuit implementation of $head_{K,L}$ we introduce a fourth handshake component, the selector. Its symbol and a gate-level realization are depicted in Figure 20.[8] The operation of the selector is as follows. Once activated along a (i.e., once a_0 has been pulled up) the sequencer "opens" the selector for a choice through b or d. That is, the environment is expected to make a choice by either pulling up b_0 or d_0. This pull-up may actually precede the pull-up of a_0. However, the selector assumes mutual exclusion of handshakes through b and d. The choice made is stored in a so-called handshake latch (see Section 4.2). The acknowledge from this latch is returned to the sequencer, which subsequently "closes" the selector. The sequencer then reads the choice of the variable and thereby starts an active handshake along either c (if b had been selected) or e (otherwise). The energy E_{sel} required for a selection equals 10 (the sequencer) + 4 (one of the two C-elements) + 8 (the handshake latch) + 4 (the OR gate), adding up to 26 transitions.[9]

The handshake circuit for $head_{K,L}$ is depicted in Figure 21. For $L = 0$ this circuit can be simplified by removing the sequencer and the mixer. Of course, for $L = 1$ and $K - L = 1$ the corresponding 1-fold repeaters can be eliminated. Handshake circuits $head_{2,0}$ and $head_{2,1}$, for example, are given in Figure 22 (adapted from [36]).

---

[7]The derivation for $K = 2$ is due to Frits Schalij.

[8]The circuit has been suggested by Ad Peeters (Eindhoven University of Technology).

[9]The handshake latch may consume two transitions fewer if the input chosen is identical to the one chosen before.

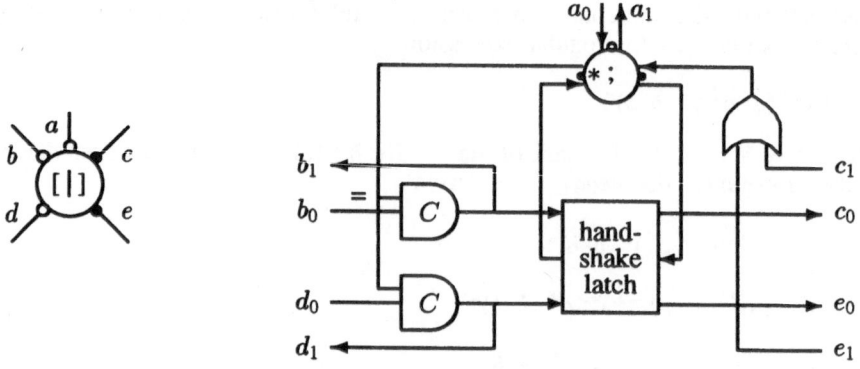

Figure 20: Symbol and circuit for a handshake selector. (Port and wire names are local to these diagrams.) A CMOS realization of the handshake latch is given in Figure 24.

For simplicity we take $A_{sel} + A_{rep} \approx 2A_{dup}$. Then we have

$$
\begin{aligned}
A_{head}(2,0) &= A_{head}(2,1) & \approx 3A_{dup} \\
A_{head}(4,0) & & \approx 4A_{dup} \\
A_{head}(4,1) &= A_{head}(4,2) = A_{head}(4,3) & \approx 5A_{dup}
\end{aligned}
$$

A modulo-N counter can be realized with N **div** K cells of type $head_{K,i}$ and a tail cell based on Equation (5). Hence, for large N, a counter based on $K = 2$ costs approximately $3 \log_2(N) A_{dup}$. For $K = 4$ the costs are about 4 to 5 times $\log_4(N) A_{dup}$, saving about 30% in the average.

The energy per cycle of the head cell depends on the selection made at the tail, i.e., the choice between $\underline{a}$ and $\underline{b}$. The most frequent case is the choice for $\underline{a}$ (N **div** K versus N **mod** K). This clearly is also the most dissipating case, with

$$
E_{head}(K, L) = E_{rep} + E_{sel} + E_{Nrep}(L) + E_{Nrep}(K - L) + M
$$

where M denotes the conditional contribution of the sequencer and the mixer:

$$
\begin{aligned}
M \ &= 0 & \text{if } L = 0 \\
&= E_{seq} + N E_{mix} & \text{if } L > 0
\end{aligned}
$$

More specifically,

$$
E_{head}(2,0) = E_{head}(2,1) = E_{rep} + E_{sel} + E_{dup} \approx 2E_{dup}
$$

And $E_{head}(4, i)$ is in the range of 4 to 5 times E_{dup}. (Recall that $E_{Nrep}(4) = 3 E_{dup}$.)

In analyzing the energy dissipation of the modulo-N counter we start with the observation that cell $head_{K,i}$ is involved in about K times as many communications along a than along $\underline{a}$. Consequently, the modulo-(N **div** K) counter in Figure 19 dissipates only about $\frac{1}{K}$-th of the energy of the head cell. Hence, the energy required

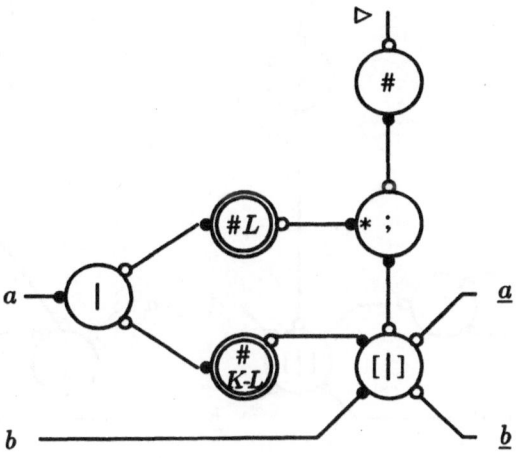

Figure 21: Handshake circuit for $head_{K,L}$.

for N ticks of the complete modulo-N counter, denoted by $E_{modN}(N, K)$, is defined by

$$E_{modN}(N, K)$$

$$\approx \frac{N}{K} E_{head}(K, N \bmod K) + E_{modN}(N \operatorname{div} K, K)$$

$$\approx \frac{N}{K} E_{head}(K, N \bmod K) + \frac{1}{K} E_{modN}(N, K)$$

Solving this equation yields

$$E_{modN}(N, K) \approx \frac{N}{K - 1} E_{head}(K, N \bmod K)$$

Hence, the energy required for $a^N\ b$ is linear in N, and the "energy per tick" is therefore independent of N. This property is called *constant power* in [36] and is not feasible for globally clocked circuits, where the power is proportional to the number of flip-flops to store the state of the repeater (at least $K + 1$). For $K = 2$ the energy dissipation of the asynchronous modulo-N counter is approximately $2E_{dup}$ per tick and for $K = 4$ at most $\frac{5}{3} E_{dup}$ per tick.[10] Modulo-N counters with similar properties have also been reported in [6].

One question remains: how to choose K? We have seen that $K = 4$ is an improvement over $K = 2$, both in terms of area and power. The advantage, however,

---

[10]As with the N-fold repeaters, leakage may become dominant when the counter operates at low frequencies.

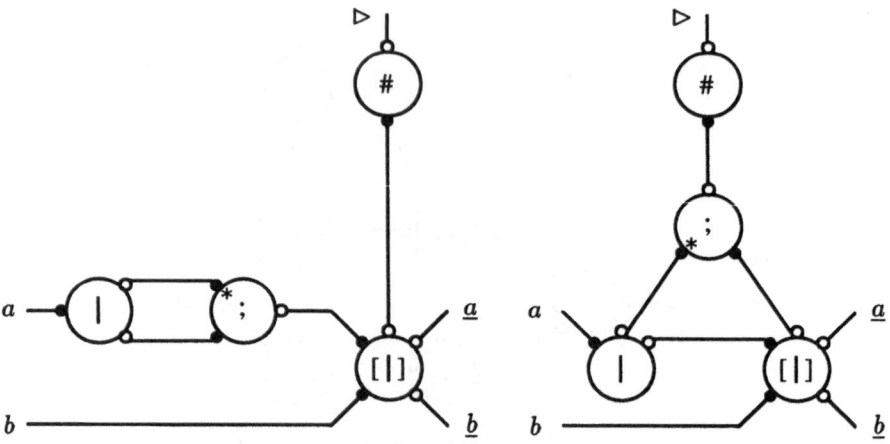

Figure 22: Handshake circuits for $head_{2,0}$ and $head_{2,1}$

depends on $L = N$ **mod** K for each cell, and tends to disappear for larger L just below a power of 2. (See also Figures 17 and 18.) Also, the jitter may increase with K, although the upperbound for the jitter has become independent of N.

The observation that the cell next to the head cell operates only at $\frac{1}{K}$-th of the speed of the head cell suggests the following practical strategy for large N. Take two head cells with small K (say, at most 4) and one tail cell (Tangram program (5)) for the remainder. Only if the jitter of the tail still "shines through" the two head cells, a third head cell should be considered or their K values should be increased. For large N this strategy combines the cost advantage of N-fold repeaters with the low jitter and high performance of modulo-N counters.

A modulo-100 counter based on cells with $K = 2$ has been realized on silicon in a 1μ CMOS process [40]. In this circuit the output wire b_0 is fed back to b_1 via an AND gate. The observed count rate exceeds 80 megahertz at a 5 volt supply voltage. Discounting the above AND gate would result in a count rate of approximately 100 megahertz. Simulation of the energy consumption of a modulo-100 counter yields 6 nanojoules for $a^{100} b$, about 60 picojoules per a handshake. These simulations have been confirmed by measurements. At a count rate of 80 megahertz this corresponds to a power consumption of less than 5 milliwatts. This 60 picojoules correspond also to $2E_{dup}$ or 52 transitions, about 1.2 picojoules per transition.

3.6 Control distribution

Consider a sequence of N actions (e.g. assignments or communications). The control circuit for executing this sequence can be realized by cascading $N - 1$ sequencers. The control for executing this sequence requires $(N - 1)E_{seq}$ transitions, amounting to a constant energy overhead per action. Recall that the energy overhead for unbounded and N-fold repetition of the same action is also constant. This is quite typical for asynchronous circuits with a high degree of control distribution. During each step typically 1 or 2 state variables change state. Furthermore state encoding and decoding is often extremely simple, involving a few transitions only.

In a centralized controller, the state comprises tens or sometimes hundreds of registers, all enabled at frequency f_c. State encoding and decoding is often quite complex and deep, often involving power-burning dynamic ROMs. Most synchronous circuits have centralized controllers. Distribution of control is possible for synchronous circuits as well. However, since it generally requires *additional* state variables (enabled at frequency f_c), it may very well increase the power consumption.

For applications with complex control structures, the natural match of control distribution with asynchronous operation may be a major source of power savings.

4 Moving data

In this section we address the asynchronous implementation of data paths. As with control, we shall build such data paths with a small set of building blocks that use handshake signaling as their only means of interaction. The communication of data by means of handshake signaling introduces the issue of the data encoding. Various alternative encodings have been applied, with large differences in circuit cost, speed, power, and design effort. This choice among encodings is orthogonal to the choice between 2-phase versus 4-phase handshake signaling as addressed in Section 3.1. Once introduced to data encoding, we will be able to address communication, storage, and manipulation of data. (The latter forms the subject of Section 5.)

First we focus on simple assignments of the form $x := y$, where source y and destination x are distinct variables. The energy required for such assignments depends linearly on the word width measured in bits, as we shall see. Using such simple assignments, some disguised as input or output statements, a large variety of data-structure-like programs can be designed. In this section shift-registers are used as vehicle to analyze power consumption of such structures and to investigate the trade-offs between power, costs, and performance.

The construction of the relevant handshake circuits requires two new handshake components: the handshake latch and the transferrer. Furthermore the mixer of Section 3 requires generalization.

4.1 Data encoding

Data can be encoded in the request phase (corresponding to active outputs and passive inputs), or in the acknowledge phase (yielding passive outputs and active inputs), or in both. Below we first consider the encoding in the request phase. Bi-directional handshake channels with data encoded in both phases are not discussed in this chapter, as they are not required for the compilation of the current version of Tangram.

The two most common data encodings [31] are *single rail* (bundled data) and *double rail*, see Figure 23. Single-rail encoding uses one wire per bit and one data-valid wire per data word containing, say, W bits. Sending a message amounts to encoding the message in the levels of the W data wires and then pulling up the data-valid wire. When the data-valid is high, the data wires must be stable. After receiving an acknowledge the data-valid wire must be pulled down. Limited control over wire delays generally requires a safety margin akin to the set-up time in clocked

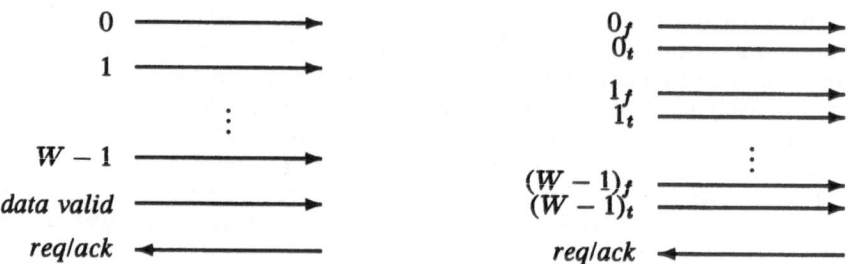

Figure 23: Single-rail data encoding (left) and double-rail encoding (right). In both encodings *req/ack* denotes the request wire if data is communicated from the passive to the active side, and the acknowledge wire otherwise.

circuits. Depending on the duration of this set-up time, more or less effort must be spent on bounding the difference among the delays of these $W + 1$ wires ("data bundling"). Single-rail encoding has been applied in [33, 1, 8].

In contrast to single rail, double-rail encoding is very robust: it is safe even when the individual wires have widely different delays. It requires two wires per bit, which are both initially low. A message can be communicated by pulling up exactly one wire per bit pair. The arrival of the message at the receiving end can easily be detected by checking if one wire per bit is high. Consistent with four-phase operation the message is erased by pulling down the high wires after an acknowledge has been received ("return-to-zero"). A double-rail handshake channel of W bits requires $2W + 1$ wires. As W wires make 2 transitions with each handshake, double-rail encoding is clearly less energy efficient than single-rail encoding. Double-rail encoding has been applied in [19, 35]. Alternative delay-insensitive encodings are presented in [43].

4.2   Energy of an assignment

A simple assignment involves the transfer of data stored in one variable to another variable. For the storage and transfer of data we introduce two new handshake components, namely the handshake latch and the transferrer. These are the two most common handshake components used in data-paths. The handshake latch corresponds to a register in synchronous circuits. The transferrer forms the prime interface with the control part of handshake circuits.

The handshake latch of Figure 24 has two passive ports, w for writing and r for reading. A write handshake starts by pulling up either wire w_0 (write "zero"), or wire w_1 (write "one"). This double-rail encoded value is stored in a conventional latch consisting of a cross-coupled pair of NOR gates. Then an acknowledge is generated by making w_a (write acknowledge) high. This indicates that the new value has arrived *and* that the content of the latch is consistent with this value ("completion detection"). The four-phase write handshake is completed by making w_0 (w_1) low again, followed by a down-going transition on w_a. Similarly, a read handshake starts by making r_r (read request) high. The handshake latch responds by making either r_0 or r_1 high, depending on the value stored in the latch. Then r_r is made low, followed by a down-going transition on r_0 or r_1. After each write

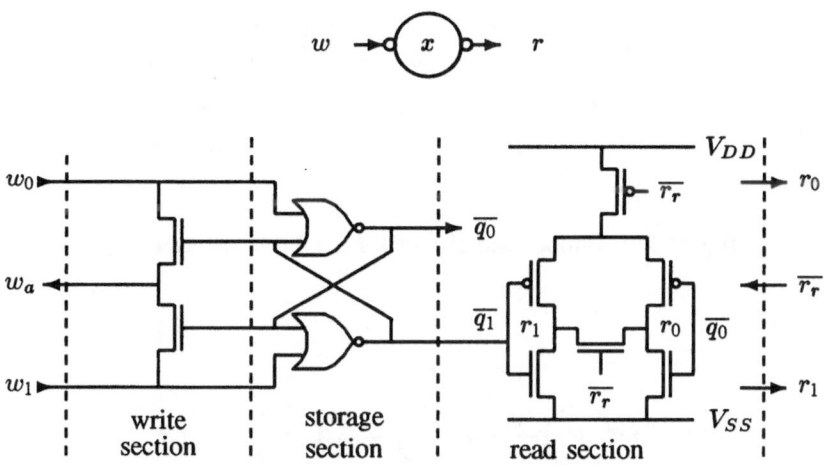

Figure 24: Handshake latch x: symbol and CMOS circuit.

or read handshake all external wires (w_0, w_1, w_a, r_r, r_0, and r_1) are low again. Read and write handshakes must be mutually exclusive, which can for example be realized by using a sequencer, as is shown below.

The energy required for a write action amounts to 2 or 4 transitions, depending on the occurrence of a state transition of the storage section. A read action requires 4 transitions, 2 for the inversion of the read request, and 2 for pulling up and down either r_0 or r_1.

The handshake latch of Figure 24 stores 1 bit. W-bit wide versions can be constructed by placing W handshake latches in parallel. The W read requests can be connected to form a collective read request. The combination of the W write acknowledge signals requires a W-input Muller-C element: its output must rise after all W latches pull up their acknowledge and its output must fall after these W signals have become low. Such a large fan-in C-element can, for instance, be constructed as a tree of binary C-elements. Such a tree requires $W - 1$ C-elements. For large W this results in an additional 4 transitions in energy per bit per write action. Worse is the impact of this so-called completion tree on performance. The generation of the completion signal requires a time proportional to the depth of the tree, which is of $O(\log W)$. A slight improvement can be expected by considering C-elements with more than 2 inputs, thereby reducing the depth of the tree.

The transferrer of Figure 25 has passive control port a, active input port b, and active output port c. A request arriving at a_r is passed without delay to b_r. Incoming data through b is simply passed through c, and an acknowledge arriving at c_a is passed to a_a. The return-to-zero cycle follows the same sequence. As its name suggests, the primary use of the transferrer is to transfer data from a source (e.g. a read port of a handshake latch) to a destination (e.g. a write port). Generalization to multiple bits is straightforward. Besides being an indispensable component, the transferrer is also a real bargain: its circuitry contains no gates, its operation introduces no delay and requires no energy!

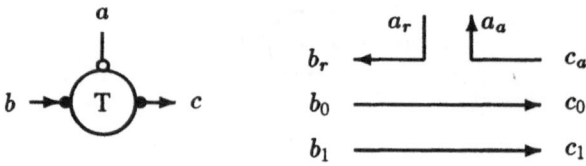

Figure 25: Symbol and circuit for a 1-bit transferrer.

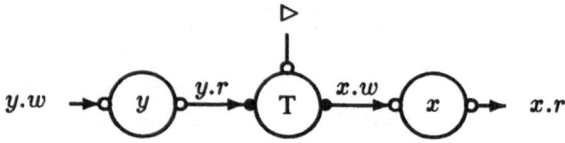

Figure 26: Handshake circuit for the assignment $x := y$.

A simple assignment involves two handshake latches and one transferrer (see Figure 26). The transfer is executed by a performing a handshake through passive port ▷ (pronounced "go"). The energy required for a simple assignment, denoted by $E_{:=}$, depends on the word width W and comprises $4W$ for reading y and at most $8W$ for writing x, corresponding to a total of at most $12W$. If we assume that about half the bits change value, a simple assignment of 8-bits wide dissipates 11×8, or 88 transitions. Assuming about 1.5 picojoules per transition, this agrees nicely with a measured energy of about 140 picojoules. In the remainder of this section we shall assume a fixed W, and use $E_{:=}$ and $T_{:=}$ as constants to denote energy and time required for an assignment. Constant A_{var} denotes the area of a W-wide handshake latch.

The handshake latch and transferrer have also prepared us for a family of practical and useful handshake circuits: ripple registers.

4.3 N-place ripple registers

A shift register has one input port and one output port of a given type T (which we shall assume to be W bits wide). We assume a storage capacity of N words, $N > 1$. Its behavior must satisfy two requirements:

1. Output and input actions must strictly alternate, starting with an output.

2. After the first N outputs, the subsequent sequence of output values must be a copy of the sequence of input values.

The first class of shift-register implementations are called *ripple registers*, because data ripple through the circuit, from input to output. A 2-place ripple register is

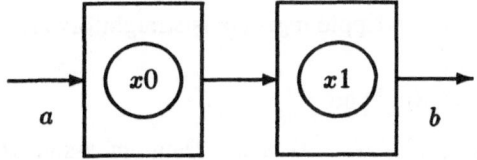

Figure 27: 2-place ripple register.

described by the following Tangram fragment.

proc $(a?T \ \& \ b!T)\cdot$
begin
   $x0, x1 :$ **var** T
| **forever do** $b!x1 ; x1 := x0 ; a?x0$ **od**
end

Here $b!x1$ denotes the output of the value stored in $x1$ through handshake port b.
Similarly, $a?x0$ stands for an input through handshake port a, where the incoming
value is stored in $x0$.

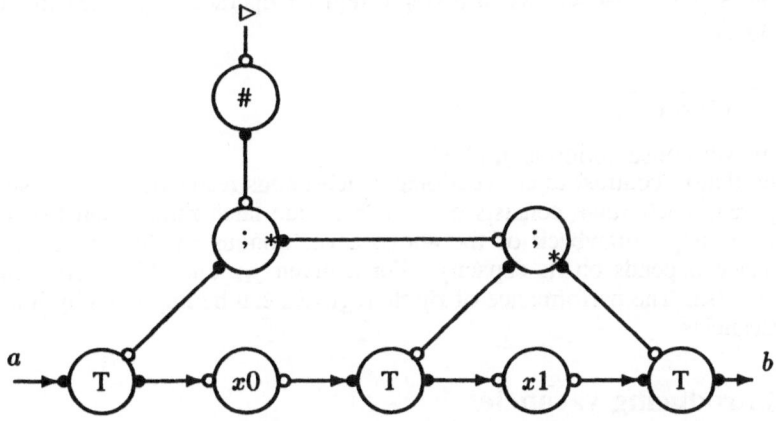

Figure 28: Handshake circuit for a 2-place ripple register.

A handshake circuit for a 2-place register is depicted in Figure 28. The data
path comprises two handshake latches ($x0$ and $x1$, both of width W), and three
transferrers to move data. These transferrers are cyclically served from right to
left under control of a small handshake circuit consisting of a repeater and two
sequencers. The operation of the ripple register is started by a request through port
▷. Comparison of the Tangram text with the handshake circuit of Figure 28 reveals
a direct structural correspondence. The repeater implements the **forever do** .. **od**
construct and the sequencers the semicolons in the program text. The transferrers
implement the input '?', the assignment ':= ', and the output action '!'.

Generalization to N-place ripple registers is straightforward. This is captured by

proc $(a?T \ \& \ b!T)\cdot$
forever do $ripple_N(a, b)$ **od**

where statement $ripple_0(a, b)$, $N > 0$ is the Tangram assignment $b := a$ and for $N > 0$, statement $ripple_N(a, b)$ is defined as

begin x : **var** T | $ripple_{N-1}(x, b)$; $x := a$ **end**

The statement $ripple_N(a, b)$ defines basically a sequence of N variables, internally connected by $N - 1$ assignments, and connected to the environment by two more assignments. It is generalized by allowing a and b to denote channels as well: if a is declared as a channel, assignment $x := a$ should be replaced by input statement $a?x$. Likewise, $b := x$ should be replaced by output statement $b!x$. (For simplicity we exclude a and b both being channels. This particular description using $ripple_N$ appears convenient when we describe other types of shift registers.) Execution of $ripple_N$ amounts to the creation of a "vacancy" by outputting a value, followed by a backward propagation of this vacancy towards the input stage (in $N - 1$ steps), and is completed by filling the vacancy with a fresh input.

The area of the ripple register is clearly proportional to the number of stages. For sufficiently large W the area and power of the control part can be ignored, and the required area is about $N A_{var}$. The minimum period is $(N+1)T_{:=}$, and the energy per cycle is $N E_{:=}$. Hence, when a ripple register operates at a fixed input-output frequency f_{io}

$$f_{io} = \frac{1}{(N + 1)T_{:=}}$$

and its power consumption is $f_{io} N E_{:=}$.

Note, that in contrast to conventional synchronous realizations, no slave latches are required: each stage consists of a W-bit wide latch rather than two of those latches. A major drawback of the above asynchronous ripple register is that its performance depends on its capacity. For a given f_{io} this effectively bounds N, and vice versa. The performance of ripple registers can be improved by introducing more vacancies.

4.4 Introducing vacancies

Let us consider the following fragment of *pseudo* Tangram.

proc $(a?T \ \& \ b!T)\cdot$
begin
  x : **var** T
| **forever do** $ripple_N(a, x) \ || \ ripple_N(x, b)$ **od**
end

The operator $||$ denotes parallel composition. Both $ripple_N$s operate in parallel: the one on the left-hand side starts by filling x, the other one ends by emptying x. Tangram forbids this form of concurrent reading and writing in the same variable, because their interference may corrupt the value being read. Given realistic timing assumptions, the circuit corresponding to Tangram program above will most likely operate correctly for $N \geq 2$, and behave as $ripple_{2N}(a, b)$.

Because of the parallelism, the ripple time is only about half that of $ripple_{2N}(a, b)$, at the expense of only one additional variable (in area) and one additional assignment (in energy).

The idea of providing an additional variable so that two holes can back propagate in parallel can be implemented correctly by introducing additional synchronization. For example,

```
proc (a?T & b!T)·
begin
   x, y : var T
|  forever
   do   (ripple_N(a, x) || ripple_{N-1}(y, b))
   ;    y := x
   od
end
```

Alternatively, a more balanced and slightly faster version is

```
proc (a?T & b!T)·
begin
   x, y, z : var T
|  forever
   do   (ripple_{N-1}(z, x) || ripple_{N-1}(y, b))
   ;    (z := a || y := x)
   od
end
```

In both alternatives, variable y may be viewed as a slave register to master x. Introducing more slave variables (combined with more parallelism) progressively increases the number of vacancies. With $2N - 1$ slaves, this results in pairing a master with each slave, approaching synchronous operation. This is shown in Figure 29.

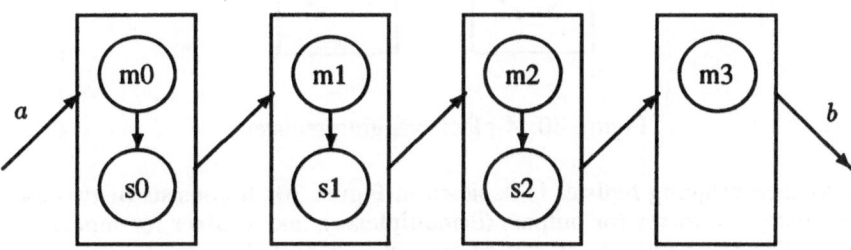

Figure 29: 4-place ripple register with 3 vacancies.

184

```
proc (a?T & b!T).
begin
    m0, s0, m1, s1, m2, s2, m3 : var T
|   forever
    do  (s0 := m0  ||  s1 := m1  ||  s2 := m2  ||  b!m3)
    ;   (a?m0       ||  m1 := s0  ||  m2 := s1  ||  m3 := s2)
    od
end
```

This ripple register clearly mimics the operation of a clocked shift register, in which the sequencer (implied by the semicolon) provides the equivalent of 2-phase non-overlapping clock signals. The circuit area, power, and cycle time are comparable to those of the corresponding synchronous version. In particular, the cycle time has become practically independent of N.

What have we achieved so far? The dramatic improvement in throughput (two assignments per cycle, more or less independent of N) came with an area and power overhead of almost 100%! Still, a simultaneous reduction of cycle time and energy per message can be achieved, as we shall see next.

4.5 Wagging registers

The key property of *ripple* registers is that outgoing values have visited all N stages, resulting in an energy dissipation proportional to N. With *wagging* registers we provide alternative, shorter paths for incoming values.

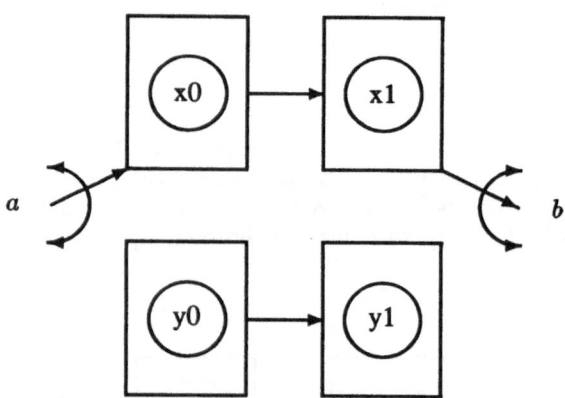

Figure 30: 4-place wagging register.

A 4-place wagging register is depicted in Figure 30. It consists of two 2-place ripple registers, a mixer for outputs (demultiplexer), and a mixer for inputs (multiplexer). (The latter two are shown as switches connected to input a and output b respectively.) Incoming messages are alternatingly passed via the upper and lower ripple register, thereby effectively making only 3 transfers, in contrast to the 5 transfers required for a 4-place ripple register.

A Tangram program for a $2N$-place wagging register is

```
proc (a?T & b!T)·
begin
  forever
    do   ripple_N(a, b)
    ;    ripple_N(a, b)
    od
end
```

The handshake circuit for a wagging register is depicted in Figure 31. The handshake components labeled with a '|' are generalizations of the mixer of Figure 13. The one connected to b transfers a message encoded in the request phase, and the other one in the acknowledge phase. Their roles are similar to those of multiplexers and demultiplexers.

The cycle time and power dissipation of a $2N$ wagging register are comparable to those of a N-place ripple register, and only are half of those of a $2N$-place ripple register. For small N, however, the contributions of (de-)multiplexers to cycle time and power consumption must not be ignored.

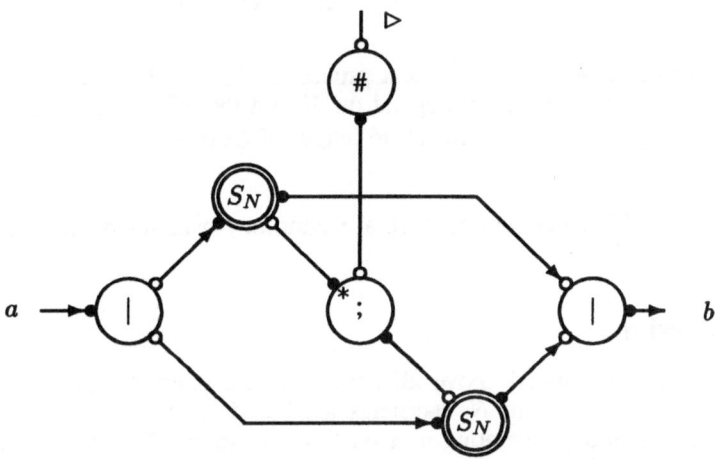

Figure 31: Handshake circuit for a $2N$-place wagging register.

For large N, one may consider to apply the wagging scheme recursively to each of the two ripple registers. This leads to trees of multiplexers and demultiplexers, and suggests an energy consumption of $O(\log N)$ per message. However, there does not seem to exist layouts for this recursive structure that bound the wire length involved in the transfer of a message from input to output by $O(\log N)$. Hence, our assumption that the average energy per transition is constant, that is, independent of N, is not valid for recursive wagging. There are schemes to layout such structures in $O(N)$ area, such as the H-tree [28], cf. Figure 32, that suggest a upper bound of $O(\sqrt{N})$ for the energy per message. A more detailed analysis of such layouts shows that the $\sqrt{N}$ contribution dominates over the $\log N$ contribution only for very large

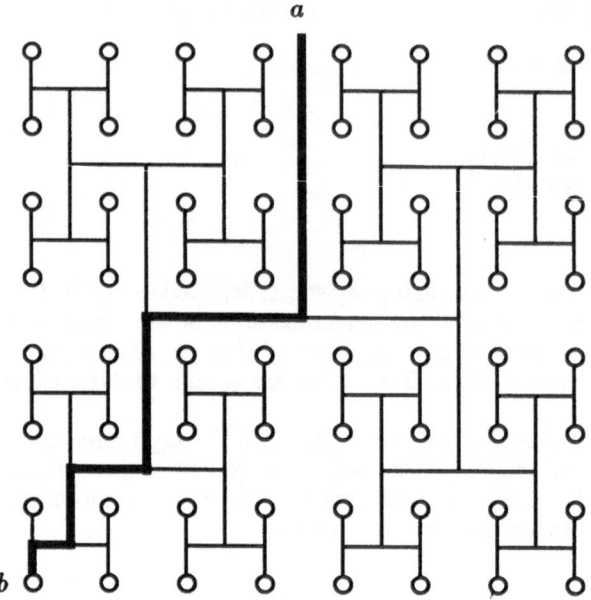

Figure 32: H-tree layout for $N = 2^6$ leaf processes (denoted by bubbles). Note that the density of leaf cells does not depend on N and that all paths from a to a leaf cell (e.g. the fat path to b) have the same length of $O(\sqrt{N})$.

N (above 100-1000, depending on several technology parameters and other design choices)

4.6 Discussion

The shift registers discussed above all have different internal transfer rates. The N-place ripple registers with N vacancies are based on two parallel assignments per I/O cycle, essentially operating in a synchronous mode. Extreme application of wagging also results in two assignments per cycle. All other solutions have higher internal rates. An advantage of asynchronous operation is that these higher rates can be chosen independently of the availability of a suitable clock. Phrased differently, clocked versions of many of these alternatives are less practical, because they require clock frequencies that depend on N.

In all alternatives the power dissipation depends linearly on the I/O rate, and can be expressed in terms of watts/hertz. This power is, of course, independent of a prevailing clock frequency. The internal rate of the ripple registers is equal to the I/O rate. This makes their power consumption proportional to the register capacity N plus the number of vacancies. In wagging registers we see internal rates that are *lower* than the I/O rate. In a synchronous context, this technique would actually lower the activity, but not the power dissipation, unless the clock is switched off for the branch not in use. This technique is also known as clock gating. Some implications of clock gating are reviewed in Section 7.

The cycle time for all shift registers is at least two assignments: the newly input value is stored in a variable, which must be emptied before the next input. Still, there is an escape that offers almost twice the throughput, by alternatingly using a different input variable. This idea [34], offers another trade-off between area, throughput, and power. Together with an alternative form of ripple registers, this type of shift registers is elaborated upon in [35]. These alternatives also allow a simple modification into FIFO queues.

The limit case of wagging also resembles the traditional solution based on register files. This type of shift registers can be based on a W by N register file (one latch per bit) and a simple control structure that cyclically addresses each word. Incoming messages are demultiplexed to the N words, using a global bus, and outgoing messages are multiplexed, using precharged bit lines. The capacitance being switched per I/O is proportional to WN, which is asymptotically equivalent to that of ripple registers. The additional power savings obtained by wagging can be imitated in these register files by splitting the bus lines using a de-multiplexer (for writing) and by splitting each bit line with an OR gate (for reading). These splits effectively reduce the capacity being switched.

5 Processing data

In this section we include some simple forms of data processing, and focus on more intricate interleavings of control and data operations. Two vehicles, namely a powerful incrementer and a parallel-to-serial converter, help us to explore asynchronous architectures that seem hard to mimic by clocked circuits.

5.1 Preliminaries

In Section 4 all assignments are simple assignments of the form $x := y$, where both x and y are variables. It is more interesting if the right hand side of the assignment is an expression involving logic and/or arithmetic. An expression is evaluated by a combinational handshake circuit [10], in a demand-driven fashion. In Figure 33 the interface of expression $f(x, y)$ is depicted. A request for an output through passive port z results in a request for inputs through active ports x and y. (These ports can be connected directly to the read ports of variables x and y.) Function f can then be applied to the incoming values and output through z. In a four-phase implementation a return-to-zero cycle returns circuit f to its initial state.

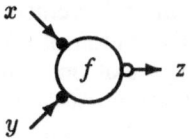

Figure 33: Handshake interface of combinational handshake circuit f.

A combinational handshake circuit can be decomposed into handshake components for elementary logic and arithmetic operators. For the implementation of asynchronous (self-timed) arithmetic the reader is also referred to [31, 20].

A Tangram assignment of the form $x := f(x)$ is called an *auto assignment*: an expression dependent on the value of x is assigned to x. Recall, however, that concurrent read and write actions on handshake latches are excluded (Section 4.2). Therefore we implement such auto assignments by introducing a local auxiliary variable y as in $y := x ; x := f(y)$. If a variable is involved in multiple auto assignments, this auxiliary variable may of course be shared. This is actually an interesting advantage over dedicated master-slave structures in which master and slave latches are paired.

5.2 $\pm 2^i$ incrementer

Incrementing and decrementing are operations commonly applied to a variety of variables, including program counters, address pointers, and filter constants. Let us first focus on an increment of an M-bit variable by one. The least significant bit (bit 0) changes with each increment, the next bit (bit 1) once every other increment, etc. Bit i changes every 2^i increments. In *average*, the total number of bits that change value approaches 2 from below for large M. Hence, a synchronous circuit with an M-bit master-slave register has an enable efficiency of only $\frac{2}{M}$. We design a generalization of this simple incrementer, that supports increments as well as decrements with an arbitrary power of 2, with an enable efficiency that is asymptotically optimal. The width is M bits.

A simple increment of variable x can be described in Tangram by the auto assignment

$$x := x + 1$$

The energy per increment is proportional to M, assuming a simple implementation of the adder without provisions for carry acceleration. In comparison with a synchronous implementation, this only saves power when the average execution rate is well below the prevailing clock rate.

Let the boolean $x.0$ represent the least significant bit of x and $x.1$ the remaining $M - 1$ most significant bits. Then the above increment can also be realized by

if $x.0 = 1$ **then** $x.1 := x.1 + 1$ **fi** ; $x.0 := -x.0$

where "$-$" denotes boolean negation. Note that the increment of $x.1$ is only executed every other invocation, yielding a very fast response and very low energy consumption when $x.0 = 0$. The same scheme can be applied to the increment of $x.1$. Recursive application results in an incrementer with an *average* response time of $O(1)$ and an *average* energy dissipation of $O(1)$, because on average only two bits are inverted. The down side is that once in every 2^M increments all M bit sections are involved, yielding a poor worst-case performance. We shall generalize the above idea to a variable that allows increment and decrement operations with an arbitrary power of two (modulo 2^M).

Let x denote a tuple variable comprising M booleans, numbered from 0 to $M - 1$. When x is interpreted as an integer, boolean $x.0$ corresponds to the least significant bit, and $x.(M - 1)$ to the most significant bit. Furthermore, let up be a boolean variable, with $up = true$ corresponding to incrementing and $up = true$ to decrementing. Variable up is global in that it is accessible to all bit-sections. An increment and decrement by 2^{M-1} amount to the same action, namely the inversion of bit $x.(M-1)$. This is captured by the Tangram procedure inc_{M-1}

inc_{M-1} : **proc** ().
   $x.(M-1) := -x.(M-1)$

For i, $0 \le i < M-1$, the increment is programmed by

inc_i : **proc** ().
   begin
      s : **var** *bool*
   | $s := x.i$
   ; $(x.i := -s \parallel$ **if** $s = up$ **then** inc_{i+1} **fi**)
   end

Procedure inc_i effectively inverts $x.i$. If $up = true$, carry propagation occurs only if $x.i$ was *true* initially. Similarly, borrow propagation occurs if both up and $x.i$ were *false* initially. Hence expression $s = up$ guards the combined carry/borrow propagation. An increment with 2^i can now be implemented by $up := true$ followed by an invocation of procedure inc_i. The sequence $up := false$; inc_i realizes a decrement by 2^i. The expansion of the auto assignment $x.i := -x.i$ has been made explicit with the assignment to local "slave" variable s. This expansion allows some concurrency in the negation of $x.i$ and the conditional invocation of inc_{i+1}

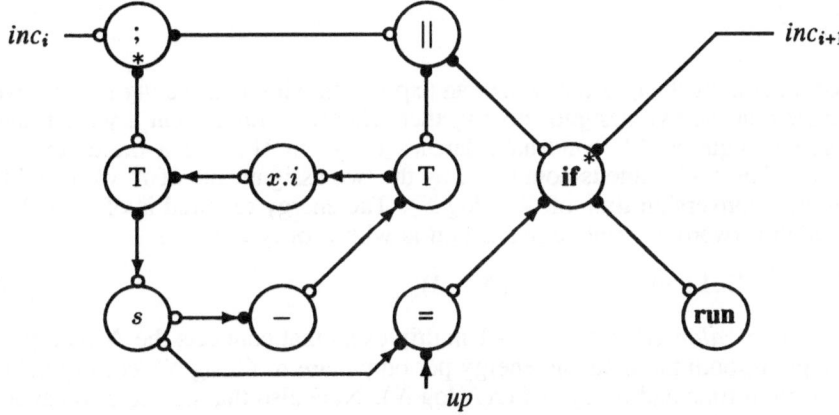

Figure 34: Handshake circuit for incrementer/decrementer cell inc_i.

The handshake circuit for the $\pm 2^i$-incrementer consists of $M-1$ identical sections that implement inc_i, and one for inc_{M-1}. The former is depicted in Figure 34. The handshake component labeled '=' computes the exclusive NOR of s and up. Component **if** implements the selection: after activation through its passive port, it requests a boolean input (the result of the evaluation of $s = up$). If this evaluates to *true* procedure inc_{i+1} is invoked. Otherwise a void handshake with **run** is executed. (The **else** case is omitted in the Tangram text.) Note that each section has access to variable up. Read access to $x.i$ can be extended by adding read ports. Write access can be extended by inserting a multiplexer at the write port of $x.i$. Note that access to cell inc_i shared the access to inc_{i+1} with the environment. This can be implemented by introducing a mixer.

By presetting boolean up and selecting among procedures inc_i a wide range of increment/decrement operations are made available. Invocation of inc_i requires $O(1)$ time and dissipates $O(1)$ energy.

5.3 Parallel-to-serial converters

Conversion of an NW-bit word to a sequence of N words each of W bits and vice versa is a standard exercise in digital circuit design. It is commonly applied to off-chip I/O in order to save on I/O pins. The standard synchronous solution is to shift word for word (N shifts in total), requiring NW master-slave latches and an energy per conversion proportional to N^2. Below we show how to improve this to $N \log N$ while keeping the required conversion time proportional to N. This combination does not seem possible for synchronous implementations.

Let B denote the set of all words of W bits and let B_i denote a tuple of i words of type B, where i is in the range $[0..N)$. Our first converter has output port a of type B and (local) variables x and y of type $\langle B, B_{N-1} \rangle$. The traditional (synchronous) solution for parallel-to-serial conversion of the content of x, with least-significant word $x.0$ first is

$$
\begin{array}{ll}
\textbf{for } N \\
\textbf{do} & (a!x.0 \,\|\, y := \langle x.1, x.0 \rangle \textbf{ cast } \langle B, B_{N-1} \rangle) \\
; & x := y \\
\textbf{od}
\end{array}
$$

The **cast** operation changes the type of an expression without affecting the bit-level value. Note that the two assignments together effectively implement a word rotate. This solution requires $2N$ handshake latches of width W, and a multiplexer of width NW. The cycle time is dominated by the two assignments of N words wide, resulting in a conversion time of $O(N \log N)$. The energy required is of $O(N^2)$.

A straightforward alternative realization is with x of type B_N is

$$
a!x.0 \,;\, a!x.1 \,;\, a!x.2; .. \,;\, a!x.(N-1) \tag{6}
$$

If we assume a *balanced* tree of $N-1$ multiplexers that connects the N read ports $x.i$ to output a, both the time and energy per output are of $O(\log N)$, corresponding to a conversion time and energy of $O(N \log N)$. Note also that this solution avoids auxiliary handshake latches.

In a third attempt we aim at an asymptotic reduction of the conversion time. The word width is restricted to powers of two, with $N = 2^M$. Let procedure $ps_i(x)$ serialize a word of type B_i, stored in variable x. An attractive realization of $ps_i(x)$ is

$$
\begin{array}{ll}
ps_1(\langle x \rangle) & = \; a!x \\
ps_{2.i}(\langle x.0, x.1 \rangle) & = \; ps_i(x.0) \,;\, x.0 := x.1 \,;\, ps_i(x.0)
\end{array}
$$

Here $x.0$ and $x.1$ are both of type B_i, denoting the least and most significant halves of variable x respectively. An alternative for $ps_{2.i}$ is

$$
ps_i(x.0) \,;\, x.0 := x.1 \,;\, ps_i(x.1)
$$

This is essentially (6) for N a power of two.

The handshake circuit for $ps_{2.i}(\langle x.0, x.1 \rangle)$ is given in Figure 35. Passive port x is effectively the write port of structured variable $\langle x.0, x.1 \rangle$. It splits an incoming value into a least- and a most-significant fraction that are of equal sizes. The former is "written into" ps_i, and the latter into handshake latch $x.1$. The component labeled '$\rangle\langle$' is called a splitter, and is simply realized by joining the incoming requests with a C-element and by splitting the incoming wiring bundles into two bundles corresponding to $x.0$ and $x.1$ respectively. Hence, the area, delay, and energy of the splitter are those of the C-element.

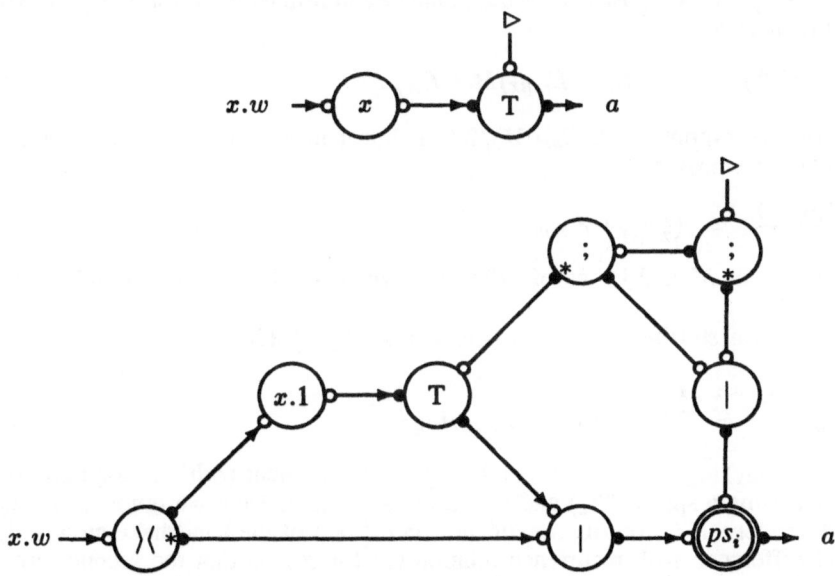

Figure 35: Handshake circuit for Parallel-to-serial converter ps_1 (top) and for $ps_{2.i}$ with $i > 0$ (bottom).

In the analysis of area, power, and cycle time we ignore the contribution of the control section of $ps_{2.i}$. A more detailed analysis shows that the constant contributions of the two sequencers and the control mixer are negligible compared to the other contributions that grow with i.

The area for a parallel-serial converter with width 2^M is denoted by $A_{ps}(2^M)$. According to the handshake circuit of Figure 35 we have

$$A_{ps}(1) = A_{var}$$
$$A_{ps}(2i) = A_{ps}(i) + i(A_{var} + A_{mux})$$

Constants A_{var} and A_{mux} denote the areas of a W-bit variable and W-bit multiplexer respectively. Solving this simple recurrence relation with $N = 2^M$ yields

$$A_{ps}(2^M) = 2^M A_{var} + (2^M - 1)A_{mux}$$

Note that this is also the area for the converter described in Tangram program (6). Compared to the classical shift-register-based converter this is about half the number of latches. Unfortunately, for both converters it turns out that wiring will dominate for large M. (See also the remark on wagging registers.) This makes it impossible to give an upper bound of the wire lengths involved, making our conclusions on energy consumption and cycle time somewhat optimistic.

The energy for shifting out 2^M words is denoted by $E_{ps}(2^M)$.

$$
\begin{aligned}
E_{ps}(1) &= E_{:=} \\
E_{ps}(2i) &= 2E_{ps}(i) + (E_{:=} + 2E_{mux})i
\end{aligned}
$$

The assignment $x.0 := x.1$ is i bits wide, contributing $E_{:=}i$ for each word, plus $E_{mux}(1 + \frac{1}{2} + \frac{1}{4}..)i \approx 2E_{mux}i$ for the branching multiplexer path into ps_i. Solving this equation yields

$$
E_{ps}(2^M) = \left((\tfrac{1}{2}E_{:=} + E_{mux})M + E_{:=}\right)2^M \tag{7}
$$

The energy per output word, $E_{ps}(M)/2^M$, is linear in M, and hence logarithmic in the number of words N:

$$
\frac{E_{ps}(2^M)}{2^M} = (\tfrac{1}{2}E_{:=} + E_{mux})M + E_{:=}
$$

This is linear in M and hence of $O(\log N)$, which is identical to parallel-to-serial converter (6).

The time for shifting out 2^i words is denoted by $T_{ps}(i)$.

$$
\begin{aligned}
T_{ps}(1) &= T_{:=} \\
T_{ps}(2i) &= 2T_{ps}(i) + (T_c + T_{mux})\log i
\end{aligned}
$$

The $(T_c + T_{mux})\log i$ term corresponds to the "assignment path" of $\log i$ sections, each comprising a splitter (T_c) and a data mixer (T_{mux}). Here we ignore a constant (i.e. independent of i) contribution of the control part of the handshake circuit. The essential difference with recurrence relation (7) for E_{ps} is that the second term is logarithmic in i, rather than linear. Solving the above equation yields

$$
T_{ps}(2^M) = (T_{:=} + 2(T_c + T_{mux}))2^M - (T_c + T_{mux})(M + 2)
$$

Hence, the average conversion time per output word $T_{ps}(2^M)/2^M$ is bound by $T_{:=} + 2(T_c + T_{mux})$, which is independent of M. This is quite interesting, because the assignment times vary logarithmically with the word widths. Apparently, the exponentially lower frequencies of these wide assignments make their contributions to the total conversion times invisible.

In a synchronous solution, the time for an assignment also grows logarithmically with the word width, because of the clock driver involved (See Section 2.2). Synchronizing all actions with a (periodic) clock would result in a clock period that grows with $\log N$.

6 Systolic computations

A systolic computation [17, 29] is a computation that is carried out by a regular network of communicating processes. The processes in the network are identical, or come in only a few varieties, and communicate with their neighbor processes only. Each process has a fixed, small number of local variables, and spends the major part of its time in communication actions. A systolic computation is thus an example of fine-grained parallelism. The order in which the processes execute their communications does not depend on the values communicated: systolic computations are data-independent.

We discuss three types of systolic computations: linear systems, wagging systems, and tree-like systems. They all solve the same problem, viz. the problem of computing window maxima, and they do so at quite high a rate. They differ, however, in area usage and energy consumption. We show that there is a trade-off involved between area and energy: energy reduction costs area.

Each system operates on an input stream A and produces an output stream B, both containing integer values. The problem is to move a window of size N (where N is a given constant, $N \geq 2$) over the input stream. For each position of the window the maximum value covered by it has to be output. Indexing all streams from 0 upwards, the functional specification of this problem may be written as

$$B(i) = (\textbf{max } j: 0 \leq j < N: A(i-j)) , i \geq N-1$$

It expresses for $i \geq N-1$ that output value $B(i)$ should be the maximum of window $A(i-N..i]$. Notice that it does not specify the values of the first $N-1$ output communications; value $B(N-1)$ should be the maximum of window $A[0..N)$, value $B(N)$ the maximum of $A[1..N+1)$, etc.

6.1 A linear system

A system consists of communicating processes. One of these processes has to output the B stream. According to the specification, that process has to perform a max-quantification over domain $\{j \mid 0 \leq j < N\}$. Since we are aiming at simple processes, it should not compute the whole quantification itself. Thereto, it receives from another process the maxima of windows of size $N-1$ and determines maxima of windows of size N. That other process receives, from yet another process, the maxima of all $(N-2)$-sized windows, etc. We thus get N processes, one for each window size in the range 1 through N.

We number the N processes from 1 upwards and give each process n ($0 < n \leq N$) an output port b_n along which it produces values given by

$$b_n(i) = (\textbf{max } j: 0 \leq j < n: A(i-j)) , i \geq n-1$$

Notice that the relation above is obtained from the system specification by replacing constant N by n. Consequently, process N produces output stream B. For process 1 we have

$$b_1(i) = A(i) , i \geq 0$$

Process 1 only has to copy the input stream. For $n \geq 2$ the relation above allows a nice recursive decomposition [27]:

$$b_n(i) = A(i) \textbf{ max } b_{n-1}(i-1)$$

for $i \geq n-1$.

Observe that in order to compute $b_n(i)$, process n needs output $b_{n-1}(i-1)$ of process $n-1$. It also needs value $A(i)$. To receive $A(i)$ we equip each process n with an input port a_n along which it receives sequence A:

$$a_n(i) = A(i) , i \geq 0$$

Process N receives these values from the environment. The other processes n receive them from process $n+1$, cf. Fig. 36. An output port in one process and an

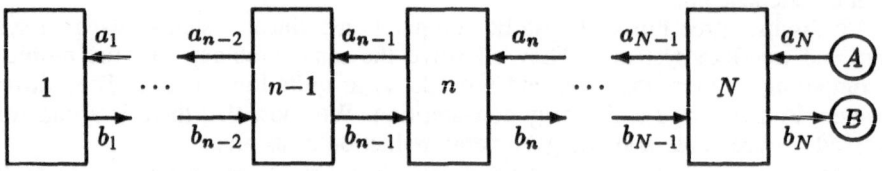

Figure 36: Linear system

input port with the same name in another process together form a channel. The processes communicate with each other by means of channels. The output action and the corresponding input action are synchronized, and together they carry out a distributed assignment.

The outputs of process n may be computed by

$$a_{n-1}(i) = a_n(i)$$
$$b_n(i+1) = a_n(i+1) \textbf{ max } b_{n-1}(i)$$

for $i \geq 0$. Each output value can be computed from the input values last received if process n executes its communications according to behavior

$$a_n \ ; (a_{n-1}, \ b_n \ ; a_n, \ b_{n-1})^*$$

in which the comma, which takes priority over the semicolon, denotes concurrent execution.

Thus process n for $n \geq 2$ becomes process $maxl(a_n, b_{n-1}, a_{n-1}, b_n)$, with $maxl$ given by the following Tangram text:

$$maxl = \textbf{proc } (ai, bi?\text{int } \& \ ao, bo!\text{int})\cdot$$
$$\textbf{begin } va, vb: \textbf{var int}$$
$$| \quad ai?va$$
$$; \quad \textbf{forever}$$
$$\textbf{do } (ao!va \parallel bo!(va \textbf{ max } vb))$$
$$; \quad (ai?va \parallel bi?vb)$$
$$\textbf{od}$$
$$\textbf{end}$$

Notice that variable vb has not been initialized; we assume that all variables have unknown values upon declaration.

Process 1 is simply $buf(a_1, b_1)$ with buf defined by

$$buf = \textbf{proc } (a?\text{int } \& \ b!\text{int})\cdot$$
$$\textbf{begin } va: \textbf{var int}$$
$$| \quad \textbf{forever do } a?va \ ; b!va \textbf{ od}$$
$$\textbf{end}$$

The whole system could be created by a statement like

$$buf(a_1, b_1) \;\|\; (\| \; n: 1 < n \leq N: maxl(a_n, b_{n-1}, a_{n-1}, b_n))$$

in which $\|$ denotes parallel composition. Input stream A equals a_N and output stream B equals b_N.

The system contains $2N-1$ integer variables. In order to assess its speed, we draw a schedule that, for $N = 5$, maps all communications in the system onto time slots. In the schedule below the communications with subscript i are listed in the row of process i only.

process\time slot		0	1	2	3	4	5	6	$\cdots$
1:	$(a_1 \,; b_1)^*$					a_1	b_1	a_1	
2:	$a_2 \,;(a_1 , b_2 \,; a_2 , b_1)^*$				a_2	b_2	a_2	b_2	
3:	$a_3 \,;(a_2 , b_3 \,; a_3 , b_2)^*$			a_3	b_3	a_3	b_3	a_3	
4:	$a_4 \,;(a_3 , b_4 \,; a_4 , b_3)^*$		a_4	b_4	a_4	b_4	a_4	b_4	
5:	$a_5 \,;(a_4 , b_5 \,; a_5 , b_4)^*$	a_5	b_5	a_5	b_5	a_5	b_5	a_5	

Process 5 starts (with input) in time slot 0. Process 4 starts one time slot later, process 3 again a time slot later, etc. We also observe that the inputs and the outputs of the system – i.e. a_5 and b_5 – occur in alternating time slots, starting in time slot 0. The throughput (number of outputs produced per time slot) is, consequently, 1/2. The latency is the time between an output and the last input on which the value of that output depends. For our problem that is the time between the ith input and the ith output: 1 time slot.

In the analysis above we have neglected the time it takes a process to carry out the **max**-operation. For a detailed speed analysis such internal computing should be taken into account as well. We refrain from doing so, since it would only slightly affect the performance characteristics.

A good approximation for the energy consumption is the number of communications, assignments – which do not occur in this program – and arithmetic operations. We count how many of these the system performs per output action. Since an output action occurs in every other time slot, we count the number of communications and **max**-operations in two consecutive time slots. From time slot $N-1$ onwards there occurs exactly one communication along each channel in every two successive time slots. The number of communications per output action, consequently, equals the total number of channels: $2N$. In $N-1$ processes the computation of the b output involves a **max**-operation. The total energy of a *maxl* system for window size N is, consequently,

$$2N + N - 1 = 3N - 1$$

per system output.

According to Section 4.2 an assignment requires on the average $11W$ transitions, where W is the word width. The number of transitions per output is, consequently, $11W(3N-1)$. Assuming that the outputs have to be produced at a frequency f, the number of transitions per second equals

$$11Wf(3N-1)$$

If, for example, $f = 10$ megahertz, $W = 8$, and $N = 16$, this amounts to 4.1×10^{10} transitions per second. In today's technology (cf. Section 2.1) the energy required for a single transition is about 1.5 picojoules, which results for the figures above in a power dissipation of 62 milliwatts for the whole circuit.

6.2 A linear system with merged processes

By making our system slightly more coarse-grained, we can, without appreciably degrading its speed, decrease both the number of variables and the energy consumption. We effectively merge pairs of processes into single processes. Rather than doing that with the system above, we consider again relation

$$b_n(i) = A(i) \text{ max } b_{n-1}(i-1)$$

Unfolding this relation once, we find for $n \geq 3$

$$b_n(i) = A(i) \text{ max } A(i-1) \text{ max } b_{n-2}(i-2)$$

Now we have a relation between the outputs of process n and those of process $n-2$. We, consequently, need only half as many processes. All except one compute their outputs by means of the relation above. The computation of the last process depends on whether N is odd or even. For odd N the last one is $buf(a_1, b_1)$ (as before); for even N the last process should compute

$$b_2(i) = a_2(i) \text{ max } a_2(i-1)$$

which may be done by a *max2* process:

```
max2 = proc (a?int & b!int)·
          begin va, ba: var int
          |   forever
              do a?va ; b!(va max ba)
              ;   a?ba ; b!(va max ba)
              od
          end
```

The other processes are $maxm(a_n, b_{n-2}, a_{n-2}, b_n)$, with *maxm* given by

```
maxm = proc (ai, bi?int & ao, bo!int)·
          begin va, vb, bb: var int
          |   ai?va
          ;   forever
              do (ao!va || bo!(va max bb))
              ;   bb := va max vb
              ;   (ai?va || bi?vb)
              od
          end
```

For odd N the system may be created by

$$buf(a_1, b_1) \parallel (\parallel n: 0 < n \leq (N-1)/2: maxm(a_{2n+1}, b_{2n-1}, a_{2n-1}, b_{2n+1}))$$

and for even N by

$$max2(a_2, b_2) \parallel (\parallel n: 1 < n \leq N/2: maxm(a_{2n}, b_{2n-2}, a_{2n-2}, b_{2n}))$$

Notice that the repetition of process *maxm* differs from that of process *maxl* in that it has an additional assignment statement. One of the communications between the amalgamated, neighboring processes has changed into an assignment; the other one − communicating A-values − has disappeared. The latter phenomenon saves a variable per process. Thus the system occupies less area: the number of variables has roughly gone down from $2N$ to $\frac{3}{2}N$. The *maxm* system is also more energy-efficient: N communications have disappeared, only half of which have become assignments. Thus, the energy (number of communications, assignments, and operations) per output has roughly decreased from $3N$ to $\frac{5}{2}N$.

6.3 A wagging system

In the previous sections linear systems of processes computed the maximum for each window. Now we apply the adage of 'divide and conquer' and use two subsystems that each do half of the work. This is most easily explained for even-sized windows, although a similar scheme may be devised for windows that are odd-sized. Let the window size equal $2M$: $N = 2M$. We use two copies of the linear system already designed − each of size M − and we add a head cell to distribute the input sequence and to collect the results, cf. Fig. 37. The subsystems establish relations

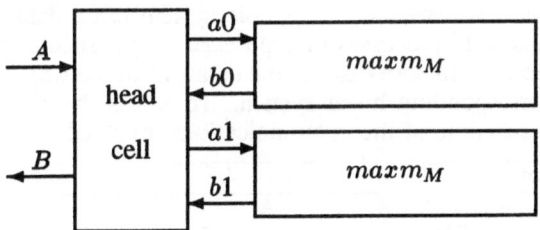

Figure 37: Wagging system

$$b0(i) = (\mathbf{max}\ j: 0 \leq j < M: a0(i-j))$$
$$b1(i) = (\mathbf{max}\ j: 0 \leq j < M: a1(i-j))$$

They do so under communication behaviors $(a0\ ;b0)^*$ and $(a1\ ;b1)^*$. The head cell distributes the input values it receives along A alternately towards $a0$ and $a1$ ($i \geq 0$):

$$a0(i) = A(2i)$$
$$a1(i) = A(2i+1)$$

This part of the head cell may be implemented by process $toggle(A, a0, a1)$, with

```
toggle = proc (a?int & b, c!int)·
         begin va: var int
         |  forever
            do a?va ; b!va
            ;  a?va ; c!va
            od
         end
```

The output values of the system depend on those of the two subsystems by means of

$$B(2i) = b0(i) \textbf{ max } b1(i-1)$$
$$B(2i+1) = b1(i) \textbf{ max } b0(i)$$

The part of the head cell that produces the outputs on B can, consequently, be carried out by process $merge\,(b0, b1, B)$ defined by

$merge$ = **proc** $(a, b?\text{int} \& c!\text{int})\cdot$
  begin va, vb: **var int**
  | **forever**
    do $a?va$; $c!(va \textbf{ max } vb)$
    ; $b?vb$; $c!(va \textbf{ max } vb)$
    od
  end

We have designed the head cell as two (unconnected) processes: $toggle\,(A, a0, a1)$ and $merge\,(b0, b1, B)$.

Since the head cell contains only 3 variables and each of the two $maxm$ subsystems contains about $\frac{3}{2}M$ variables, the whole system has a data size of about $\frac{3}{2}N$ variables. If we had used processes of type $maxl$ in our subsystems, the total data size would be about $2N$. In both cases is the data size of the wagging system 3 more than that of the corresponding linear system. To assess the speed of the wagging system we draw a schedule of the communications involved:

process\time slot		0	1	2	3	4	5	6	7	8	...
$toggle$:	$(A\,;a0\,;A\,;a1)^*$	A			A		A		A		A	
$merge$:	$(b0\,;B\,;b1\,;B)^*$				B		B		B			
$maxm_M$:	$(a0\,;b0)^*$		$a0$	$b0$				$a0$	$b0$			
$maxm_M$:	$(a1\,;b1)^*$				$a1$	$b1$				$a1$	$b1$	

The system's throughput is still 1/2 and its latency is now 3. The system's energy, however, has been halved. Per output of each subsystem there occur two outputs of the whole system. Therefore, the energy consumed in each subsystem per system output is only half of what we found above: $\frac{5}{4}M$. The whole system thus has an energy consumption per output of $\frac{5}{2}M$, which equals $\frac{5}{4}N$. If we had used $maxl$ rather than $maxm$ subsystems, the energy per output would be about $\frac{3}{2}N$.

Without significantly increasing the energy per output action, we can easily increase the speed of the system. We do that by changing the two processes in the head cell in such a way that they perform input and output concurrently. Thereto we change process $toggle$ into

$toggle1$ = **proc** $(a?\text{int} \& b, c!\text{int})\cdot$
  begin vb, vc: **var int**
  | $a?vb$
  ; **forever**
    do $(a?vc \parallel b!vb)$
    ; $(a?vb \parallel c!vc)$
    od
  end

and process *merge* into

```
merge1 = proc (a, b?int & c!int)·
         begin va, vb, vc: var int
         |  a?va
         ;  forever
              do (b?vb || c!(va max vc))
              ;  ⟨vb, vc⟩ := ⟨va, vb⟩
              ;  (a?va || c!(vc max vb))
              od
         end
```

The schedule above now becomes:

process\time slot		0	1	2	3	4	5	6	...
toggle1:	A ;$(A, a0$;$A, a1)^*$	A	A	A	A	A	A	A	
merge1:	$b0$;$(b1, B$;$b0, B)^*$				B	B	B	B	
$maxm_M$:	$(a0$;$b0)^*$		$a0$	$b0$	$a0$	$b0$	$a0$	$b0$	
$maxm_M$:	$(a1$;$b1)^*$			$a1$	$b1$	$a1$	$b1$	$a1$	

With only two additional variables the throughput has increased to 1. The energy per output action increases with only one assignment. The new processes do not really affect the energy per output, but they allow the circuit to be operated at a higher speed.

6.4 A tree-like system

If the wagging system halves the energy consumption, can we decrease the energy consumption even more by repeatedly applying the wagging scheme to ever smaller windows?

Consider the maximum computation for windows of size $N = 2^K$ for $K \geq 1$. For $K = 1$ the specification reads

$$B(i) = A(i-1) \textbf{ max } A(i) , i \geq 1$$

This is exactly the computation carried out by process *max2*. For $K \geq 2$ we split the computation into two subsystems using process *toggle* to send the input values to the subsystems, and process *merge* to produce the maximum of their output values. Fig. 38 shows the system thus designed for $N = 8$. The figure may look more like two binary trees with amalgamated leaves, but by flipping the lower part up it becomes a tree with bidirectional edges. The system may be created by statement

$$(\| \, n: 1 \leq n < N/2: toggle\,(a_n, a_{2n}, a_{2n+1}))$$
$$\| \, (\| \, n: N/2 \leq n < N: max2\,(a_n, b_n))$$
$$\| \, (\| \, n: 1 \leq n < N/2: merge\,(b_{2n}, b_{2n+1}, b_n))$$

The tree has a data size of about $\frac{5}{2}N$ variables, as opposed to $\frac{3}{2}N$ for the wagging *maxm* system. To get an impression of its performance we draw a schedule for $N = 8$. The ten processes are shown in the order of the statement above: first three *toggle* processes, then four *max2* processes, and finally three *merge* processes:

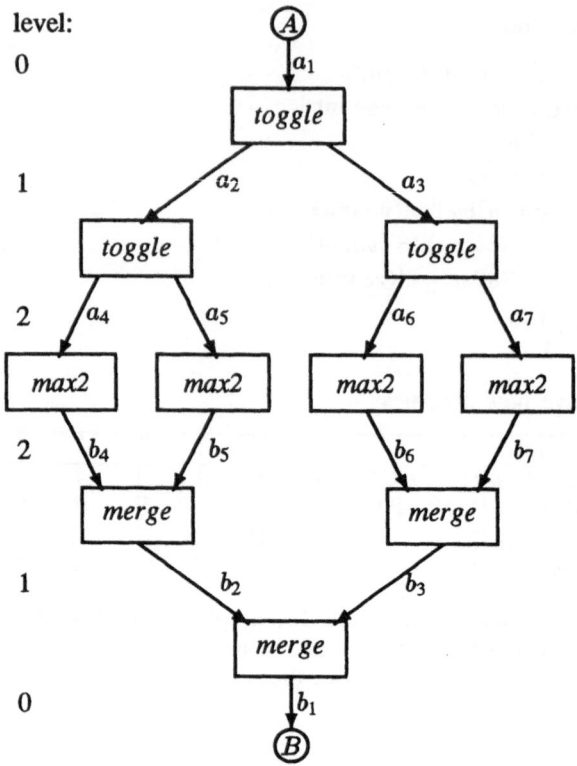

level:

Figure 38: Tree-like system for computing window maxima ($N = 8$)

	0	1	2	3	4	5	6	7	8	9	10	11	12	...
1	a_1		a_1		a_1		a_1		a_1		a_1		a_1	
2		a_2				a_2				a_2				
3				a_3				a_3				a_3		
4			a_4	b_4							a_4	b_4		
5						a_5	b_5							
6					a_6	b_6							a_6	
7									a_7	b_7				
8					b_2				b_2				b_2	
9							b_3				b_3			
10						b_1		b_1		b_1		b_1		

We observe that the throughput is 1/2. The latency is $2K-1$, which equals $2\log_2 N - 1$. The $2K-1$ can be attributed to the different types of processes as: $K-1$ for the *toggle* processes, 1 for the *max2* process, and another $K-1$ for the *merge* processes.

The schedule above shows that the processes operate at a rather relaxed speed, which is indicative of a low power dissipation. The edges in the tree come in levels. There are K levels of a edges and K levels of b edges: $2K$ levels for the whole system. Every two time slots one output is produced. During these two time slots there occurs exactly one communication in each of the $2K$ levels. The K communications in the lower levels are each accompanied by one **max**-operation. The number of communications and operations – there are no assignments – per output action is consequently $3K$, which equals $3\log_2 N$. This is a dramatic reduction indeed. Thus far the most energy-efficient solution was the wagging *maxm* system with an an energy per output of about $\frac{5}{4}N$. When operated (as in Section 6.1) at 10 megahertz, with 8-bit integers and a window size of 16, the circuit has a power dissipation of only 15 milliwatts, which is a factor of 4 ($=\log_2 N$) less than the *maxl* system. Of course, the caveat with respect to tree-structured layouts, expressed at the end of Section 4.5, applies to this example as well.

Just as in Section 6.3, the throughput of the system can be increased to 1 by replacing processes *toggle* and *merge* by processes *toggle1* and *merge1*. There is no need to replace all of them. Actually, only the top *toggle* and the bottom *merge* have to be replaced. As in Section 6.3, this speed-up increases the number of variables by 2 and the energy per output by 1. It does not affect the latency of the circuit.

6.5 No best solution

VLSI Programming is designing functionally correct programs while keeping control over such circuit characteristics as area, speed, and power. In this chapter we have focused our attention in particular on energy and power. It turns out to be possible to address these circuit characteristics at a rather high abstraction level, viz. by considering different Tangram programs for the same functional specification.

Fig. 39 shows the energy consumption against the area (number of variables) for the five systems we have designed to compute window maxima. The *maxm* systems ($\ominus$ and $\oslash$) are better – both in area and energy – than the corresponding *maxl* systems ($\oplus$ and $\otimes$). There is, consequently, no reason to choose the most fine-grained solution; merging pairs of processes yields better systems.

Among $\ominus$, $\oslash$, and $\odot$ there is no best solution. Wagging roughly halves the energy, but it costs some area. The tree is the system we obtain when the wagging is repeated in the subsystems until the window size becomes 2. We could, actually, have repeated this another step, until windows of size 1. Then the tree consists of *toggle* and *merge* processes only. This system, indicated in Fig. 39 by •, is inferior to the original tree. Just like linear systems, trees should not be too fine-grained. There are also solutions between $\oslash$ and $\odot$: we can do the wagging until, for example, the window size becomes 4 and then use *maxm* systems. This system is plotted as o in Fig. 39: indeed it nicely fits between the wagging *maxm* system and the tree. And then there are, of course, also the throughput-1 systems.

In short, there is a wide variety of solutions, exemplifying the different trade-offs that can be made between (in this case) area and energy. Which solution is to be preferred depends on characteristics of the environment (the application) in which the circuit will be used. This is typical of VLSI programming. By virtue of the fact that a VLSI programming language allows investigating all these solutions at a high level of abstraction, we can with relative ease explore the whole solution space and select the system that suits our purposes best.

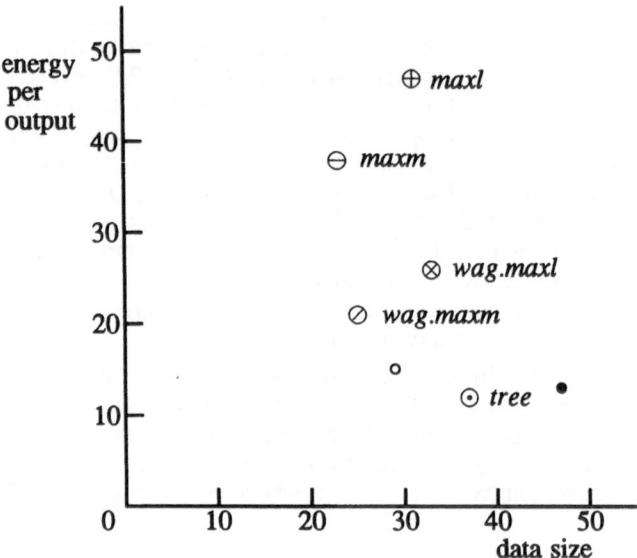

Figure 39: Energy per output versus data size of five throughput-1/2 systems for window size 16. The energy is measured in number of assignments (the unit is about 1.3×10^{-10} joules for 8-bit integers); the data size is the number of variables used; $\otimes$ and $\oslash$ are the wagging versions of *maxl* and *maxm*.

7 Conclusion

7.1 In practice

A part of a graphics processor, programmed in Tangram, was successfully compiled onto silicon in early 1987 [34]. A well-documented IC [40], that includes a modulo-100 counter based on the design of Section 3.5 (with $K = 2$), confirms our models for timing and power consumption.

A large-scale application of VLSI programming for low-power are two ICs that, together with a commercially available 256k DRAM, form a DCC error corrector [14, 39, 38]. The power dissipation of the two ICs is only one fifth of that of their synchronous counter parts. Both ICs use 4-phase control and double-rail data encoding, and their operation is quasi delay insensitive [2]. A major source for power savings is the very low activity in audio ICs: approximately 10%. Most sample streams have frequencies of 100 to 200 kilohertz, whereas 6 megahertz was used as clock frequency for the synchronous version. In addition, the complex control is highly distributed, avoiding high-frequent accesses to status registers and an instruction ROM. Another contribution is rather specific for error correctors. Error detection generally consists of two phases: syndrome computation followed by the computation of error values and locations. If the syndrome equals zero, the code word is correct and the second phase can be omitted. The second phase requires more than twice the amount of energy of the first phase. Because over 95% of all code words are correct, the *average* energy consumption the detector is close to the energy required for syndrome computation. After establishing the correctness of a code word the detector simply waits until the next code word is offered, consuming negligible power only (leakage). Hence, most VLSI-programming effort was spent on minimizing the energy requirements for the first phase. The architectural differences between the two versions make it difficult to factorize the factor 5 over the different contributions. A similar, but much simpler error corrector, offering a comparable reduction in power dissipation, is in [15].

The last three ICs also validate the simple idea of counting transitions to predict the energy consumption. Using 1.5 picojoules per transition the accuracy was generally within 20%. These three ICs were used for the histograms of Figure 4.

7.2 Tuning V_{DD}

So far we have treated the supply voltage V_{DD} as a constant. As we have seen in Section 2.1, the energy per transition reduces more than quadratically with the supply voltage. It therefore pays off to reduce the supply voltage from 5 volts to, say, 2.5 volts, saving about 80% of the energy per transition. The main drawback is a substantial increase in computation times. The following idea [24, 23] takes advantage of the fact that most ICs are designed very conservatively with regard to performance (often with a 100% margin), and that the work load may be highly data dependent. The idea is referred to as "just-in-time processing" or as "adaptive scaling of the supply voltage", and turns out to be particularly attractive for asynchronous processors. The main application is in signal processing, where an incoming stream of data samples is processed into an outgoing stream. The data rates may differ.

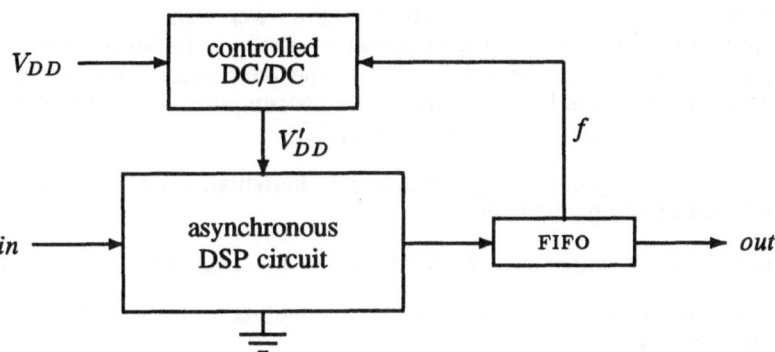

Figure 40: Just-in-time DSP. The supply voltage V_{DD} is regulated such that FIFO is kept about half full.

A typical "just-in-time" configuration is shown in Figure 40. The asynchronous signal processor operates on a supply voltage V'_{DD} This supply voltage is derived from a constant source V_{DD} by a so-called DC/DC converter,[11] under control of

---

[11] Such converters can be realized with conversion efficiencies of more than 90% [9].

signal f. Signal f is a measure for the degree of filling of a FIFO buffer that is connected to the output of the asynchronous signal processor. If the processor is operating too fast, the buffer becomes more than half full, and the DC/DC converter is requested to reduce V_{DD}. Similarly, if the processor is running too slow, V_{DD} is increased.

This form of adaptive scaling of the supply voltage tends to keep the supply voltage and the power consumption as low as possible. Even with a constant work load, V_{DD}' tends to be lower because most circuits are designed to operate according to a specified performance at a lower voltage, a high temperature, and after worst case IC-processing. If the work load varies, as for instance in the case of error correctors, the advantage may easily become as high as 50-80%. For a detailed quantitative analysis of the power savings that can be obtained by this technique see [23].

In [18] a synchronous version of adaptive scaling of the supply voltage was proposed. However, their feed-back loop does not contain a FIFO buffer, and can therefore only handle limited variations in workload. They control V_{DD}' by means of an "equivalent critical path", a circuit with delay-versus-V_{DD} properties identical to those of the actual critical path. When full CMOS gates are used, the design of such an equivalent critical path is not that hard, although some margin must be allowed for spread in device parameters and on-chip temperature gradients. When pass gates are used or forms of dynamic circuitry, additional margins are required to guarantee that set-up and hold times are not violated.

7.3 Low-power synchronous circuits

Power consumption of CMOS ICs has not been a real concern in most applications until recently. With less complex circuits and fewer battery-powered applications there was no real need to pay much attention to the power consumption of synchronous circuits. Also, the transition from high-power NMOS to low-power CMOS during the eighties made power consumption a seemingly irrelevant issue. This all is changing rapidly, as we have indicated in the introductory section. Considerable attention is being paid to reducing power consumption of synchronous CMOS circuits. We therefore can expect (cf. Section 2.2):

- lower clock frequencies, by considering alternative algorithms and architectures with more parallelism;

- multiple clock frequencies, each adapted to the local processing needs;

- extensive use of clock gating, that is, decoupling a unit's clock wiring when that unit is standing by.

These techniques aim at increasing the circuit's activity and at improving the enable efficiency. Although clock gating may be quite simple to apply in a coarse grain size (both in time and in circuit size), it becomes increasingly more complex to apply to finer grain sizes. Nevertheless, when applied to an extreme degree, similar advantages can be expected as with asynchronous circuits. However, multiple clock frequencies and clock gating introduce some new problems with regard to clock distribution and scan-based testing. More importantly, few or no CAD tools yet exist that support these techniques. Furthermore, clock gating does little to the power overhead spent in the clock pre-driver, and does offers no prospects for reducing EMC problems. Also, synchronous operation complicates application of just-in-time processing, because the clock frequency must be scaled together with the supply voltage. Especially, it seems more difficult to take advantage of variations in work load, corresponding to variations in the number of required clock cycles per sample.

7.4 Low-power asynchronous circuits

In this report we have argued that asynchronous circuits have an intrinsic power advantage relative to synchronous circuits. In many cases we could even show an asymptotic advantage in terms of a functional parameter of the circuit. In a practical context this is not too convincing. What is needed is a quantified (measured) advantage for practical (industrially relevant) applications. In practice it turns out that such a detailed comparison is very complicated, because differences in technology, cell libraries, CAD tools (for layout), requirements on testability, and, most importantly, differences in architecture, tend to pollute and confuse these comparisons. (Simplifying the specification by leaving out the difficult parts make a comparison even less interesting.) One of the few attempts at making a fair and detailed comparison is [32].

Even if an apples-to-apples comparison can be made, this only provides a single instance for that particular application, comparing a specific synchronous style with a specific asynchronous style. In particular, the field of asynchronous circuit design is still immature, and for many questions the answers are not yet available:

- Which is more power efficient: 2-phase or 4-phase handshake signaling?

- Which data encoding dissipates less: single-rail or double-rail?

- How much additional power does quasi delay-insensitive operation cost?

(Be warned, answers are less obvious than they may seem!) Interestingly, in analyzing and illustrating the power advantage of asynchronous circuits, we were able to abstract from these choices, using handshake circuits as architecture.

Few asynchronous ICs have been designed that allow a meaningful comparison. The DCC ICs mentioned above suffer to some extent from the apples-to-pears comparison problem. Architectural differences preclude a precise factoring of the factor 5 advantage. Also, it would be very interesting to compare a 2-phase, single rail asynchronous implementation with a highly clock-gated synchronous version! One of the most impressive asynchronous circuits to date that allows a detailed comparison is the AMULET microprocessor [25] by Manchester University also described in this volume. In [21] attention is drawn to the low power per MIPS (million instructions per second) of a fully asynchronous (QDI) 16-bit microprocessor. Unfortunately, a comparable synchronous version is not available.

7.5 VLSI programming for low power

In this report we approach the design of VLSI circuits as a programming activity. For large and complex applications such an approach heavily relies on automatic compilation and analysis tools. VLSI programming allows the VLSI designer to explore many algorithmic and architectural variations for a given specification. As we have seen in Sections 4 and 6, these alternatives are vastly different in terms of

IC area, performance, and power dissipation. These differences are often much more significant than the differences between the various styles of asynchronous design.

In order to explore these alternatives it is of utmost importance that a VLSI programmer can predict whether a particular architectural modification results in a win or in a loss in power dissipation. For this reason we consider a *transparent* translation from Tangram to gate-level circuits as essential. We are able to achieve such transparency by choosing handshake circuits as a highly uniform intermediate architecture.

VLSI programming for low power also requires analysis and simulation tools that quantify power consumption on the basis of the Tangram text. These tools predict power consumption with a reasonable accuracy (within about 20%) for our double-rail asynchronous circuits. More accurate feedback requires the availability of the actual layout of the circuit.

By using a highly transparent translation and powerful power-analysis tools, we have transformed the design of low-power digital ICs into a programming task. Considerable effort in computer science has been spent on designing programs that consume minimal time and resources (memory). Unfortunately, few techniques, tactics, and methods are available today to assist the designer in VLSI programming for low power.

Acknowledgements

Tangram, the tools (compiler, simulators, test generation, fault simulation), and the ICs are the result of team work at Philips Research Laboratories. The current Tangram team consists of Ronan Burgess, Joep Kessels, Marly Roncken, Frits Schalij, and the first author. Many students also contributed to the above results, in particular Jaco Haans, Ad Peeters, and Rik van de Wiel (all EUT, Eindhoven). The project also benefits from advice by Cees Niessen (Philips Research) and the second author. Presentations at the Eindhoven VLSI Club (EUT) and at the ACiD workshop on asynchronous circuit design in Lyngby (Denmark) helped in organizing and improving this chapter. Joep Kessels, Rudolf Mak, and Tom Verhoeff provided valuable feedback on a draft version of this text. Part of the work has been funded by the European Commission under ESPRIT contract 6143 (EXACT).

A   Cheapest unary expression

The problem of the cheapest N-fold repeater exclusively constructed from binary mixers and binary sequencers (see Section 3.4) can also be formulated as follows. The set of unary expressions UExp is given by the following syntax:

$$
\begin{array}{lll}
\text{UExp} & ::= & \text{``1''} \\
& | & \text{``(''} \; \text{UExp} \; \text{``+''} \; \text{UExp} \; \text{``)''} \\
& | & \text{UExp} \; \text{``} \times \text{''} \; \text{UExp}
\end{array}
$$

The *value* of a unary expression can be obtained by evaluating it using the standard arithmetic of natural numbers. For instance, $((1 + 1) * (1 + 1) + 1)$ evaluates to 5. So does $(((1 + 1) + (1 + 1)) + 1)$. The *costs* of a unary expression is its number of "+" symbols; multiplication is for free[12]. The costs of the above expressions with value

---

[12]A unit cost corresponds to A_{dup} in the construction of N-fold repeaters.

5 are 3 and 5 respectively. The problem is how, for given N, $N \geq 1$, to construct a unary expression with value N and lowest costs.

The strategy of "binary division" applied in Subsection 3.4 does not always yield the cheapest unary expression. For instance,

$$15 = ((1 + 1) \times ((1 + 1) \times ((1 + 1) + 1) + 1) + 1)$$

costs 6, whereas

$$15 = 3 \times 5 = ((1 + 1) + 1) \times ((1 + 1) \times (1 + 1) + 1)$$

costs only 5. The latter expression is based on "factorization followed by binary division of the factors", which seems attractive since multiplication comes for free. However

$$33 = 3 \times 11 = ((1 + 1) + 1) \times ((1 + 1) \times ((1 + 1) \times (1 + 1) + 1) + 1)$$

costs 7, whereas

$$33 = 2^5 + 1 = ((1 + 1) \times (1 + 1) \times (1 + 1) \times (1 + 1) \times (1 + 1) + 1)$$

costs only 6, resulting in a win for binary division. Of course, one may enumerate all possible unary expressions with value N. But, does there exist a more elegant and more efficient solution to the problem of the cheapest unary expression?

References

[1] Erik Brunvand. *Translating Concurrent Communicating Programs into Asynchronous Circuits*. PhD thesis, Carnegie Mellon University, 1991.

[2] Steven M. Burns. *Performance Analysis and Optimization of Asynchronous Circuits*. PhD thesis, California Institute of Technology, 1991.

[3] James B. Burr and John Scott. A 200mv Self-Testing Encoder/Decoder using Stanford Ultra-Low-Power CMOS. In *ISSCC 1994 Digest of Technical Papers*, pages 84--85, 1994.

[4] Anantha P. Chandrakasan, Samuel Sheng, and Robert W. Broderson. Low-Power CMOS Digital Design. *IEEE Journal of Solid-State Circuits*, 27(4):473--483, 1992.

[5] T.R. Crompton. *Battery Reference Book*. Butterworths, 1990.

[6] Jo C. Ebergen and Ad M. G. Peeters. Modulo-N counters: Design and analysis of delay-insensitive circuits. In J. Staunstrup and R. Sharp, editors, *2nd Workshop on Designing Correct Circuits, Lyngby*, pages 27--46. Elsevier, North Holland, 1992.

[7] Daniel W. Dobberpuhl *et al.* A 200-MHz 64-b Dual-Issue CMOS Microprocessor. *IEEE Journal of Solid-State Circuits*, 27(11):1555--1565, 1992.

[8] S. B. Furber, P. Day, J. D. Garside, N. C. Paver, and J. V. Woods. A micropipelined ARM. In *Proceedings of VLSI 93*, pages 5.4.1 --5.4.10, September 1993.

[9] Frank Goodenough. Off-Line And One-Cell IC Converters Up Efficiency. *Electronic Design*, pages 55--64, June 27, 1994.

[10] Jaco Haans, Kees van Berkel, Ad Peeters, and Frits Schalij. Asynchronous Multipliers as Combinational Handshake Circuits. In S. Furber and M. Edwards, editors, *Proceedings of the IFIP WG 10.5 Working Conference on Asynchronous Design Methodologies, Manchester*, pages 149--163. Elsevier Science Publishers B.V., 1993.

[11] Nils Hedenstierna and Kjell O. Jeppson. CMOS Circuit Speed and Buffer Optimization. *IEEE Transactions on Computer-Aided Design*, 6(2):270--281, 1987.

[12] C.A.R. Hoare. Communicating Sequential Processes. *Communications of the ACM*, 21(8):666--677, 1978.

[13] Gordon M. Jacobs and Robert W. Brodersen. A fully asynchronous digital signal processor using self-timed circuits. *IEEE Journal of Solid-State Circuits*, 25(6):1526--1537, December 1990.

[14] Joep Kessels, Kees van Berkel, Ronan Burgess, Ad Peeters, Marly Roncken, and Frits Schalij. VLSI Programming of a Low-Power Asynchronous Error Corrector for the DCC Player. Technical Report TN 023/94, Philips Research Laboratories Eindhoven, Prof. Holstlaan 4, 5656 AA Eindhoven, The Netherlands, 1994.

[15] Joep Kessels, Kees van Berkel, Ronan Burgess, Marly Roncken, and Frits Schalij. An Error Decoder for the Compact Disc Player as an Example of VLSI Programming. In *Proceedings of the European Conference on Design Automation*, pages 69--74, 1992.

[16] Joep L.W. Kessels. Calculational Derivation of a Counter with Bounded Response Time and Bounded Power Dissipation. *Distributed Computing*, 8(3), 1994.

[17] H.T. Kung and C.E. Leiserson. Systolic Arrays (for) VLSI). In *Sparse Matrix Proc. 1978*, pages 256--282, 1979.

[18] Peter Macken, Marc Degrauwe, Mark van Paemel, and Henr Oguey. A Voltage Reduction Technique for Digital Systems. In *ISSCC 1990 Digest of Technical Papers*, pages 238--239, 1990.

[19] Alain J. Martin. Programming in VLSI: From Communicating Processes to Delay-Insensitive Circuits. In C.A.R. Hoare, editor, *UT Year of Programming; Institute on Concurrent Programming*. Addison-Wesley, 1989.

[20] Alain J. Martin. Asynchronous Datapaths and the Design of an Asynchronous Adder. *Formal Methods in System Design*, 1(1):119--137, 1992.

[21] Alain J. Martin. Tomorrow's digital hardware will be asynchronous and verified. In J. van Leeuwen, editor, *Information Processing 92, Vol. I: Algorithms, Software, Architecture*, volume A-12 of *IFIP Transactions*, pages 684--695. Elsevier Science Publishers, 1992.

[22] C. Mead and L. Conway. *Introduction to VLSI Systems*. Addison-Wesley, 1980.

[23] Lars Skovby Nielsen, Cees Niessen, Jens Sparso, and Kees van Berkel. Low-Power Operation Using Self-Timed Circuits and Adaptive Scaling of the Supply Voltage. *IEEE Transactions on VLSI Systems*, 2(4):7, 1994.

[24] Cornelis Niessen and Cornelis Hermanus van Berkel. An apparatus featuring a feedback signal for controlling a powering voltage for asynchronous electronic circuitry therein. European Patent Application 92203949.0, N.V. Philips' Gloei-lampenfabrieken, 1993.

[25] N. C. Paver. *The Design and Implementation of an Asynchronous Microprocessor*. PhD thesis, Department of Computer Science, University of Manchester, June 1994.

[26] Ad Peeters. The Asynchronous Bibliography. Available for anonymous ftp on Internet. Uniform Resource Locator (URL) `ftp:-//ftp.win.tue.nl/pub/tex/async.bib.Z`. Corresponding e-mail address: `async-bib@win.tue.nl`.

[27] M. Rem. Lecture Notes in Parallel Computing. Technical report, Eindhoven University of Technology, 1994.

[28] Martin Rem. Mathematical aspects of VLSI design. In Charles L. Seitz, editor, *Proc. Caltech Conf. on VLSI*, pages 55--63. Computer Science Dept., California Institute of Technology, 1979.

[29] Martin Rem. Trace Theory and Systolic Computations. In J.W. de Bakker, A.J. Nijman, and P.C. Treleaven, editors, *PARLE Parallel Architectures and Languages in Europe*, volume 258 of *Lecture Notes in Computer Science*, pages 14--33. Springer-Verlag, 1987.

[30] Frits Schalij. Tangram manual. Technical Report UR 008/93, Philips Research Laboratories Eindhoven, P.O. Box 80.000, 5600 JA Eindhoven, The Netherlands, 1993.

[31] Charles L. Seitz. System Timing. In C.A. Mead and L.A. Conway, editors, *Introduction to VLSI Systems*, chapter 7. Addison-Wesley, 1980.

[32] J. Sparso, C. D. Nielsen, L. S. Nielsen, and J. Staunstrup. Design of self-timed multipliers: A comparison. In S. Furber and M. Edwards, editors, *Asynchronous Design Methodologies*, volume A-28 of *IFIP Transactions*, pages 165--179. Elsevier Science Publishers, 1993.

[33] Ivan Sutherland. Micropipelines. *Communications of the ACM*, 32(6):720--738, 1989.

[34] C. H. (Kees) van Berkel, Cees Niessen, Martin Rem, and Ronald W. J. J. Saeijs. VLSI programming and silicon compilation. In *Proceedings ICCD '88*, pages 150--166, Rye Brook, New York, 1988. IEEE Computer Society Press.

[35] Kees van Berkel. *Handshake Circuits. An asynchronous architecture for VLSI programming*. International Series on Parallel Computation 5. Cambridge University Press, 1993.

[36] Kees van Berkel. VLSI programming of a modulo-N counter with constant response time and constant power. In S. Furber and M. Edwards, editors, *Proceedings of the IFIP WG 10.5 Working Conference on Asynchronous Design Methodologies, Manchester*, pages 1--11. Elsevier Science Publishers B.V., 1993.

[37] Kees van Berkel, Ronan Burgess, Joep Kessels, Ad Peeters, Marly Roncken, and Frits Schalij. A Fully-Asynchronous Low-Power Error Corrector for the DCC Player. In *ISSCC 1994 Digest of Technical Papers*, pages 88--89, 1994.

[38] Kees van Berkel, Ronan Burgess, Joep Kessels, Ad Peeters, Marly Roncken, and Frits Schalij. A Fully-Asynchronous Low-Power Error Corrector for the DCC Player. *IEEE Journal of Solid-State Circuits*, 29(12):11, 1994.

[39] Kees van Berkel, Ronan Burgess, Joep Kessels, Ad Peeters, Marly Roncken, and Frits Schalij. Asynchronous Circuits for Low Power: a DCC Error Corrector. *IEEE Design & Test*, 11(2):22--32, June 1994.

[40] Kees van Berkel, Ronan Burgess, Joep Kessels, Marly Roncken, and Frits Schalij. Characterization and evaluation of a compiled asynchronous IC. In S. Furber and M. Edwards, editors, *Proceedings of the IFIP WG 10.5 Working Conference on Asynchronous Design Methodologies, Manchester*, pages 209--221. Elsevier Science Publishers B.V., 1993.

[41] Kees van Berkel, Joep Kessels, Marly Roncken, Ronald Saeijs, and Frits Schalij. The VLSI-programming language Tangram and its translation into handshake circuits. In *Proceedings of the European Design Automation Conference*, pages 384--389, 1991.

[42] Harry J.M. Veendrick. Short-Circuit Dissipation of Static CMOS Circuitry and Its Impact on the Design of Buffer Circuits. *IEEE Journal of Solid-State Circuits*, SC-19(4):468--473, 1984.

[43] Tom Verhoeff. Delay-insensitive codes: an overview. *Distributed Computing*, 3:1--8, 1988.

[44] Eric A. Vittoz. Low-Power Design: Ways to Approach the Limits. In *ISSCC 1994 Digest of Technical Papers*, pages 14--18, 1994.

[45] Neil H.E. Weste and Kamran Eshraghian. *Principles of CMOS VLSI Design; A System Perspective;* SECOND EDITION. Addison-Wesley Publ. Company, 1993.

Computing without Clocks: Micropipelining the ARM Processor

Steve Furber
Department of Computer Science, University of Manchester
Oxford Road, Manchester M13 9PL, England.

Abstract

High-performance VLSI microprocessors are becoming very power hungry; this presents an increasing problem of heat removal in desk-top machines and of battery life in portable machines. Asynchronous operation is proposed as a route to more energy efficient computing. In his 1988 Turing Award Lecture, Ivan Sutherland proposed a modular approach to asynchronous design based on "Micropipelines". The AMULET group at Manchester University has developed an asynchronous implementation of the ARM microprocessor based on micropipelines as part of a broad investigation into low power techniques. The design is described in detail, the rationale for the work is presented and the characteristics of the chip described. The first silicon from the design arrived in April 1994 and an evaluation of it is presented here.

1 Motivation

The motivation for the work described in this chapter stems from the need for approaches to the design of VLSI devices which result in chips with lower power consumption, and from the potential which asynchronous logic appears to have to meet this need.

1.1 Power Trends

VLSI techniques with lower power consumption are needed for at least two different reasons.

Firstly, the portable equipment market is growing rapidly. Products such as lap-top computers and personal digital assistants rely on batteries for power, hence battery life is an important specification point. Mobile telephones are increasingly using significant computational functions, and again battery life is a key feature in the marketplace.

Secondly, the trend in CMOS VLSI is towards very high dissipation. Chips like the DEC Alpha and the TI Viking SPARC illustrate the trend towards unmanageable power consumption where 20 to 30 watts is not unusual for a high-performance CMOS processor today. If current design practice were to continue, by the year 2000 we would expect to see a 0.1μm 5V CMOS processor dissipating 2kW! Of course, there are trends towards lower supply voltages which reduce the power somewhat, but 3v (and later 2v) operation only reduces power by a factor of 3 (and then 6). More is needed if the technology of the next century is not to have its performance potential compromised by power issues.

The desirability of lower power consumption is now widely recognised and many conferences have sessions and keynote speeches on the subject. There have been workshops dedicated to low power, and manufacturers of commercial processors are incorporating power management features into their products. The goal of more power-efficient design is not contentious; what is contentious is the proposition that asynchronous logic may have a role to play in achieving this goal.

1.2 Asynchronous Logic

There has recently been a resurgence of interest in asynchronous logic design techniques, which had for two previous decades been largely neglected by designers. One reason for this renewed interest is the observation that synchronous logic is beginning to run into serious limits. As VLSI devices incorporate increasing numbers of transistors it is becoming increasingly difficult to maintain the global synchrony on which synchronous designs depend. Clock skew is already a problem at board level, and is increasingly becoming a problem on a single chip. Most notable amongst recent victories over clock skew is the DEC Alpha [1], where the clocking system is a major feat of engineering and the statistics of the clock drivers (such as the proportion of silicon area they occupy, the sizes of the transistors, the peak currents, etc.) are viewed with some awe by most 'normal' designers. But the fact that these feats were needed on the Alpha serves here to demonstrate the point that synchronous design is beginning to approach its limits.

Asynchronous logic abandons global synchrony in favour of locally generated timing signals and is therefore unaffected by clock skew. It also displays a number of specific advantages over clocked logic, particularly when low power consumption is an objective:

- Power is only used to do useful work. In synchronous design the clocks are applied to all units whether or not they are doing anything useful. Recent developments in power management may gate off clocks from some areas of the design, but these typically are applied at a coarse granularity. Straightforward asynchronous design only activates units when they are required to do useful work.

- Designs can be optimised for typical conditions. When timing is defined by a fixed clock, all operations must complete within a fixed period. Synchronous designs therefore expend considerable silicon resource on making rare worst-case operations fast. This usually results in complex circuits which use more power than is necessary under typical conditions. An asynchronous design can allow worst case operations to proceed more slowly and focus the use of resource on operations which occur frequently.

Therefore asynchronous design appears to have considerable potential for power-efficient design. However, asynchronous design is not new; in the early days of computing many designs used asynchronous approaches, but these were later abandoned in favour of synchronous styles because of the inherent difficulties of asynchronous operation. Are these difficulties still an obstacle to a wide acceptance of asynchronous logic?

This question remains to be answered. The following observations may go some way towards addressing it:

- Synchronous designs are getting increasingly complex and, where power-efficiency is important, clock gating is used with increasingly fine granularity. In the limit, clock gated synchronous circuits look very like fully asynchronous circuits, so perhaps this complexity is unavoidable where power-efficiency is a goal.

- The clock in a synchronous circuit has a role similar to a global variable in a computer program. Object-oriented programming styles are increasingly accepted as ways to increase programmer productivity by precluding global variables and encapsulating data. Asynchronous logic styles with well defined interfaces have many of the properties of object-oriented programming styles and may show the same benefits in designer productivity for systems above a certain level of complexity.

- Mathematical techniques are being developed and applied with increasing success to address some of the difficulties inherent in asynchronous design, such as proving circuits are free from deadlock.

Whilst the above arguments do not constitute conclusive proof that asynchronous design has advantages over clocked design sufficient to justify an immediate abandoning of synchronous styles by all designers, they do at least suggest that asynchronous design is worth investigating again to see to what extent its potential for power-efficiency can be realised. The AMULET group at the University of Manchester was established late in 1990 to carry out such an investigation. This chapter documents the first work of this group, the design and evaluation of AMULET1, an asynchronous implementation of the ARM microprocessor.

The chapter begins with a description of Sutherland's micropipelines, which were used as the basis for the asynchronous design style employed on AMULET1. Details of the event control cells which manage the interactions between micropipelines are presented along with discussions of various low-level VLSI design issues. The ARM processor is introduced and the asynchronous organisation of AMULET1 described. The results of the evaluation of the first silicon are presented, and the chapter closes with a discussion of the conclusions which may be drawn from the work so far and suggested directions for future work.

2 Sutherland's Micropipelines

The asynchronous design style used in AMULET1 is based on Sutherland's Micropipelines [2] which employ a 2-phase bundled data interface for sending data between functional units. 2-phase (or transition) signalling uses both rising and falling edges in turn to signal the same event; rising and falling edges are equivalent and carry the

214

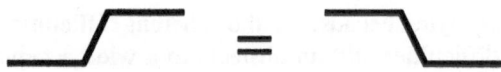

Figure 1: Transition signalling

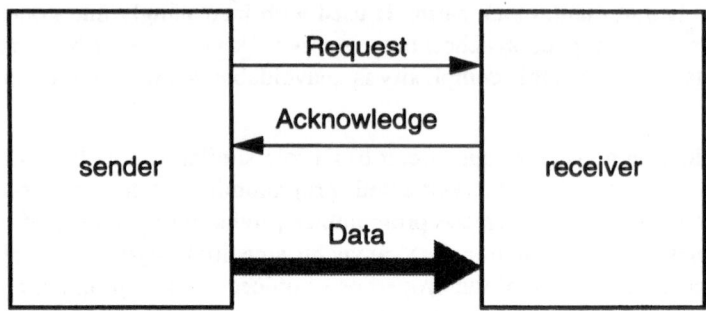

Figure 2: A bundled data interface

same information (figure 1).

A bundled data interface passes a binary value encoded conventionally on a bus from sender to receiver along with a *Request* wire which indicates when the data is valid and an *Acknowledge* wire which indicates when the data has been received (figure 2).

The communication protocol is illustrated by the timing diagram in figure 3. A valid data value is placed on a conventional bus by the sender which then indicates the availability of the data by causing a transition on the *Request* wire. The receiver senses this transition, accepts the data and then causes a transition on the *Acknowledge* wire, completing the transfer. The sender may then issue another data value in a similar manner. Note that only the order of these events is significant; the delays between them are arbitrary (though long delays will, of course, reduce performance). Also note that the Request and Acknowledge wires use 2-phase signalling; rising and falling edges are both significant and have the same meaning.

In order to design circuits based on micropipelines, a few event control blocks are required. The basic set of event control blocks proposed by Sutherland is illustrated in figure 4.

The *Muller C-gate* performs the rendezvous function for events; it waits until it has received an event on both of its inputs before issuing an event on its output.

The *XOR* (exclusive-OR) gate performs the merge function for events; a transition on either input results in a transition on its output.

The *toggle* cell transfers an event from its input to its two outputs alternately; the first event to arrive is issued to the output marked with a dot, the second to the unmarked output, and so on.

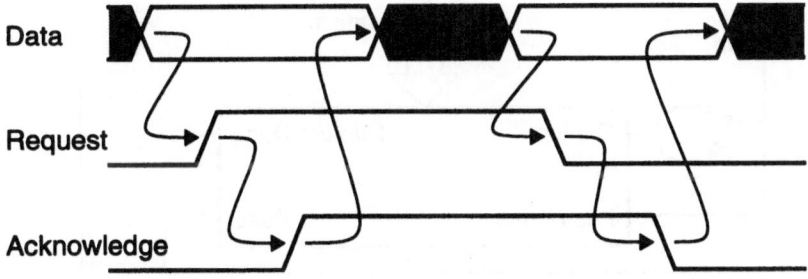

Figure 3: The 2-phase bundled data convention

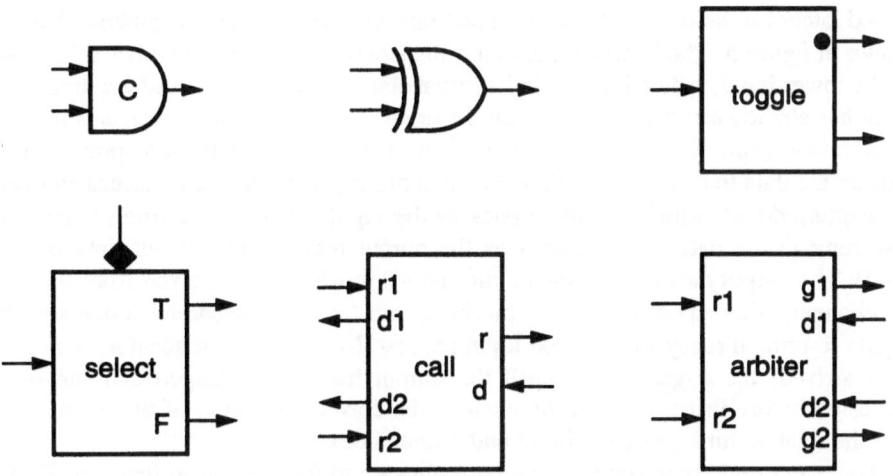

Figure 4: Event logic modules

The *select* block allows a boolean, usually derived from the binary data, to direct the input event to the True or False output.

The *call* block allows two independent processes to share a common subprocess; requests on r1 and r2 are routed through to r, and the completion signal on d is routed back to the caller. The calling processes must be mutually exclusive; if they are not, they must access the call block through an arbiter.

The *arbiter* accepts asynchronous requests on r1 and r2 and grants only one of them at a time on g1 or g2. When the shared resource is given up (signalled by an event on d1 or d2) the arbiter will then grant the other request if it is pending. The arbiter must be designed to allow for metastability in its internal circuitry, but if this is done correctly reliable operation is possible.

In order to construct a micropipeline circuit, it is necessary to employ event con-

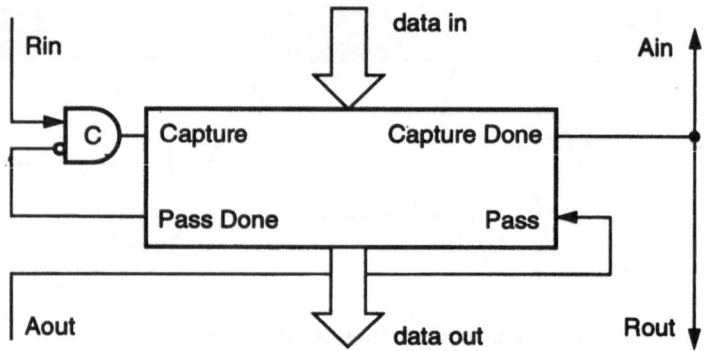

Figure 5: A micropipeline event driven register

trolled latches to hold the data stable. The high-level view of a micropipeline latch is shown in figure 5. The latch begins in a transparent state. The C-gate has a 'bubble' on the lower input, indicating that in its initial state this input is primed (as though an event has already arrived). An event on the input request wire (Rin) indicates that the input data is valid. This event passes through the C-gate to the latch 'capture' input, causing the data to be latched. When the latch has captured the data it issues an event on 'capture done', which is copied back as the input acknowledge (the sender may now remove the data) and forward as the output request (the output data is now valid). The output data is held stable until an acknowledge is received from the output channel, whereupon the latch is put back into 'pass' (transparent) mode and the C-gate re-primed ready for the next input request. Note that this request may already have arrived; the C-gate waits until the output has acknowledged and the input requested before firing and thereby ensures the correct operation of the latch whatever the relative timings on the input and output sides.

The event register in figure 5 can be replicated to form a first-in first-out (FIFO) buffer as shown in figure 6. A data value can be placed into the left event register by signalling an event on Rin, whereupon it will be passed along to the right register at a speed determined only by the properties of the logic technology from which it is constructed. The right register will then issue an event on Rout and hold the data stable until it receives an acknowledge on Aout. In the meantime further values may be entered from the left until the FIFO is full (i.e. each event register holds a data value), whereupon the first stage will not acknowledge the last value entered and the input process will stall until a value is removed by the output process.

The FIFO has two important parameters which define its performance:

- Latency - the speed at which new data passes through an empty FIFO. This parameter may not be significant in some applications where a steady stream of values flows through the FIFO. But when micropipelines are applied to the design of a microprocessor, the FIFO is likely to be flushed from time to time and then the latency will influence the speed with which a new processing thread may be established.

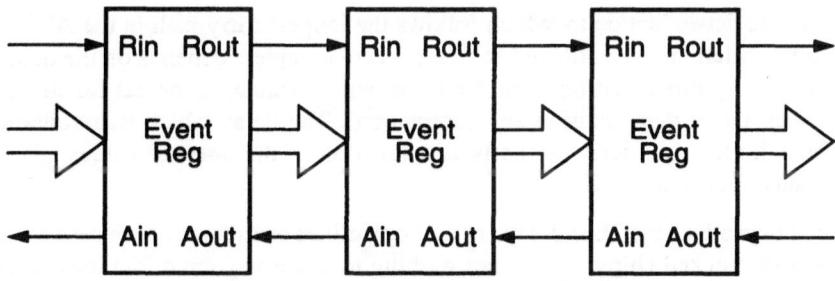

Figure 6: A FIFO - the canonical micropipeline

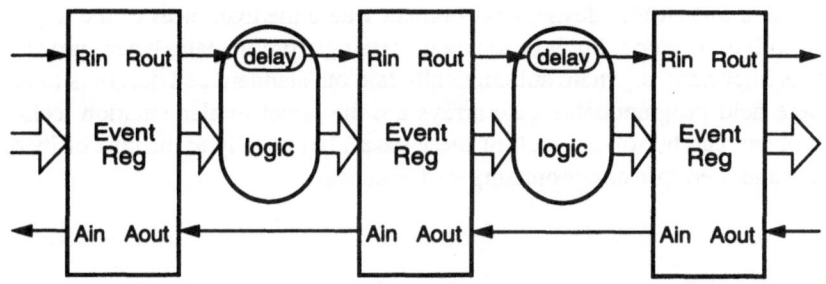

Figure 7: A FIFO with processing

- Throughput - the maximum sustainable data rate. This parameter plays the same role as the maximum clock rate in a synchronous pipeline.

One of the features of the micropipeline FIFO is its elasticity. Data can be entered at any rate up to a maximum defined by the throughput parameter and limited only by the FIFO becoming full, and it can be removed at any rate up to a similar maximum limited only by the FIFO becoming empty. Within these limits the structure will buffer a variable number of data values and pass new input values to the output at a speed limited only by the process technology. Compare this with a clocked pipeline where the input and output are regimented by the clock, and the input to output delay is a fixed number of clock cycles.

The micropipeline FIFO is fine as a buffer, but performs no logical processing on the data that flows through it. By interspersing logic between the event registers in a FIFO, a micropipeline with processing functions can be built (figure 7). The logic in the data route incurs some delay in the arrival of the input data at the next register, so a corresponding delay must be introduced into the request line to ensure that the bundling constraint is met at the input of the next register. This delay element may be implemented in a number of ways; examples on AMULET1 include:

- A 33rd register read bit which always makes a transition whenever a register is read. This 33rd bit is used to signal completion of the register read process.

218

- A data dependent delay which follows the longest carry path in the ALU. The time which must be allowed for the ALU to complete depends on the data values being processed (and the function being evaluated; logical functions are much faster than addition and subtraction). The delay which is introduced to match the ALU delay similarly is a function of the operand values, giving a data dependent delay.

The need to produce matched delays in micropipelines is a source of concern to many designers of clocked chips. Whilst not wishing to underplay the difficulties inherent in this approach, it should be noted that such delay matching is not uncommon in current synchronous designs. CMOS PLAs frequently use self-timed paths to allow dynamic operation and to eliminate DC currents, and SRAMs use self-timed paths to turn off sense amplifiers in order to save power. Delay matching is not particularly difficult in a full-custom design environment where the tolerances of the target process are well understood, but it becomes increasingly problematic where process portability is important, or where automatically laid out standard cell design is to be used, or where field programmable gate arrays are the target implementation technology. The problem can be solved in all of these cases, but usually at the cost of increasing margins and therefore compromising performance.

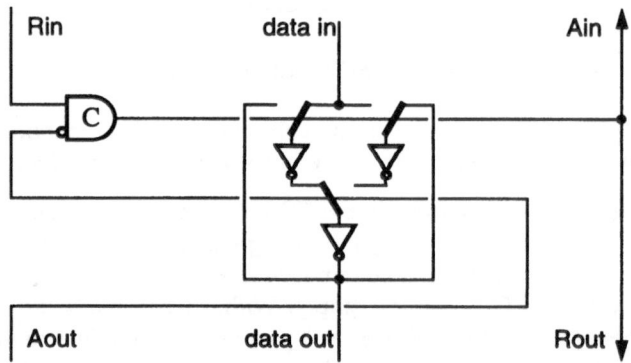

Figure 8: Sutherland's 'capture-pass' latch

3 Micropipeline Latch Structures

Sutherland proposed a latch for use in micropipelines which operates directly from the 2-phase capture and pass signals. The principle of this latch is illustrated in figure 8. Here the latch is shown in its initial, transparent, condition. An event on the capture wire (the output of the C-gate) switches the upper two multiplexers over, holding the data in a feedback loop and isolating the output from the input. A subsequent event on the 'pass' wire (from Aout) switches the output multiplexer over, reconnecting the input through to the output.

Note in this figure how the event wires pass across the latches so that the capacitive load of the latches affects the speed of operation of the control circuits. This is a common factor in all the latch designs presented here and allows the event paths to 'measure' the latch loading to ensure correct operation under all conditions by preventing the control circuits from operating faster than the latch control wires can switch.

AMULET1 does not use the Sutherland capture-pass latch. Instead, conventional level-sensitive transparent latches are used with a more complex control circuit as shown in figure 9. Here the XOR gate and toggle operate as a 2-phase to 4-phase conversion circuit. After initialisation the latch is transparent. A capture event from the C-gate closes the latch and the toggle steers that event to Rout and Ain. The pass event from Aout puts the latch back into its transparent state, then the toggle steers the 'pass done' event back to re-arm the C-gate. Note again how the events which change the state of the latch are 'measured' to ensure that the control circuits wait for the latch to complete its operation.

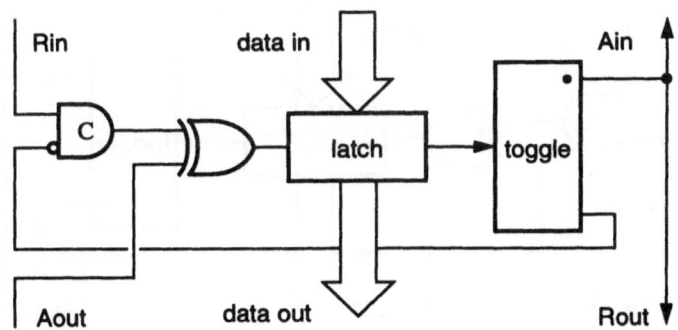

Figure 9: An event latch based on a conventional transparent latch

This latch was used in preference to the 'capture-pass' latch mainly because it is more area efficient for wide buses than Sutherland's latch. Its operation is very safe with respect to bundling constraints. Input data which changes very close to its request will be latched with a good set-up time into the latches since the C-gate, XOR and latch enable line delays all increase the input margins. Rout is delayed further by the latch enable and toggle delays so the output bundle is well margined.

The latch has a latency (Rin to Rout time) of around 10ns on a 5V 1μm CMOS process (worst case temperature, power supply and process parameters) and when several such latches are built into a FIFO structure the bandwidth corresponds to a latch cycle time of around 30ns.

The latch circuit shown in figure 9 is widely used on AMULET1. There are some places, however, where its latency is too high. One such place is the instruction prefetch buffer. When a branch is taken this buffer is flushed and the branch target instruction must pass through 5 latch stages in this buffer before it can begin execution, which at 10ns per stage would add significantly to the branch cost. In such cases a slightly modified 'fast forward' form of the latch is used. This latch control struc-

220

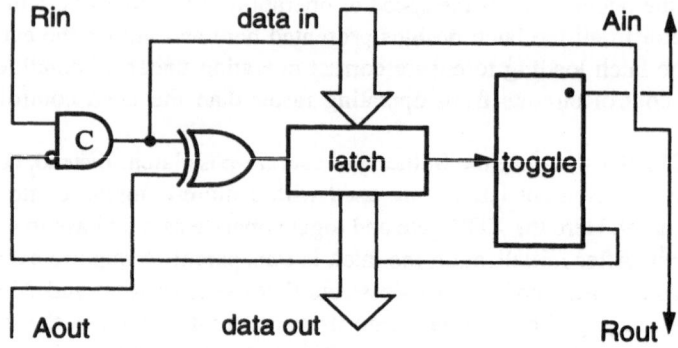

Figure 10: A 'fast forward' event latch

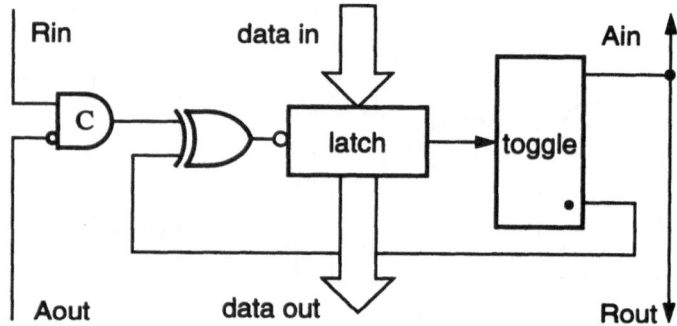

Figure 11: A blocking (normally closed) event latch

ture is shown in figure 10.

The change to the control circuit is to take Rout directly from the C-gate output rather than from the toggle. This is allowable (under certain conditions) since Rout indicates merely that the output data is valid; it need not be delayed until the latches have closed. This latch has a faster forward propagation time (around 2ns Rin to Rout) than the standard latch, giving the 5-stage instruction buffer a refill latency of 10ns rather than the 50ns it would have using standard latches.

The fast-forward latch is less safe on bundling constraints and requires the latch input to output time to be no slower than the C-gate delay. It also requires the Rout to Aout delay to be above a minimum value. This is readily satisfied when several of these latches are composed into a FIFO; under these conditions the latch operates safely and has a cycle time around 20ns.

All the above latches are transparent when empty, so transients on the input data buses will propagate through the latches, causing power to be dissipated uselessly down the pipeline. In some cases it is highly desirable to eliminate this effect in order to save power, and indeed this may be achieved by inserting a 'blocking' (normally closed) latch at the top of the pipeline. A control circuit for a blocking latch is shown in figure 11.

The blocking latch uses the first event from the C-gate to open the latches, letting the data through to the output. The toggle returns this event to the XOR, where it then closes the latch and is passed through the toggle to Rout and Ain. Aout simply re-primes the C-gate. This latch is very safe with respect to the bundling constraints, but is rather slower than the previous latches with an Rin to Rout time of around 20ns on 1μm CMOS (worst case) and around 40ns cycle time when built into a FIFO.

4 Event Control Elements

The basic library of event control elements proposed by Sutherland was introduced in figure 4. In this section the transistor structures used for these elements in AMULET1 will be introduced and relevant design issues raised.

The circuit used for the 2-input Muller C-gate is shown in figure 12. Series transistor stacks pull the internal node high or low when both inputs are at the same logic level and the weak feedback inverter retains the previous state when the inputs are at differing levels. Since the input levels are unknown at initialisation, reset circuitry (connected to *Cdn*, the active-low 'clear-down' signal) is required to force the output to zero independently of the inputs.

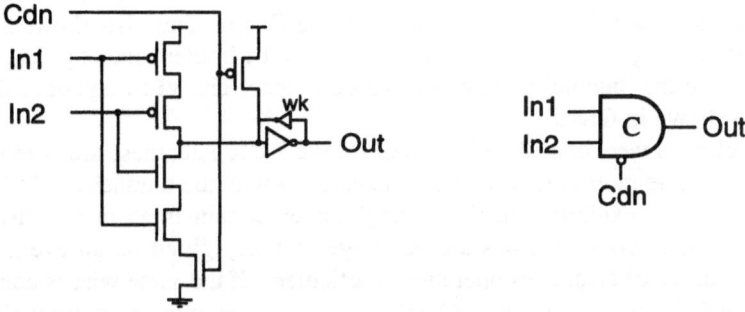

Figure 12: Muller C-gate schematic and symbol

This circuit is similar to Sutherland's dynamic C-gate, with the addition of the weak feedback inverter to give fully static operation. It should be noted that this circuit has unusual logic thresholds on its inputs since when one input is switching the other input will turn off one of the series transistor stacks. The switching threshold for a rising input is therefore V_{tn}, and for a falling input V_{tp}. The C-gate therefore switches early on the transition of the input and may detect the transition significantly before another gate connected to the same signal. To avoid this causing any problems it is advisable to ensure that all event signals have fast rising and falling edges.

222

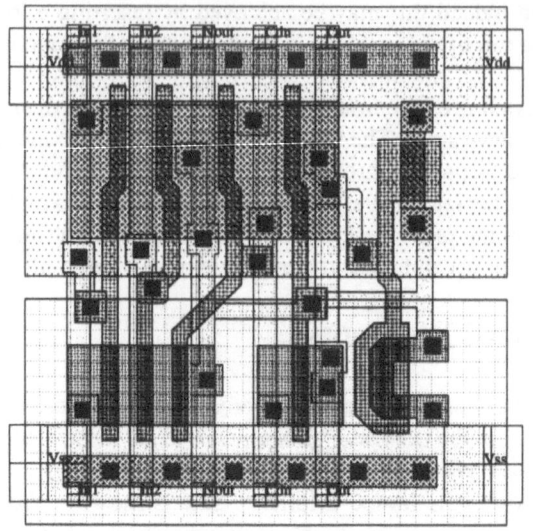

Figure 13: Muller C-gate layout

The observant reader will have noted that whilst all the introduction to micropipe-lines referred to transitions rather than levels, the implementation of the C-gate is clearly level sensitive. This is best understood by observing that provided the C-gate inputs are initialised to the same level (zero), they must thereafter each make the same number of transitions and must therefore retain the same level relationship, allowing a level-sensitive implementation of the C-gate. This also allows the primed version of the C-gate (as used in all the latches) to be implemented by simply adding an inverter to the 'bubble' input of a standard C-gate. The VLSI layout of the Muller C-gate is shown in figure 13

Two circuits were considered in detail for the XOR gate; these are both illustrated in figure 14. The most compact layout is achieved with the 6 transistor design. How-ever this design exhibits 'charge-sharing' under certain input conditions, causing glitches on In1. These glitches are not large, but any glitch on an event wire is a potential source of erroneous operation (particularly if the same wire is connected to a low threshold device such as a C-gate) so all glitches must be thoroughly investi-gated. There is a danger that, in using circuits which are known to produce glitches, the designer will neglect to investigate other glitches which are the source of real problems. Therefore the 6 transistor circuit was rejected in favour of the larger 8 tran-sistor design which is not prone to glitching (but which also requires the inverse of both its inputs).

A circuit which does not appear in Sutherland's cell library is a transparent latch for events. It is used as a building block for other event control elements where its basic role is to block events, and as it may be closed at initialisation it requires reset circuitry to ensure the output is initialised to zero. (Transparent latches for data are simpler as they do not require reset circuitry; they will be introduced later.)

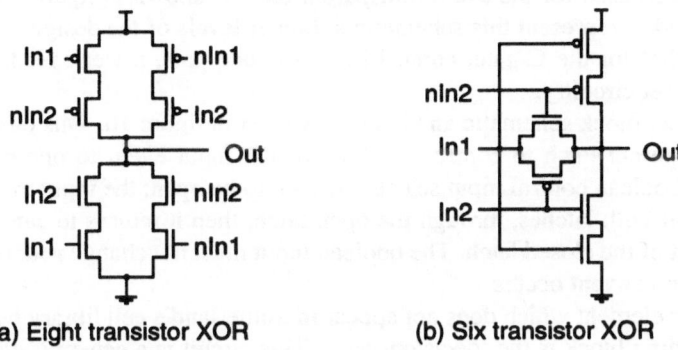

(a) Eight transistor XOR (b) Six transistor XOR

Figure 14: Alternative XOR schematics

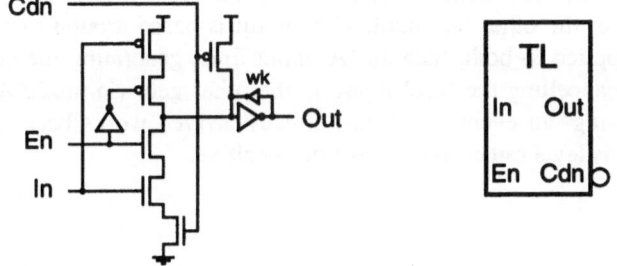

Figure 15: Transparent latch schematic and icon

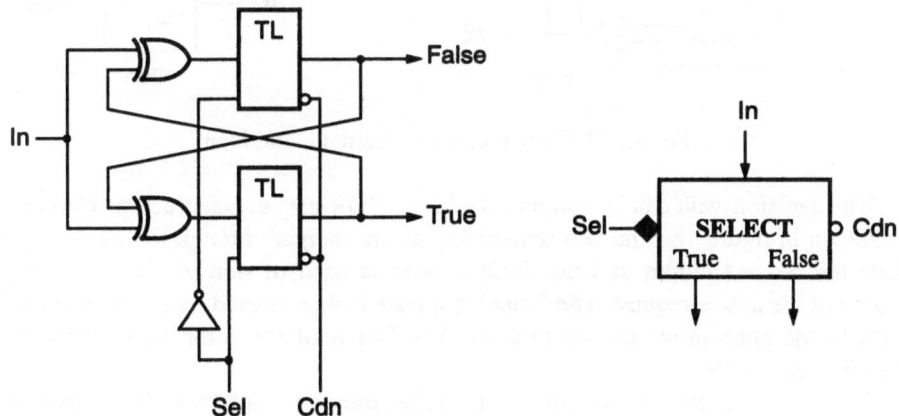

Figure 16: Select block schematic and icon

224

The circuit used for the event transparent latch is shown in figure 15 along with the icon used to represent this schematic at higher levels of the design. The circuit is similar to that for the C-gate, comprising a tri-state gate, a weak feedback storage node and reset circuitry.

The select block schematic and icon are shown in figure 16. This circuit uses the event transparent latch to control the flow of the input event to one output or the other. The boolean control input selects the path to be open; the input event passes to the inputs of both latches, through the open latch, then it returns to cancel the event on the input of the closed latch. The boolean input must not change near to the time at which an input event occurs.

Another element which does not appear in Sutherland's cell library but which is a useful building block is the 'decision-wait'. This circuit is a generalisation of the C-gate, performing a rendezvous between one event line and either of a pair of event lines. A circuit schematic and icon are shown in figure 17. Events can arrive in any order on the 'fire' input and either 'A1' or 'A2' (but not both), then an output event is generated on 'Z1' or 'Z2' according to which 'A' input made a transition. Once the output has fired the circuit is ready to perform the next rendezvous between 'fire' and either the same or the other 'A' input. The circuit is based around two C-gates; the 'fire' input is applied to both, then an 'A' input fires, generating the corresponding 'Z' output and cancelling the 'fire' input on the other gate via the XOR. This technique of 'removing' an event which has already arrived works because the C-gate implementation is level rather than transition sensitive.

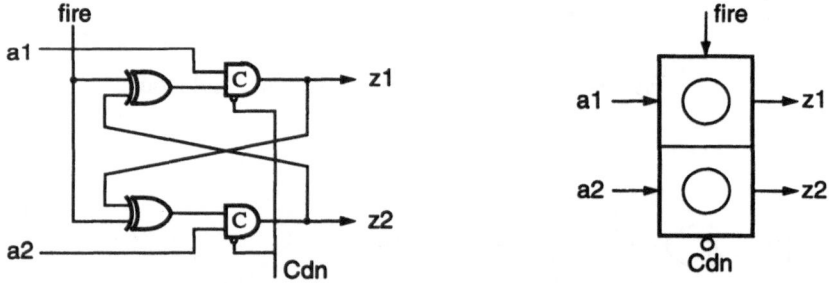

Figure 17: Decision-wait schematic and icon

The decision-wait can be composed with an XOR gate to construct the Call block as shown in figure 18. The two request inputs are merged through the XOR to produce the request output and the decision-wait is used to store the identity of the source of the active request. The 'done' response is then steered by the decision-wait back to the appropriate calling process. The layout of the complete Call block is shown in figure 19

The toggle appears, in principle, to be quite simple to implement. In practice, this was the most problematic of all the event control blocks because the circuit contains an inherent race hazard. The principle of the toggle implementation is shown in figure 20. The input event alternately opens and closes two latches in anti-phase, allow-

225

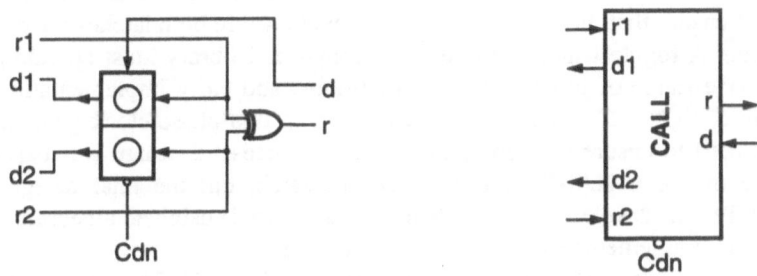

Figure 18: Call block schematic and icon

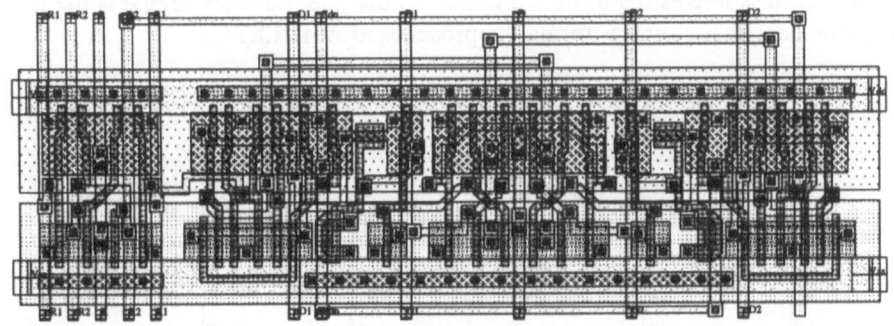

Figure 19: Call block layout

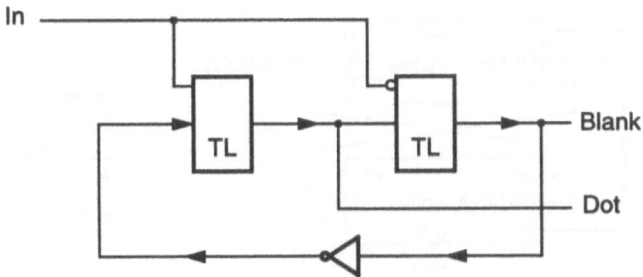

Figure 20: Toggle principle

ing a boolean value to go round a closed loop with one inversion each time round the loop. An implementation must take into account the finite rise and fall time of the input and ensure that there is no point in the cycle where both latches are open at the same time. A toggle which is to be used from a cell library must operate correctly over a wide range of input and load conditions, and early implementations of the principle of figure 20 required a non-overlapping two-phase clock generator (built into the cell) to ensure safe operation. A more successful design was based on the principle of the 6 NAND gate TTL D-type latch, but the final design used in AMULET1 was developed by Jay Yantchev (at Oxford) using an algebraic approach to derive an implementation from the specification.

The circuit of the Yantchev toggle is shown in figure 21. This circuit follows the principle illustrated in figure 20 closely, but succeeds in minimising the race problem by carefully interlocking the two latches. Even so, the circuit can fail if the input and its complement are not carefully aligned to each other, but alignment can be ensured by building the inverter which generates the complement into the cell itself rather than allowing the external circuit to produce it; under these conditions the circuit is very robust. It is interesting to note, in passing, that the Yantchev toggle is the most significant contribution from 'formal' approaches to AMULET1.

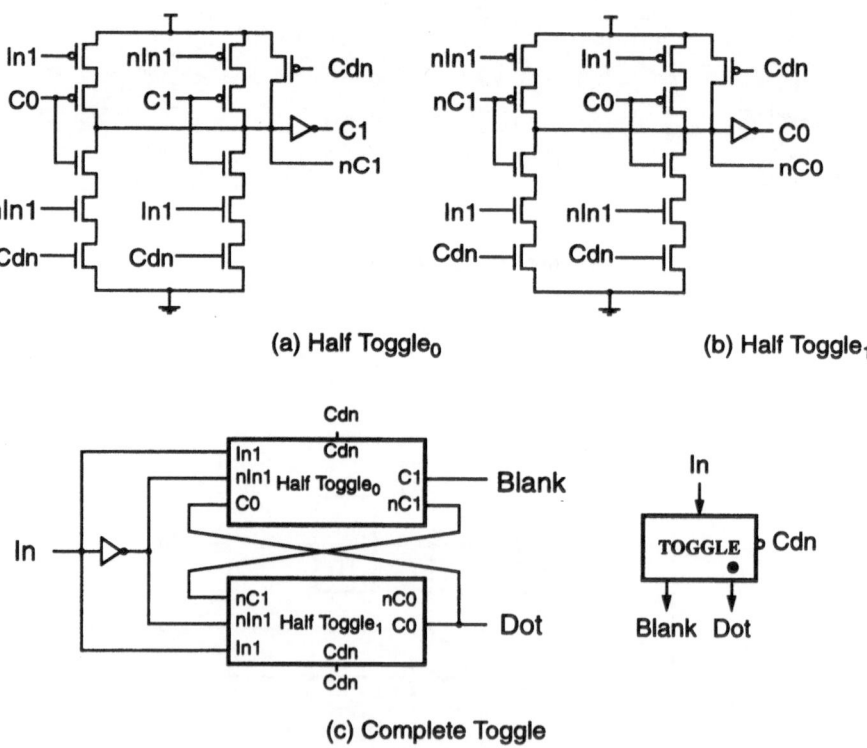

Figure 21: The Yantchev toggle circuit

The arbiter is required to ensure the mutual exclusivity of the grants to two asynchronous requesting processes. As this amounts to making a discrete decision on a continuous set of possible input conditions, it is fundamentally impossible to guarantee that the decision will be made within a bounded time. The circuit must be designed to cope with internal metastability under a range of input conditions. A suitable circuit schematic is shown in figure 22. The basic arbitration unit is a 4-phase level-sensitive mutual exclusion element with a cross-coupled pair of NAND gates forming an R-S flip-flop. The request inputs are normally low, but switch high to indicate an active request. If both switch high at the same time, the R-S flip-flop may go metastable. The output circuit will not, however, issue either grant until the difference between the internal nodes i1 and i2 exceeds V_{tp}, by which time the R-S flip-flop has exited from its metastable state.

The micropipeline (2-phase) arbiter is constructed from the mutual exclusion element by adding suitable 2-phase to 4-phase conversion circuits. The circuit shown in figure 22 is suitable provided there is a reasonable delay between the 'done' input and a subsequent 'request' on the same input.

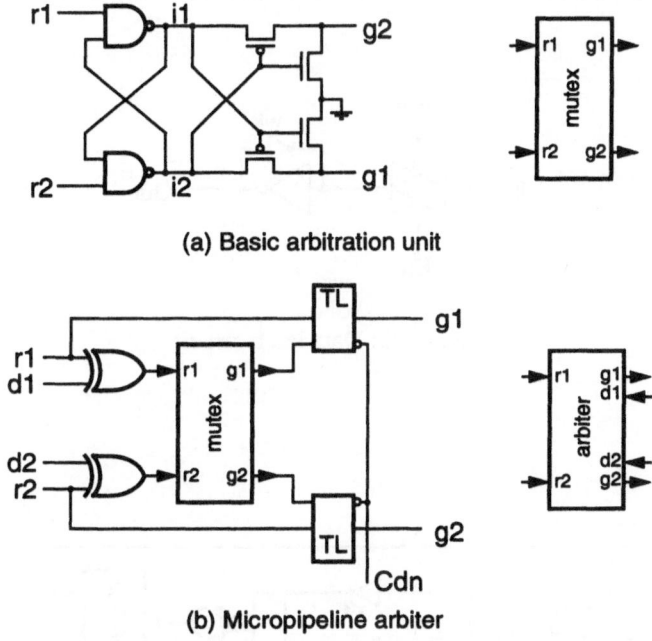

(a) Basic arbitration unit

(b) Micropipeline arbiter

Figure 22: Arbiter schematics

It is worth noting here a fundamental difference between synchronous and asynchronous circuits. In a synchronous circuit the designer must accept some residual probability of failure whenever an asynchronous input enters the clocked domain of the circuit. This probability can be made very small, at the cost of reduced performance, but can never be zero. The designer of an asynchronous circuit can, on the other hand, use an arbiter which will never fail, though it may take a very long time to make a decision. If the system in question is, for example, a flight control computer, it would make little difference to the unfortunate passengers whether a failure is due to synchronization failure in a clocked system or terminal indecision in an asynchronous system. But note that the synchronous system must accept the worst-case cost of synchronizing to the desired level of reliability on every asynchronous sampling process, whereas the asynchronous system typically incurs the average cost and only very rarely comes near to the worst-case cost.

The circuit used to construct pipeline data latches on AMULET1 is shown in figure 23. This circuit is the same as that used on ARM6 for data latches, comprising a transmission gate which, when enabled, overdrives the weak feedback which at other times retains the state on the forward inverter. A practical control circuit is illustrated in figure 24, which shows the buffer circuits used to drive 32 latch loads and a C-gate used to detect that both the latch enable and its complementary signal have switched before firing the toggle.

To illustrate the use of some of the event control cells that have been introduced, circuits are shown below for a conditional pipeline fork (figure 25) and a conditional pipeline join (figure 26). The first of these shows the last stage of one pipeline which

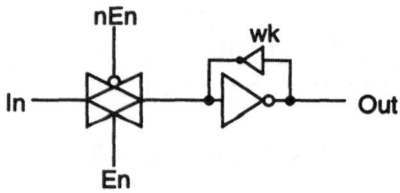

Figure 23: Data latch circuit

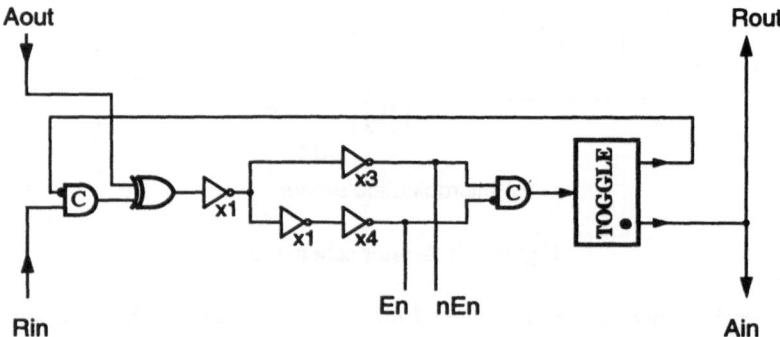

Figure 24: A practical event register control circuit

always sends data down the left output pipeline, but only sends data down the right output pipeline if a boolean in the current data value indicates that it should. The select block implements the decision by steering the event appropriately. Note how the 'False' select output event just bypasses the request to the right pipeline and that both output pipelines are called in parallel, with a C-gate used to wait until both have completed. (A simpler, but slower, circuit could be used to call first the left pipeline and then, conditionally, the second. This would save the cost of the C-gate.)

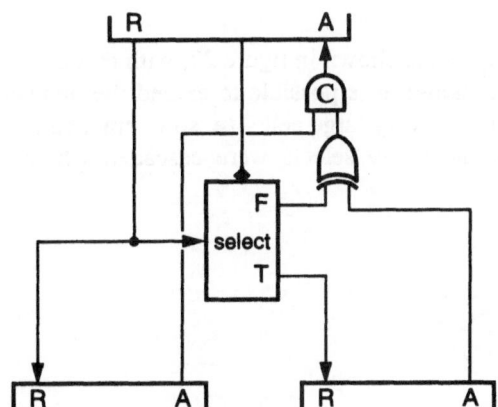

Figure 25: Conditional pipeline fork circuit

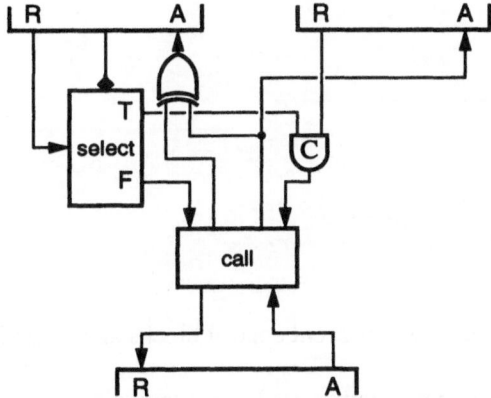

Figure 26: Conditional pipeline join circuit

The conditional pipeline join circuit has a boolean in the left input pipeline which determines whether or not there should be a rendezvous with the right input pipeline before the output pipeline is called. Again, a select block makes the decision by steering its input event. One output uses a C-gate to rendezvous with the second input

230

pipeline, the other bypasses the rendezvous. These two events could now be merged through an XOR to form the request to the output pipeline, but that would leave the problem of handling the acknowledge properly, which may or may not need to go to the right hand input pipeline. A select block on the acknowledge path could be used to achieve the required effect, but rather than taking the same decision again the circuit in figure 26 uses a call block to remember the original decision and steers the acknowledge accordingly.

4.1 Cell Areas

The areas of various cells are shown in figure 27, with that of a standard inverter for reference. Note that although it is possible to extend the number of outputs from a select block, this leads to a very large cell size, so in practice selects were limited to two outputs and several 2-way selects were cascaded where more outputs were required.

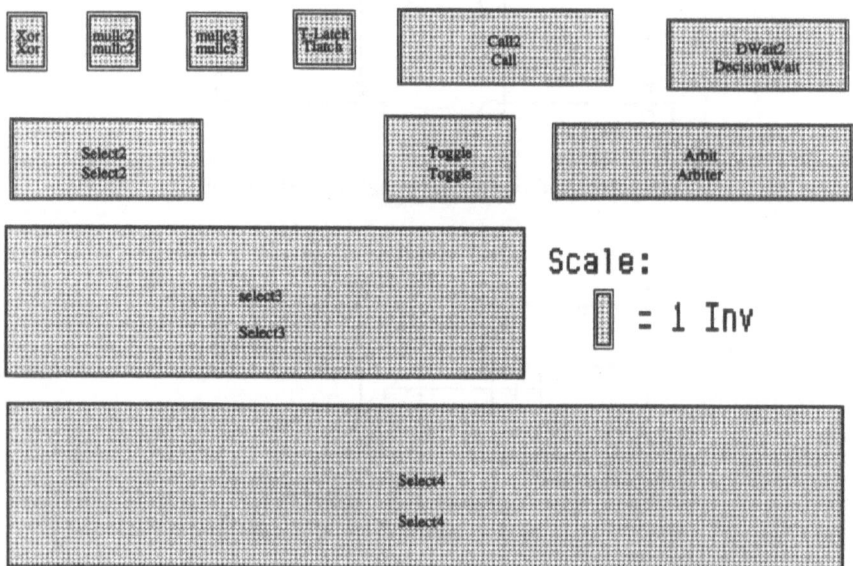

Figure 27: Layout areas of event control blocks and a standard inverter.

Earlier it was stated that capture-pass latches took significantly more silicon area than standard transparent latches, though the control for standard latches is larger. We can now look at this in more detail. The layout areas for 24 data latches and the control circuits are shown in figure 28. From these areas it can be shown that standard latches do, indeed, save on silicon area whenever there are more than 8 data bits in the latch [3]. On AMULET1 latches typically are 32 bits wide, so the area saving from using standard latches is significant.

This concludes the introduction to micropipelines and related engineering issues.

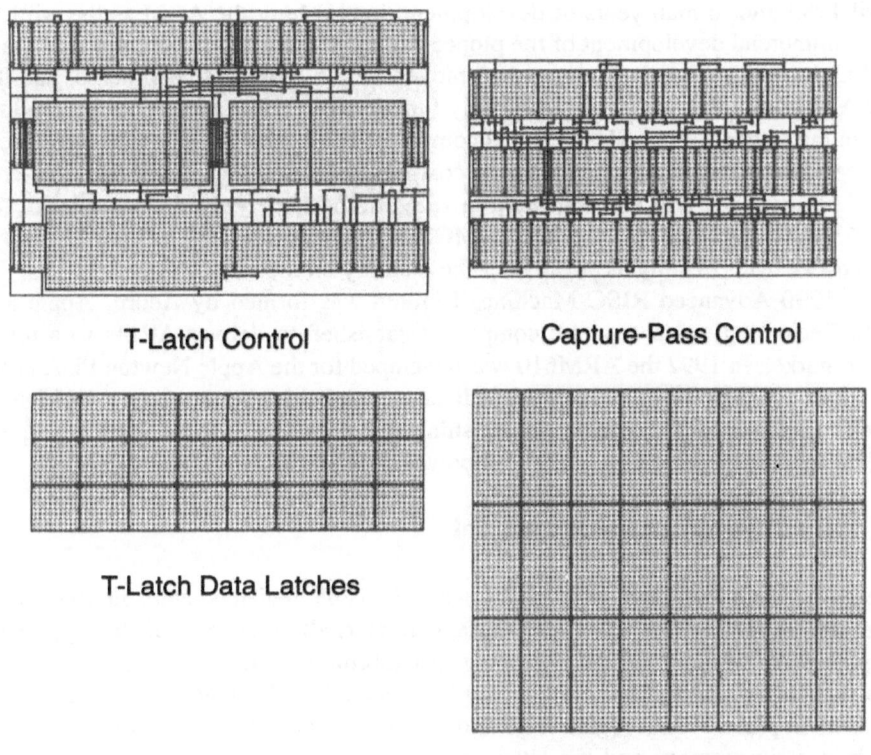

T-Latch Control

Capture-Pass Control

T-Latch Data Latches

Capture-Pass Data Latches

Figure 28: Latch area comparisons

We will now go on to see how micropipelines may be applied to the design of an asynchronous microprocessor. The next section gives an introduction to the ARM microprocessor, then subsequent sections describe the micropipeline implementation of the ARM, the results from the test silicon and the conclusions that can be drawn from the work so far.

5 The ARM Processor

AMULET1 is an asynchronous implementation of the ARM microprocessor [4]. This section gives some background on the ARM and the company which was responsible for developing it, Acorn Computers Limited, which is based in Cambridge, England.

In 1982 Acorn launched the BBC microcomputer, based on the 8-bit 6502 microprocessor, which established Acorn's education market in the UK and certain other countries. (Over 1.5 million BBC machines have been sold worldwide.)

Design of the 32-bit ARM1 was started in 1983. This was the first implementation of the ARM, based on a 3μm CMOS process. The first ARM1 samples worked in

April 1985 after 6 man-years of development, establishing the ARM as the world's first commercial development of the pioneering RISC ideas from Berkeley and Stanford Universities. The design was then moved onto a 2μm CMOS process, and the first ARM2 samples were delivered, fully functional, in 1987 and went into production in the Acorn Archimedes personal computer. The Archimedes sold in reasonable numbers, but real volume came with the cost-reduced A3000 which brought low-cost RISC performance into UK education in 1989. In the same year, first samples of the ARM3 were delivered, a 1.5 micron CMOS design incorporating an ARM2 macrocell and 4k bytes of fully associative cache memory on the same chip.

In 1990 Advanced RISC Machines Limited was formed by Acorn, Apple and VLSI Technology as a separate company established to deliver ARMs to a much wider market. In 1992 the ARM610 was developed for the Apple Newton PDA product, and since then the number of ARM licensees and chips incorporating ARM macrocells has increased regularly, establishing the ARM as a world standard architecture for a range of low-cost and low-power applications.

5.1 ARM6 Programmers' Model

The ARM has a load/store architecture with 16 visible registers available to the programmer [4,5]. Register 15 is the program counter; all other registers being general purpose with the only special use being that subroutine return addresses are placed into register 14. Exceptions are handled in protected modes which switch in private registers in place of user registers 13 and 14, the former usually pointing to a private stack in main memory and the latter containing the exception return address. Fast interrupt mode also has some private work registers. The register organization is shown in figure 29.

In addition to the general registers the ARM also has a Current Program Status Register (CPSR) visible in every mode, and a Saved Program Status Register (SPSR) for each non-user mode (figure 30). Each of these registers is used to save processor mode, interrupt enable status and condition code flag bits (figure 31).

The ARM6 instruction set is summarised in figure 32. In common with other RISC processors, ARM separates those instructions which perform data processing functions from those which move data between memory and registers. The most unusual features of the instructions set include:

- All instructions are conditionally executed.

- Load and store multiple register instructions are included which transfer any subset of the currently visible registers.

The former reduces the number of branches which are required, allowing, for example, some if-then-else clauses to be compiled without a branch instruction. The latter improves the efficiency of procedure entry and exit, and increases the rate at which data block moves can be carried out.

This is the architecture that is re-implemented on AMULET1 using a fully asynchronous design style based on micropipelines. The AMULET1 chip has the functionality of the ARM6 macrocell, omitting only the coprocessor instructions and

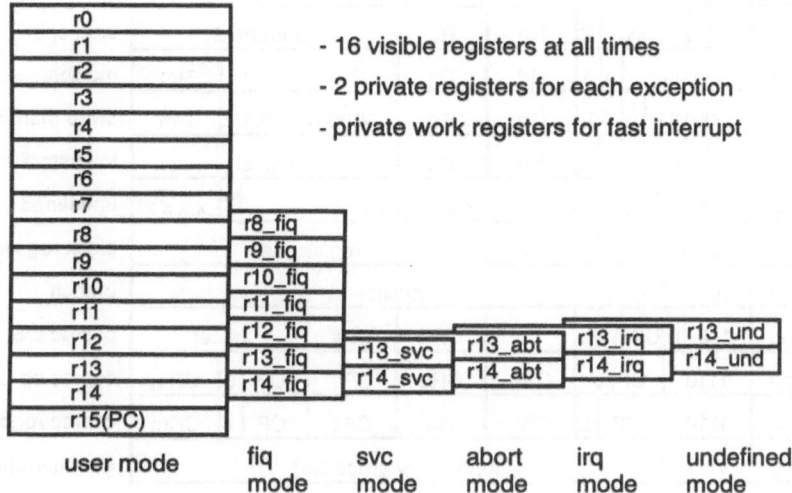

Figure 29: ARM register organization

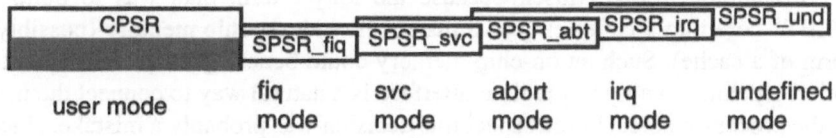

Figure 30: ARM program status registers

Figure 31: ARM PSR format

support for the 26-bit address space modes which ARM6 has for compatibility with the older ARM chips.

6 A Micropipelined ARM

The design of AMULET1 [3,6,7] begins from a consideration of the environment in which the processor will be used and the interfaces through which the processor should communicate with that environment. For instance, should the chip be

Cond	00	I	Opcd	S	Rn	Rd	Operand 2				data ops
Cond	000000			A	S	Rd	Rn	Rs	1001	Rm	multiply
Cond	00010		B	00	Rn	Rd	0000	1001	Rm		swap mem/reg
Cond	01	i	P U B W L	Rn	Rd	Offset					load/store reg
Cond	011	x x x x x x x x x x x x x x x x x x x x					1	x x x x			undefined
Cond	100	P U S W L	Rn	register list							block reg xfer
Cond	101	L	offset								branch
Cond	110	P U N W L	Rn	CRd	CP#	offset					coproc L/S
Cond	1110	CPop	CRn	CRd	CP#	CP	0	CRm			coproc op
Cond	1110	CP	L	CRn	Rd	CP#	CP	1	CRm		coproc reg xfer
Cond	1111	ignored by processor									s/w interrupt

Figure 32: ARM6 instruction set

designed to use an existing interface so that it can plug into an existing environment?

It was decided to use a micropipeline interface between the chip and its environment. This approach was chosen because the longer term plan was to build the AMULET1 core into a larger chip as a macrocell with on-chip memory (possibly in the form of a cache). Such an on-chip memory could benefit greatly from operating as a micropipeline, so a micropipeline interface is a natural way to connect the memory to the processor core. (In retrospect this decision was probably a mistake, since it made testing the AMULET1 prototypes rather difficult. Two-phase control circuits are easy to implement in VLSI but are much harder to work with at board-level.)

The top level interface is shown in figure 33. The MMU and memory are not part of the AMULET1 design task, but are shown here to illustrate the environment in which the processor will be employed. The processor produces one output bundle containing the memory address, the 'write' data (if any) and control bits. The control bits include a read enable, a write enable, an opcode fetch bit (to indicate whether an instruction or a data item is to be fetched), a privilege mode bit and two bits which hint at sequential address behaviour. A second (input) bundle is used to transfer read data back to the processor.

Note that at this stage no assumption is made about the pipeline depth of the memory subsystem. Indeed, if the memory includes a cache, the effective pipeline depth may depend on whether the cache is hit or not. The memory must, however, return results in the same order as the requests were issued.

As the processor handles memory faults as exact exceptions, it must internally prevent any state change after issuing a request for data from memory until it knows that the request will succeed. Therefore a fault/no fault response from the MMU is time critical on data transfers, and the design employs a dual-rail encoded abort response signal from the MMU. A transition on one wire signifies 'abort', causing exception entry, whereas a transition on the other wire signifies 'no abort', allowing

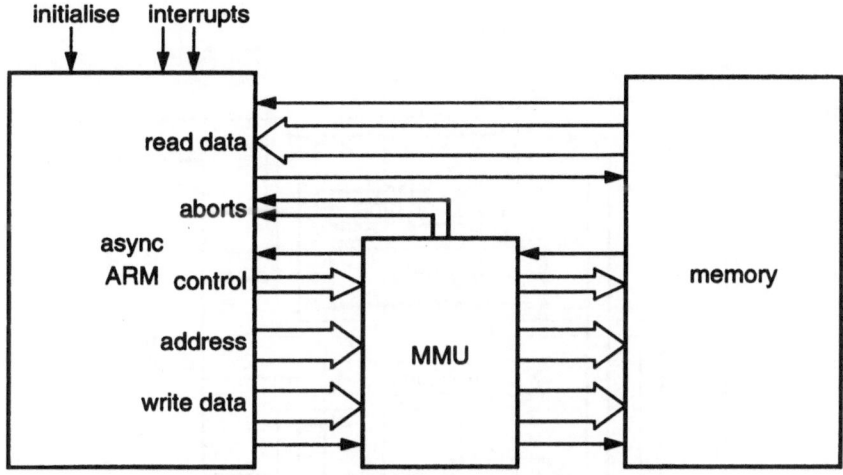

Figure 33: AMULET1 top-level interface

the processor to proceed. (For instruction prefetches a boolean flag returned with the instruction is sufficient to indicate an unsuccessful access; the trap is then initiated when the instruction enters the decode stage.) Conventional level-sensitive interrupt inputs are provided for compatibility with existing peripheral chips, though this is not really a satisfactory model for an asynchronous system as there is no control on the timing of the release of an interrupt line. A system initialisation input completes the top level interface.

6.1 Processor Organisation

The internal organisation of the processor is shown in figure 34. The processor state is held in the register bank, which has two read ports for operand access. The operands are processed by the execution unit and the result written back through a write port. The write port also allows data to be written into the register bank from memory. The third input port is used to allow the PC (which resides in the address interface unit) to be available as register 15 as required by the ARM instruction set.

The address interface unit produces sequential addresses autonomously and only requires input from the execution unit when the address sequence is changed; this may be a temporary change for a data access or a permanent change for a branch.

Write data are copied from one of the register bank read operand buses and then synchronised with the appropriate address for issue to memory. Values returned from memory are split into data and instruction streams and processed accordingly.

6.2 The Address Interface

The address interface includes a PC word-increment loop (see figure 35). The ARM instruction set specifies that (in most cases) R15 has the value PC+8 (exposing the

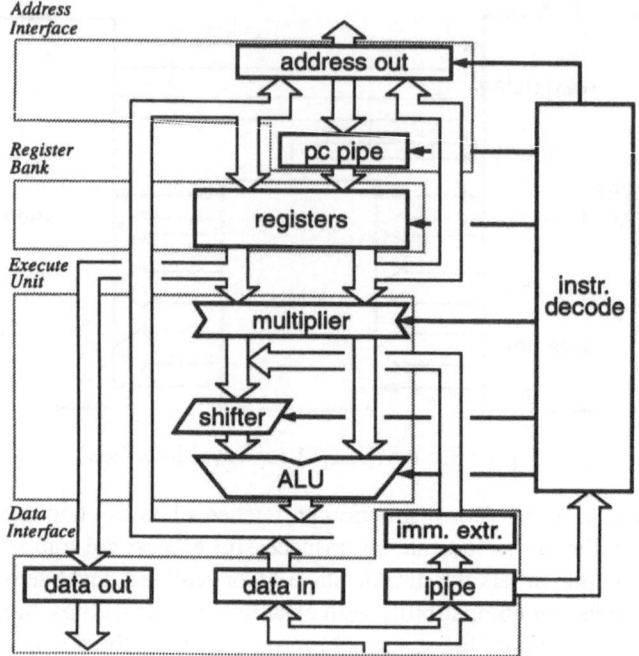

Figure 34: AMULET1 internal organization

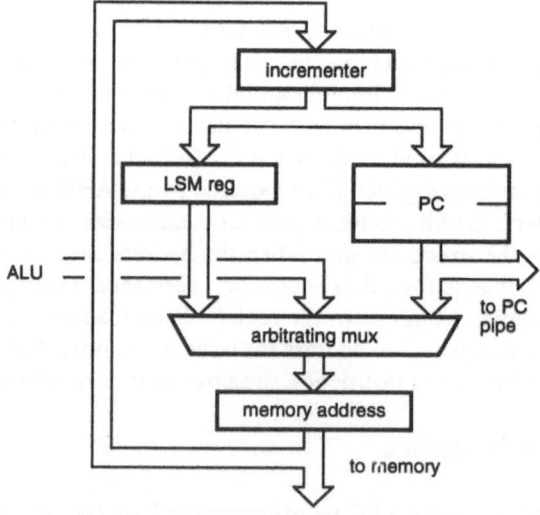

Figure 35: The address interface

depth of the synchronous pipeline implementation in the original ARM), so for code compatibility that behaviour is copied here. The first address, 0, is produced in the memory address register. After passing through the incrementer, the first value offered to the PC pipeline (which buffers PC values until they are needed as R15 during execution of the fetched instruction; see figure 46) is therefore 4. However the first value is 'thrown away', so the first value actually copied into the PC pipeline is 8. This skew between the PC pipeline and the fetched instruction stream persists and maintains the alignment of all future instructions with PC+8. The skew causes incorrect pairing of the instructions immediately following a taken branch, but as these are not executed this is of no consequence. Correct pairing is re-established by the time the new instruction stream begins execution.

The incrementer loop operates autonomously so a new address from the ALU arrives asynchronously. An arbiter is required to ensure safe interruption of the incrementer loop, and the precise point where the loop is interrupted is therefore non-deterministic, resulting in the non-deterministic depth of prefetching beyond a branch. The interruption may be transient (for a data access) or permanent (for a branch). In the former case the PC value is preserved in the PC register; in the latter case the old value is thrown away and the new one takes over the loop.

The ARM instruction set includes multiple register loads and stores of arbitrary subsets of the visible registers. Here the memory addresses are sequential, so the incrementer loop is temporarily usurped to produce these addresses. In this case the PC is held in the PC register, and the looping value passes through the load/store multiple (LSM) register.

6.3 The Register Bank

A high-level view of the operation of the register bank is shown in figure 36. An instruction specifies two source register addresses (a and b) and the destination regis-

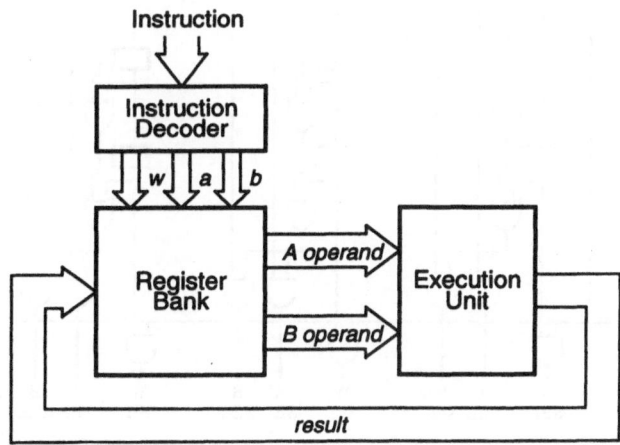

Figure 36: High-level view of register bank operation

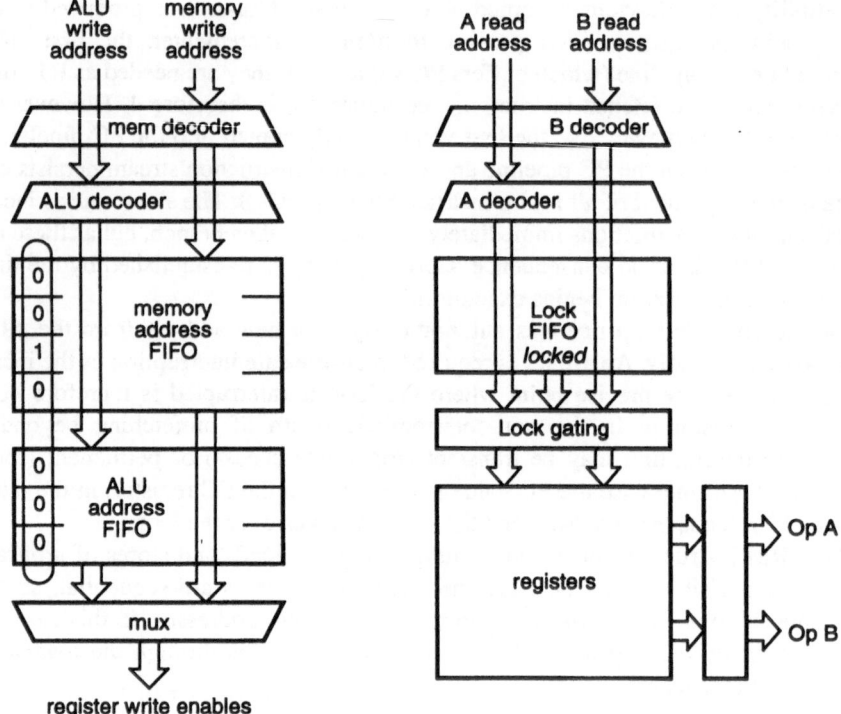

Figure 37: Register lock FIFO and read logic

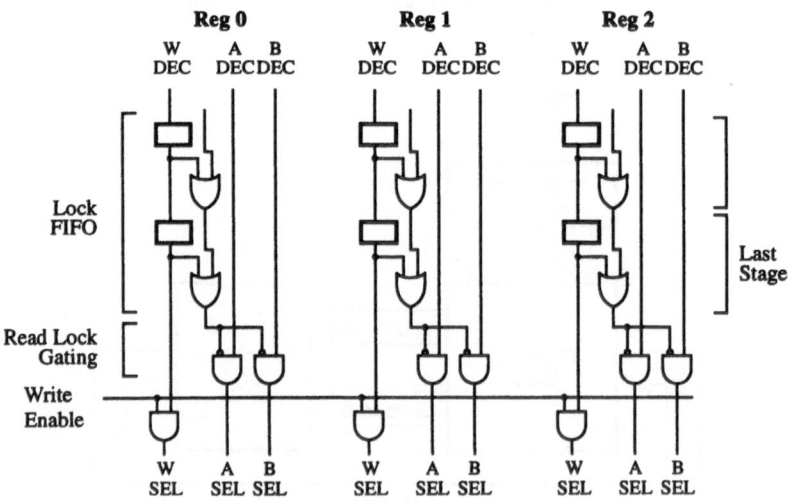

Figure 38: Register read lock gating logic

ter address (w). The register bank then produces the source operands and waits for the result to be returned before placing it in the destination. In pipelined operation, however, the register bank may produce source operands for several instructions before the result from the first one returns. The unit must therefore handle multiple pending write operations register locking to prevent access to stale register values and the asynchronous interaction between read and write operations. All these issues

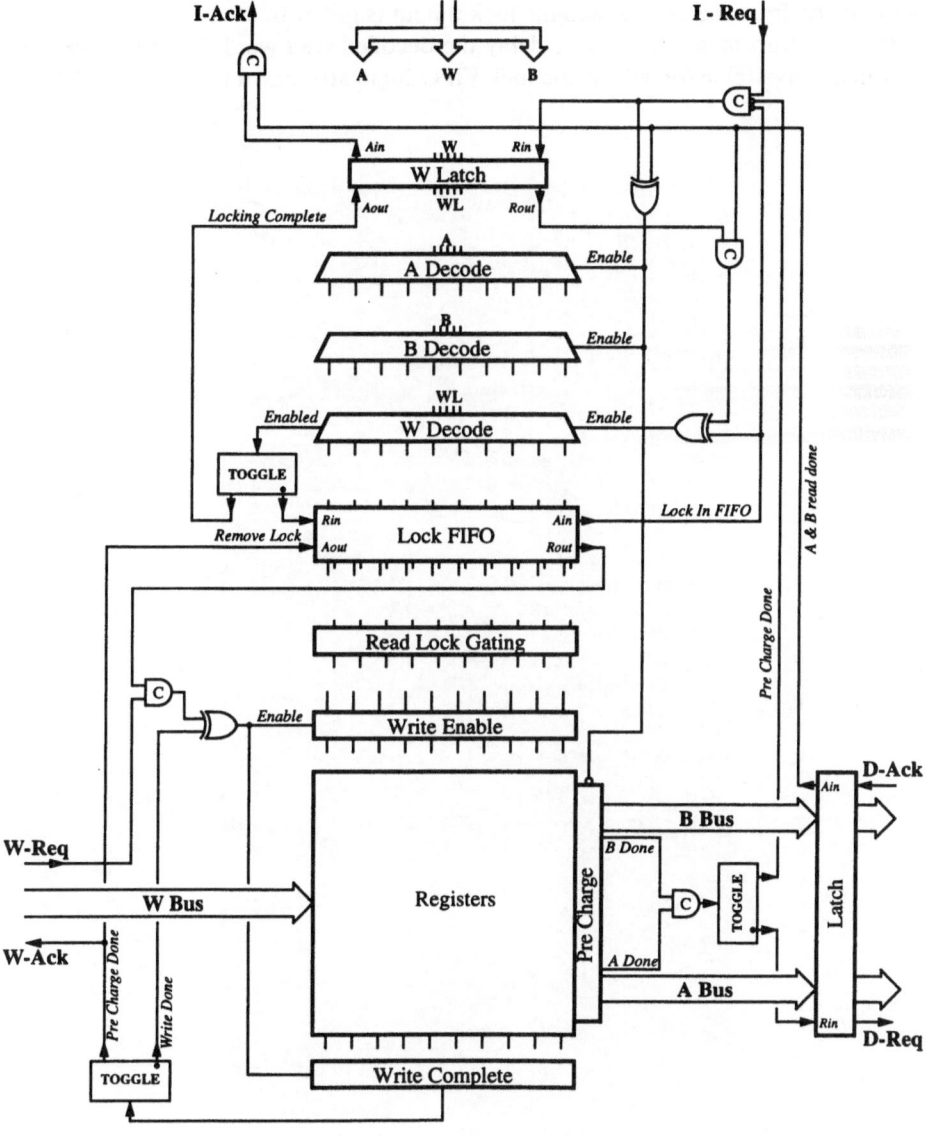

Figure 39: Register bank organization

are resolved in a single regular structure, the register lock FIFO [8], as illustrated in figure 37. (Note that in our design there are two FIFOs to allow internal results to overtake data from - potentially much slower - memory.) Here write addresses are queued in a fully decoded form, so each stage of the FIFO contains at most one '1'. A column of the FIFO contains all the information about pending writes to a particular register and a logical OR of the column provides the lock control. The OR function is not normally permissible across such an asynchronous structure but is possible here because, as data propagates through the FIFO, a '1' is copied to the new latch before it is removed from the old one and the lock output is glitch-free.

The lock information is used to delay the decoded read word lines until the correct value is available (details of the lock FIFO logic are shown in figure 38). Multi-

Figure 40: The register bank layout

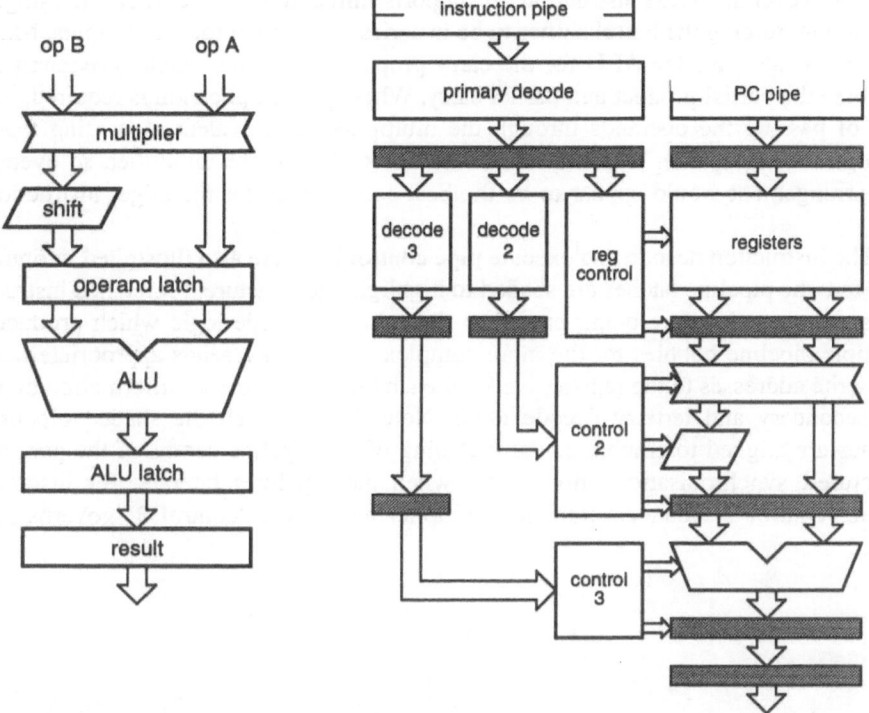

Figure 41: The execution pipeline and control structures

ple locks on a single register are handled correctly and no arbiter is required to manage the asynchronous interaction between reads and writes; these proceed independently when there is no interdependency and the lock mechanism synchronises them when a dependency occurs.

The complete organization of the register bank is shown in figure 39 and the resulting layout is shown in figure 40.

6.4 The Execution Unit

The functional units in the execution pipeline are shown in figure 41. The register operands first pass through a multiplier, which either passes them on immediately or replaces them by partial product and partial carry outputs from a carry-save multiplication unit. A barrel-shifter then modifies one of the operands before both are placed into a pipeline latch. The operands are then combined in the ALU which has a data-dependent delay and a latch to allow a dynamic structure to operate with static external behaviour. A result latch passes the output to its next destination (either a register or the address unit).

The sequential positioning of the function units is perhaps not ideal for perform-

242

ance. However the ARM instruction set supports shift and ALU operations in a single instruction, forcing the barrel shifter to be in series with one of the ALU inputs. Multiplications also use the ALU for the carry propagate addition which is required to combine the partial product and partial carry. When no multiplication is required, the cost of passing the operands through the multiplier is equivalent to passing them through the multiplexer which would be needed to bypass the multiplier, so overall this arrangement would appear to be the best compromise for the target instruction set.

The instruction decode and execute pipe control logic are also illustrated in figure 41. Here the pipeline latches are shaded to highlight the structure. Prefetched instructions are queued before being passed to the primary decode logic which produces multiple pipeline bubbles for the more complex instructions, sends appropriate read and write addresses to the register bank for each bubble and passes information on to the secondary and tertiary decode logic. Note that although the shaded pipeline latches are aligned to emphasise the matching of the pipeline depths of the parallel structures, synchronisation only occurs when the pipelines interact, for instance where 'control 2' connects into the multiplier and where 'control 3' governs the ALU.

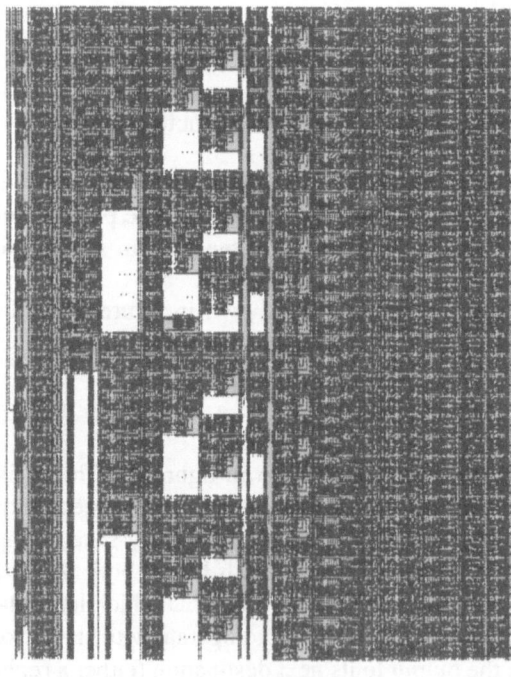

ARM6 ALU   AMULET1 ALU

Figure 42: ALU size comparisons

As an example of an area where asynchronous logic appears to offer particular advantages, consider the ALU [9]. Clocked ALUs (such as that on the ARM6) must ensure that the worst case data operands can be processed within the clock cycle time and considerable logic is added to the ALU to make these rare worst cases complete as fast as possible. An asynchronous ALU can allow rare worst cases to take longer and can therefore dedicate the logic resource to making typical cases go fast. A comparison of cell size between the AMULET1 and ARM6 ALUs is shown in figure 42.

6.5 Data Operation Datapath Activity

The operation of the datapath during a simple data processing instruction is illustrated in figure 43. The active buses are shaded for register-register and register-immediate instructions. Note that in these figures the instruction does not occupy all the shaded buses at the same time; the execution is pipelined, and at any one time different resources may be in use for several different instructions.

The only difference between the two illustrations in figure 43 is the source of the second ALU operand, which comes either from the register bank or from the immediate field extractor.

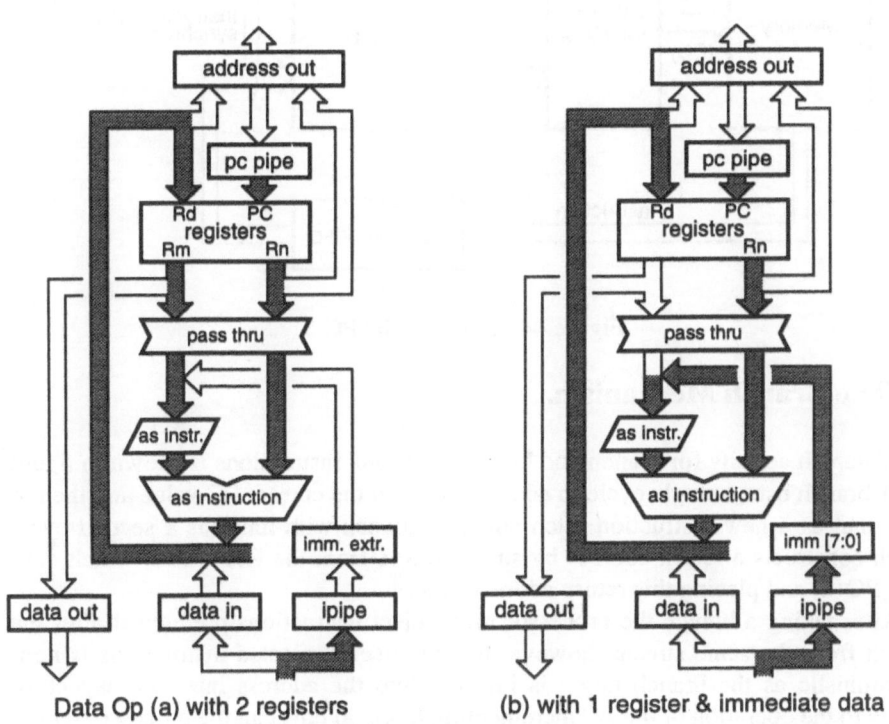

Figure 43: Data operation datapath activity

6.6 Tracking the PC

A high-level view of the mechanism which gives each instruction access to the correct PC+8 value in R15 is shown in figure 44. PC values are autonomously generated in the address interface and issued to memory as addresses for instruction fetches. As they are issued, the PC+8 value associated with each instruction is copied into the PC pipeline. Instructions which return from memory are placed in the instruction FIFO and as they are removed at the end of the FIFO to be decoded each instruction is associated with its PC+8 value. The decoded instruction first passes its register addresses to the register bank and the PC+8 value is passed with these to be used as the value to be read if R15 is accessed as one of the source registers.

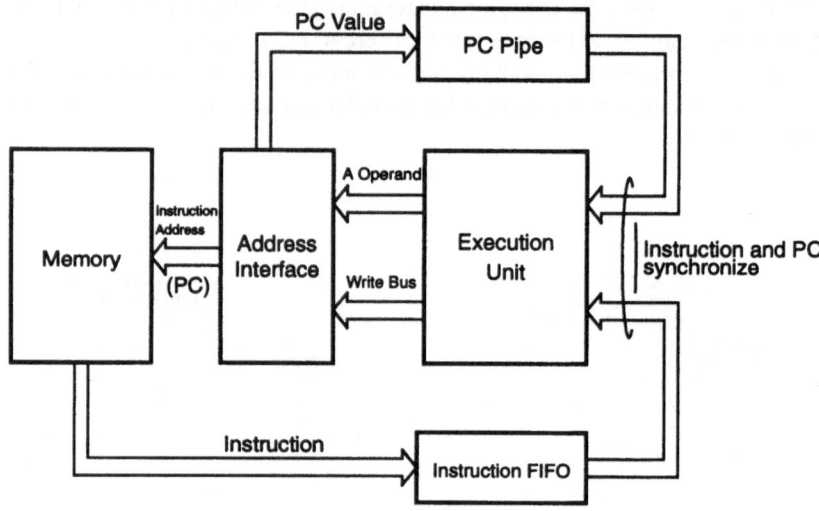

Figure 44: Tracking the PC

6.7 The Branch Mechanism.

The datapath activity for branch and branch-with-link instructions is shown in figure 45. A branch uses a single cycle to add the offset to the current PC value and then to issue that as a new instruction fetch address. Branch-with-link has a second cycle which constructs a return address by subtracting 4 from the R15 value, which contains PC+8, and placing this return address in R14.

After taking a branch the processor must reject instructions prefetched after the branch from the same stream, however the number of rejected instructions is non-deterministic as the branch target is injected into the address interface asynchronously to the operation of the PC incrementing loop. So how can the processor recognise which instructions are to be rejected and which instructions come from the branch target?

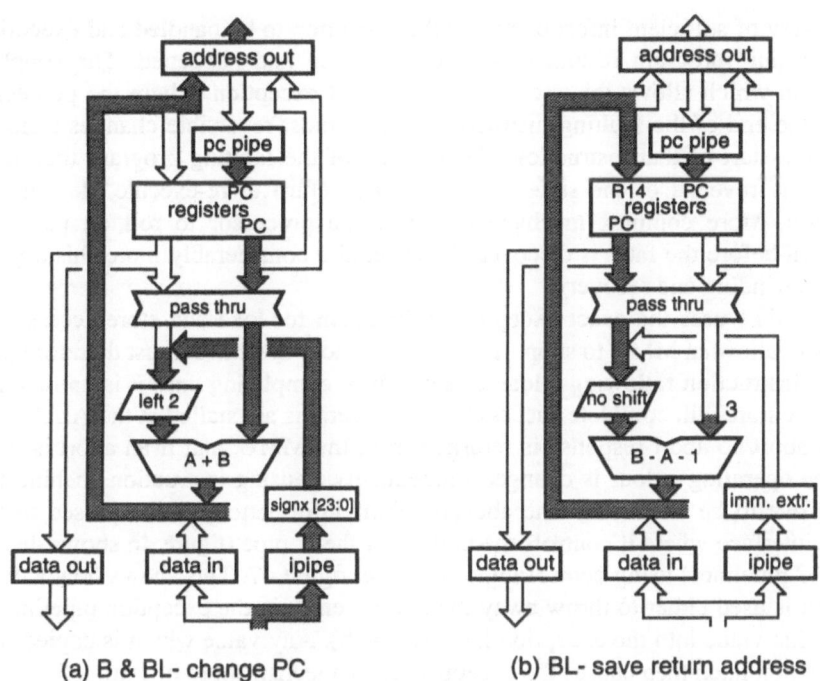

(a) B & BL- change PC   (b) BL- save return address

Figure 45: Branch datapath activity

The mechanism used is described as instruction stream 'colour'. Every instruction fetch is issued with a particular colour corresponding to the current operating colour of the processor. If the fetch colour of an instruction which comes to be executed does not match the current operating colour the instruction is rejected. Every branch changes the fetch and operating colours, so the operating colour changes immediately and the fetch colour changes from the branch target instruction.

The colour is checked at two positions inside the processor. It is checked at the ALU result stage for precise operation, and at the entry to primary decode for efficient operation. Rejecting an instruction early is efficient, since it saves the time and power that instruction would otherwise consume, but to reject instructions only at the decode stage would require every instruction to be held up there until it was known that the preceding instruction would not change the operating colour. This would compromise performance. Instead, instructions are allowed past the decode stage speculatively if their colour is correct at that point. If a preceding instruction then changes the operating colour after this instruction has entered decode, the instruction will be rejected at the ALU result stage.

6.8 Exact Exceptions

The most difficult aspect of the design of most processors is the provision for exceptions which arise during the execution of an instruction. The processor must allow for the recovery of sufficient information for the exception to be handled and execution of the original program resumed as though nothing had happened. The simplest mechanism which allows this recovery is the exact exception, where the processor stops at the end of the faulting instruction with at most reversible changes from its state at the start of that instruction. Resumption of the faulting program then only requires the reversal of the state changes and a return to re-execute the faulting instruction. More complex mechanisms allow the processor to roll on past this instruction before the fault is discovered and require considerably more 'history' to support rewinding and recovery.

AMULET1 uses the exact exception mechanism for load and store accesses to memory to allow an MMU to support a virtual memory system. It must therefore prevent any instruction following a load or store from completing until it is known that the load or store will complete successfully. Instructions are stalled at the ALU stage until an abort/no abort response is returned from the MMU, and if an abort is indicated the operating colour is changed immediately, causing instructions behind the faulting one to be discarded. The abort/no abort information is also passed to the address interface where it controls the bottom of the X pipe (figure 46 shows details of the PC pipelines) using control logic shown in figure 47. This shows how a decision-wait is used either to throw away the bottom entry in the exception pipeline, or to copy that value into the exception latch (X latch). Any value which is copied into the exception latch then causes the exception entry mechanism to be initiated.

The logic which allows the exception latch to break into the instruction stream to cause an exception entry is shown in figure 48. The normal instruction stream enters from the top of this figure. The C-gate waits until the interrupt arbiter, the PC pipeline

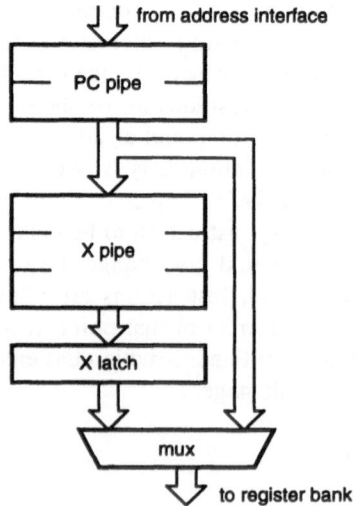

Figure 46: The PC pipelines

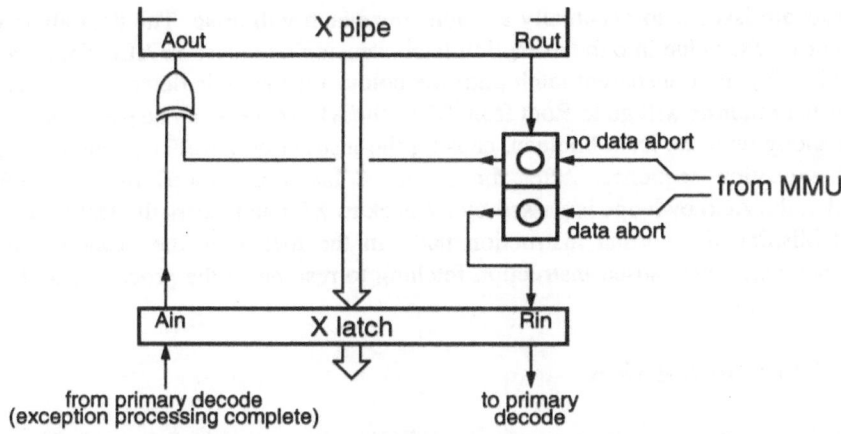

Figure 47: The Exception Latch control logic

and the instruction pipeline are all ready (ARBrq, PCrq and IPrq respectively) and then the instruction colour (Icol) and the current operating colour (OPcol) are checked to see if they are the same. If not, the select block rejects the instruction by sending the request event straight back to the acknowledge (IPack). When a data abort happens, the operating colour is changed but no further instruction fetch

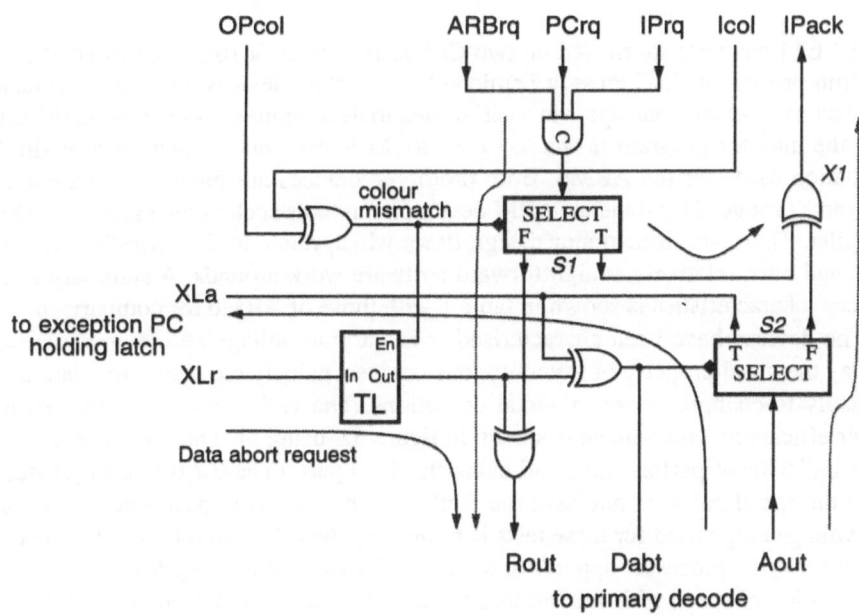

Figure 48: The Data Abort exception entry logic

requests are issued, so eventually a colour mismatch will arise. The data abort will also copy a PC value into the exception latch, causing a request on XLr. This request is held up by the transparent latch until the colour mismatch is detected, whereafter no further requests will go to Rout from S1 so the XLr request can be passed safely to Rout, along with the Dabt boolean, causing the instruction decoder to enter the data abort exception sequence. After the sequence has been issued by the primary decoder, the Acknowledge is passed by S2 back to XLa and clears the Dabt boolean, re-establishing the normal instruction path. In the meantime the exception entry sequence will have caused instruction fetching to resume so the processor will continue normally.

6.9 Chip Composition

The complete pipeline structure of AMULET1 is illustrated in figure 49. The chip was implemented with full-custom datapath components and compiled standard cell control circuits using tools from Compass Design Automation. The floorplan of the chip is shown in figure 50. The control circuits included two PLAs which were generated using a tool supplied by ARM Limited and modified to give access to external circuitry to the completion signal (which already existed inside the ARM design where it controlled power-down and precharging). The VLSI layout of the chip is shown in figure 51.

7 Results

AMULET1 has been fabricated on two CMOS processes: a 1μm process at ES2 and a 0.7μm process at GEC Plessey Semiconductors. Both devices have been evaluated on a test card which connects, via a serial line, to development tools from ARM Limited; the monitor program in the test card ROM is the same as that used in similar evaluation cards for the ARM6. Both prototype devices are functional and execute programs produced by standard ARM development tools such as the assembler and C compiler. There are three minor design flaws which relate to the operation of interrupts and have relatively straightforward software work-arounds. A summary of the devices' characteristics is shown in table 1 with those of ARM6 for comparison.

The devices have been characterised over varying voltage and temperature and display the usual property of asynchronous devices, namely the ability to adapt automatically to changing environmental conditions. The variation of performance and power-efficiency with voltage is shown in figure 52, using the Dhrystone benchmark as an indicator of performance and using the 1μm part. (The 0.7μm part operates at twice the speed but does not have the facility to measure core power consumption.) The voltage range used for these tests is limited by the other circuitry on the test card below 3.5V; the processor appears to operate in isolation down to 2.5V.

Variation of speed with temperature has also been measured. Here the test devices display a normal increase of their delays of 0.3% per °C and operate correctly between -50°C and 120°C.

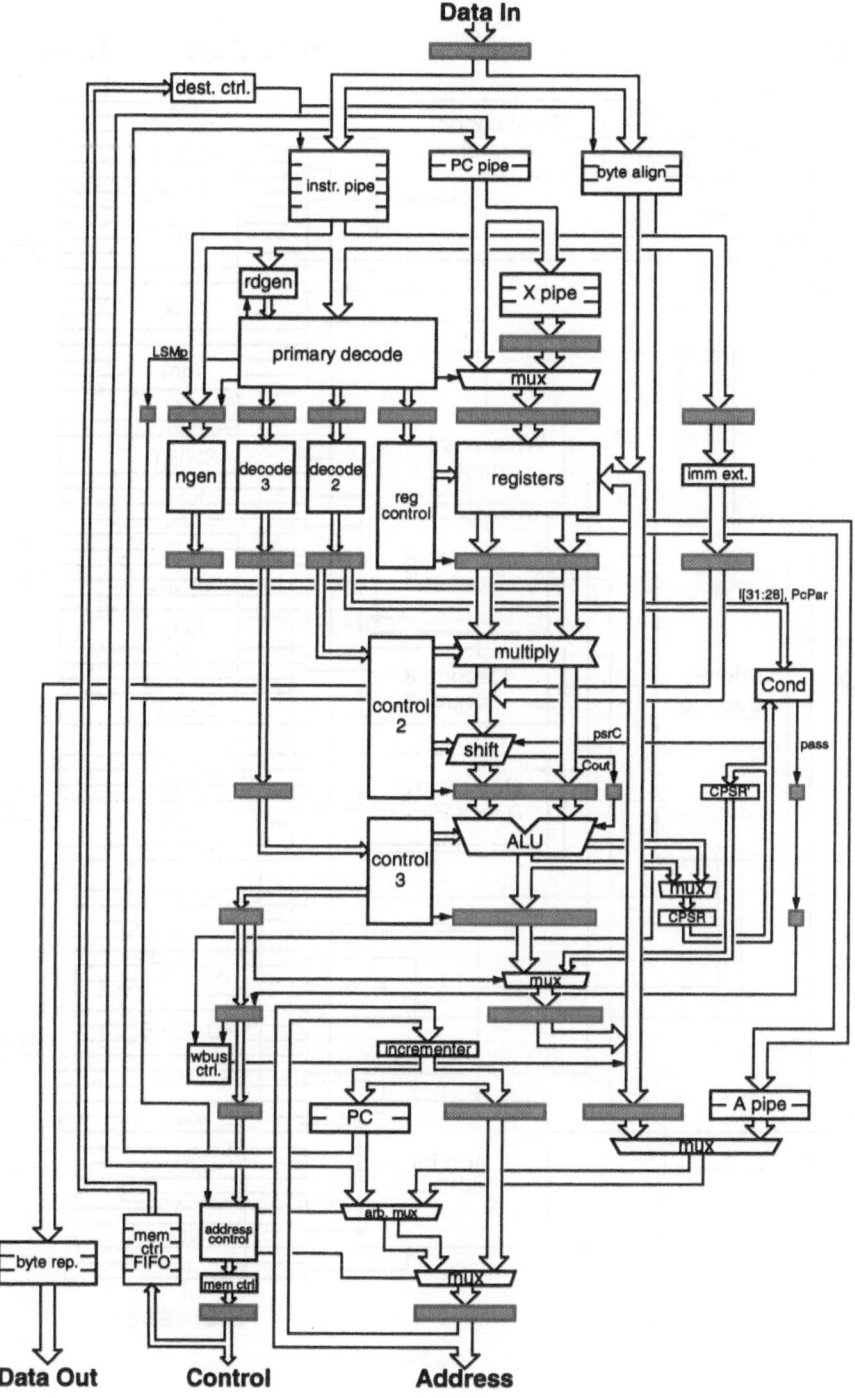

Figure 49: AMULET1 complete pipeline organisation

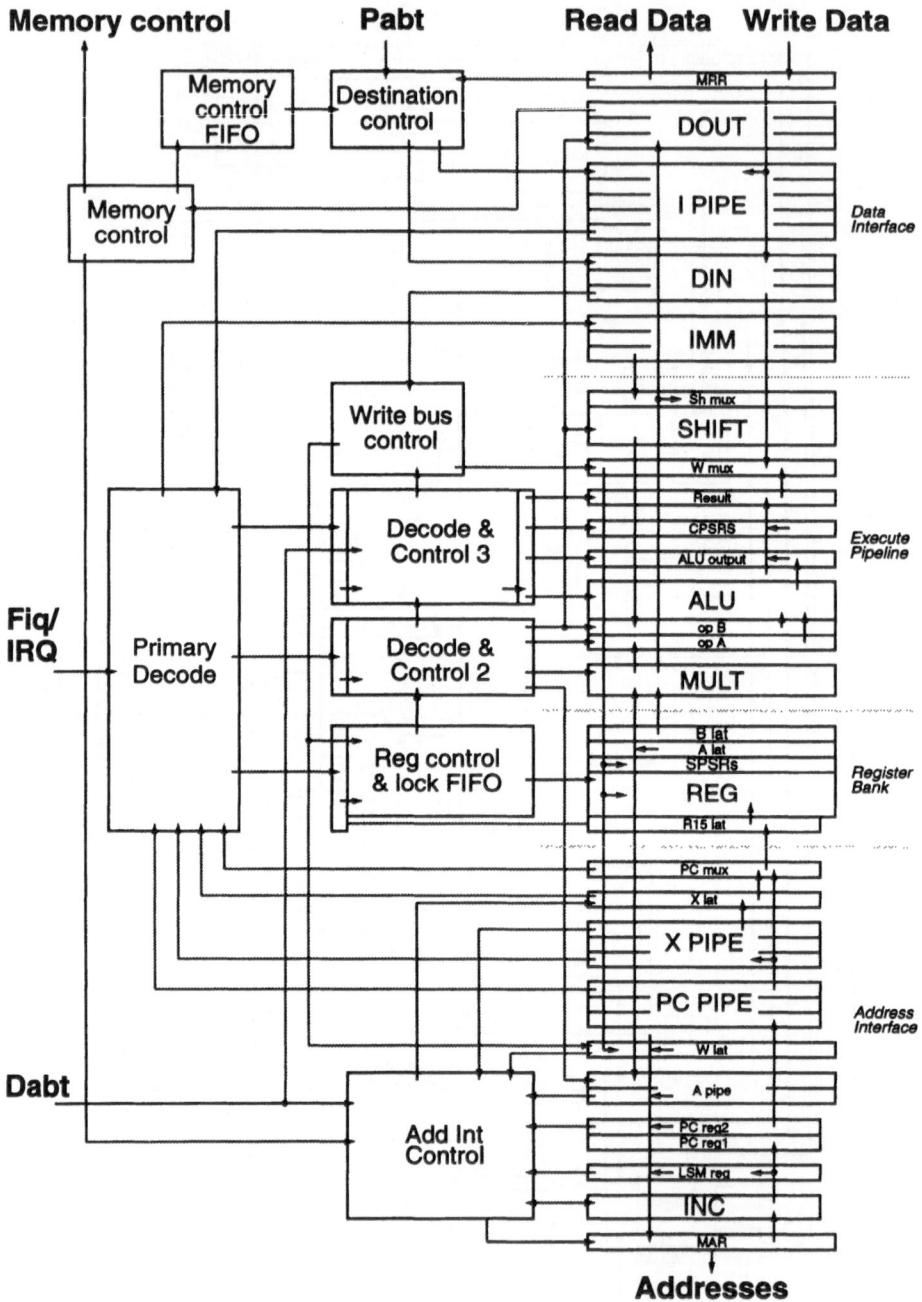

Figure 50: AMULET1 chip floorplan

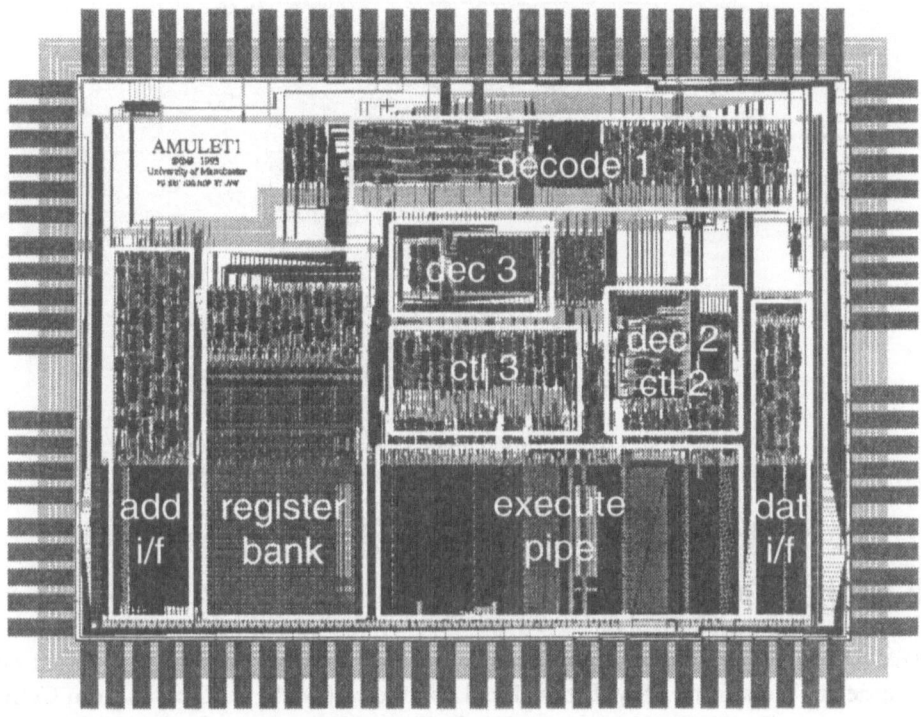

Figure 51: AMULET1 VLSI layout

	AMULET1/ES2	AMULET1/GPS	ARM6
Process	1μm	0.7μm	1μm
Area (mm^2)	5.5 x 4.1	3.9 x 2.9	4.1 x 2.7
Transistors	58,374	58,374	33,494
Performance	20.5 kDhry.	~40 kDhry.[1]	31 kDhry.
Multiplier	5.3ns/bit	3ns/bit	25ns/bit
Conditions	5V, 20°C	5V, 20°C	5V, 20MHz
Power	152mW	N/A[2]	148mW
MIPS/W	77	N/A	120

Table 1: Characteristics of AMULET1 and ARM6

1. estimated maximum performance.
2. the GPS part does not have separate core supplies for power measurement

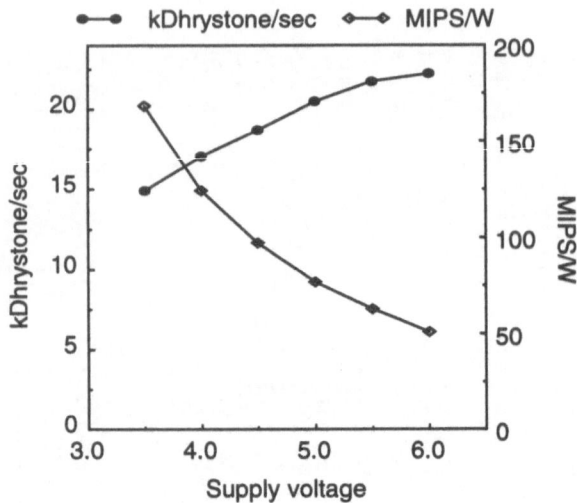

Figure 52: 1μm AMULET1 performance and power-efficiency against voltage

8 Design Tools

The AMULET1 design was developed using a conventional design flow based on proprietary tools from ARM Limited and commercial VLSI CAD tools from Compass Design Automation (see figure 53). In line with standard practice, simulations were run on transistor-level netlists extracted from the physical layout in all the process corners to cover all four combinations of fast and slow n- and p-transistors. Any serious difference between matched paths in their dependence on rising and falling delays is exposed by these skewed simulations, and on AMULET1 a few paths were adjusted it compensate for asymmetries of this sort.

Although these tools allowed the design to be developed successfully and working silicon to be produced, some aspects of asynchronous designs make conventional simulation inadequate for verification. Asynchronous circuits are prone to deadlock and simulation only establishes that deadlock does not arise under the simulated conditions; it cannot prove that liveness is guaranteed under all conditions.

In order to increase confidence in the liveness properties of the design, random delays were added into all the C-gate models in some simulation runs to simulate different time orderings of events. This only serves to increase confidence and still cannot approach an exhaustive test of all possible event orderings.

It should be noted in passing that the tendency for incorrect circuits to deadlock was a significant help during debugging! The behavioural simulator could be set up to maintain a circular buffer of past events. Then, when the model deadlocked, this buffer would contain enough history to identify the source of the problem. When similar problems arise with a clocked design all that may emerge is the wrong answer at the end of a long run, and it can be much harder to identify the time at which the

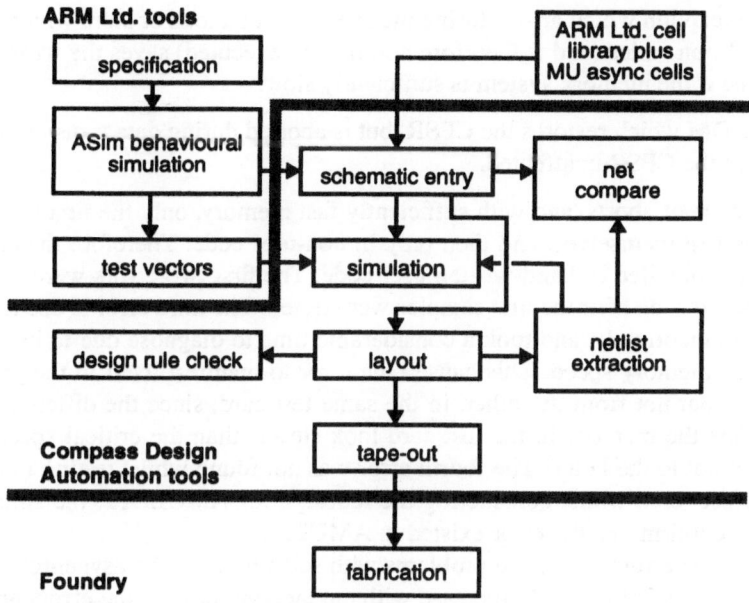

Figure 53: The design flow used to develop AMULET1

error arose and then the source of the error.

In addition to the standard tools, a tool was built to check that bundled data values never changed within the Request-Acknowledge period (with added margins) to ensure that the bundling constraints were never violated. Again, only the timing under the simulated conditions could be checked, not all possible timings under all possible conditions.

Though the above procedures increase confidence, they are not sufficient for general asynchronous design work; the present design is very conservative, and as margins are reduced better tools will be needed to achieve satisfactory confidence levels that the design will operate correctly.

8.1 Design Flaws

Though the first AMULET1 silicon is functional, there are some errors in the design which it may be instructive to review.

- an MSR or MRS executed in a non-user mode immediately after a STM accesses the wrong SPSR, due to an error in the design of logic which allows supervisor code to store the user's registers.

- if an interrupt is disabled and the flag is re-written (again to disable the interrupt) the interrupt may be transiently enabled and entered.

- an exception which arises during the execution of a load or store which fails its
  condition codes (and is therefore not, in fact, executed) saves the wrong SPSR
  value if the memory system is sufficiently slow.

- an LDM which restores the CPSR, but is aborted during data transfer, does not
  leave the CPSR unaffected.

In the absence of aborts, and with sufficiently fast memory, only the first two design
errors manifest themselves, and then only in non-user code. Therefore the chip will
run general compiled or hand-written user code. The first two errors were identified
and diagnosed soon after the first samples were tested. The third error is due to a latch
being wired incorrectly, and took a considerable time to diagnose due to its dependency on the memory speed. This caused the error to manifest itself in the part from
one source, but not from the other, in the same test card, since the different silicon
speeds made the memory in the test card look slower than the critical speed to the
former and not to the latter! The fourth error was not found whilst testing the silicon
but was discovered whilst considering the redesign for AMULET2; the silicon was
checked to confirm that the error existed in AMULET1.

None of the errors are due to problems with self-timing or the asynchronous control structure. As is frequently the case with engineering design, the errors arose not
in the areas the designers focussed on but in the peripheral detail. More thorough testing of the design under simulation would have revealed these errors before fabrication.

9 Future Enhancements

The AMULET1 design was shipped for fabrication at the end of February, 1993.
Since then work has been underway to enhance the design with respect to both its
performance and its power efficiency. Significant improvements have been made in
many areas of the design. These improvements cover all aspects of the design including the transistor structures used for data latches, the approach to event control, architectural features to reduce pipeline stalls, through to the compiler.

9.1 Compiler Optimization

As some compiler improvements also enhance the performance of AMULET1, these
will be described first.

The compiler used to evaluate AMULET1 is the standard ARM Limited C compiler. As the ARM6 displays no sensitivity to code order, the compiler makes no
attempt to separate dependent instructions. AMULET1, on the other hand, shows a
strong dependency of performance on code order. The register bank lock FIFO
ensures that instructions wait until their operands are available, but whenever an
instruction is forced to wait, the pipeline is stalled and performance is lost.

For example, this is the code for a standard string comparison which loads and
sign extends the bytes for each string:

```
strcmp
        LDRB    a3,[a1],#1
        MOV     a3,a3,LSL #24
        MOV     a3,a3,ASR #24
        LDRB    a4,[a2],#1
        MOV     a4,a4,LSL #24
        MOV     a4,a4,ASR #24
        . . .
```

This code causes register locks between the first and second instruction while the data load accesses memory and a shorter stall between the second and third instructions. The same delays are incurred again in the second group of three instructions. Effectively there are two independent threads in these six instructions which can be interleaved almost to double the execution speed. This interleaving fits the second three instructions in the gaps left by stalls during the execution of the first three.

```
strcmp
        LDRB    a3,[a1],#1
        LDRB    a4,[a2],#1
        MOV     a3,a3,LSL #24
        MOV     a4,a4,LSL #24
        MOV     a3,a3,ASR #24
        MOV     a4,a4,ASR #24
        . . .
```

There are other optimizations which improve the speed of the code running on AMULET1 (generally without impacting the speed on ARM6), such as replacing a single register load multiple (which passes four cycles down the execution pipeline) with a single register load (which passes one) and being careful about which instructions are left in a branch shadow (where some instructions may take several cycles in the execution pipeline before being annulled by the colour checking process). Producing optimised code for AMULET1 is harder than for a clocked processor because instruction dependencies have cost functions which are not constrained to discrete multiples of a clock cycle and individual instruction costs can be data-dependent.

Note that the increased inter-instruction dependencies (compared with ARM6) introduce other pressures on the compiler. Interleaving independent threads may increase register allocation pressure, so register allocation and code ordering interact, and ARM does not have very many registers compared with most modern RISC architectures.

Compiler optimization for asynchronous architectures allows for some simple improvements as outlined above, but it is not clear how far optimization can go. The lack of a discrete cost function makes this a new research area!

9.2 Improved Latch Mechanisms

The data latch used on AMULET1 is the same as that used on ARM6. Investigations into alternative latches suggest that there may be potential improvements to be gained in both performance and power-efficiency from switching to latches which do not require complementary latch enable signals. An example of such a latch is the Svensson latch as used on the DEC Alpha processor [10]. The Alpha uses a dynamic form of the latch; the transistor schematic of a variant of the latch with weak feedback to ensure fully static operation is shown in figure 54. The advantage of the single enable signal may be seen by comparing the latch control circuit for a conventional latch with complementary enable signals (figure 55) with that for a Svensson latch (figure 56).

The two forms of latch are compared in table 2. A Svensson latch with carefully sized transistors can give twice the speed of operation at one third of the energy per cycle of the ARM6 latch currently used on AMULET1. It achieves this by removing the need for the inverted enable signal (saving a C-gate and reducing the buffer loads) and by minimising the load each bit places on the remaining enable line.

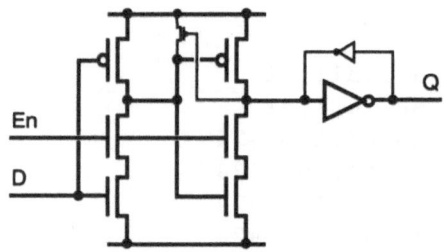

Figure 54: A fully static version of the Svensson latch

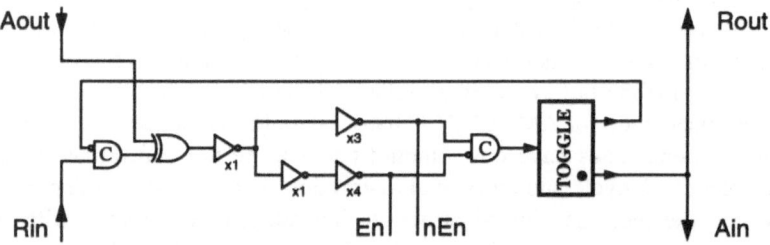

Figure 55: Conventional transparent latch control circuit

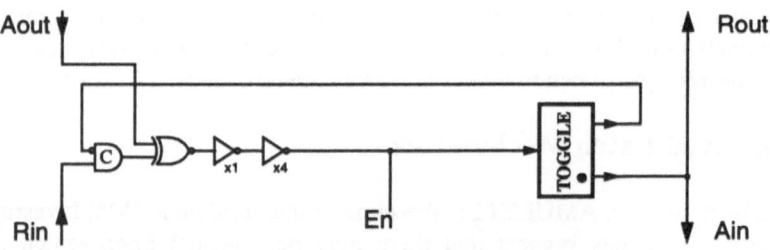

Figure 56: Single phase latch control circuit

	Svensson	AMULET1
Transistors per latch bit	11	6
Inversions in datapath	3	1
Clock load	2N	1N+1P
clock capacitance	1	3
Invs Rin -> Rout	8	11
FIFO cycle time	12ns	25ns

Table 2: Comparison of Svensson and AMULET1 latches

9.3 Four-Phase Control

All the latches described here (apart from the Sutherland capture-pass latch) use 2- to 4-phase conversion control circuits to interface the 2-phase (transition event) control environment to the level-sensitive latches. Clearly the control environment itself could be converted to 4-phase (level-sensitive) operation. Although 4-phase circuitry may appear conceptually simpler at first sight, it is in practice considerably harder to design efficient 4-phase control circuits because the return-to-zero events have no specific meaning so the communication protocol does not define when they should happen. If the return-to-zero event follows the same path around the control circuitry as the active event, the circuit tends to be very slow. Therefore the designer has to expend considerable effort in deciding the best way to handle the recovery phase within each part of the circuit.

Despite these difficulties, the performance advantages of 4-phase control seem to be significant. A 4-phase control circuit suitable for use with Svensson style latches is shown in figure 57. (Note that the C-gates do not require symmetric n- and p-transistor stacks in this circuit.)

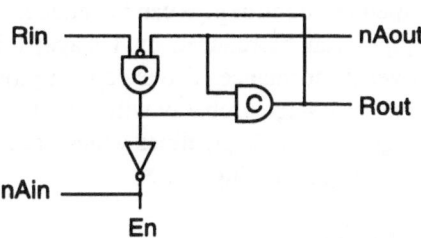

Figure 57: 4-phase pipeline latch control circuit

9.4 Engineering Margins

The self-timed circuits used on AMULET1 raise the question of what margin should be allowed in a matched signal path. If the engineering margin is too small incorrect operation could result. If the margins are too large, performance will be compromised. In order to ensure the best matching between the data route and the matching event path, the paths should be identical, close together on the chip and have the same physical orientation.

On AMULET1 the timing margins were around 20-30% on identical layout (e.g. the 33rd register bit) and 100% on standard cells paths with the same nominal delay. Since no failures on the sample chips are attributable to mismatched paths, these margins appear acceptable, at least for prototype devices. Self-timed paths on clocked chips often use much smaller margins (e.g. below 10%), but here there are usually far fewer paths on a chip and a correspondingly lower probability of failure due to one of them being out of tolerance.

The factors which need to be assessed are process, voltage and temperature variations, volume production yields and the number of potential failure points in the design.

9.5 Restructuring the Pipeline

The AMULET1 pipeline is now considered to be deeper than is optimal. To reduce the latency we propose to bypass the shifter in most cycles and to sideline the multiplier except when it is needed. The revised pipeline is illustrated in figure 58 along with the existing pipeline for comparison.

To be effective the new pipeline organization requires the shifter to be used less frequently than on AMULET1. This may be achieved by extending the immediate operand extraction logic to align all immediate operands and by causing the control logic to detect when the shifter is required for register operands, bypassing it in other cases.

The benefits of this change are lower execution unit latency and reduced power (the shifter will be activated less often). The costs are more datapath logic (a partial shifter in the immediate extraction block) and more complex control logic.

There are other places in AMULET1 where the pipeline can be slimmed without loss of performance. The memory control pipe can be reduced from 5 stages to 2 or 3. The PC and instruction pipes can be reduced by 1 stage; this reduces prefetching (saving power) and improves performance. The exception pipe is longer than is useful, and reducing the A pipe by 1 stage probably will not affect performance.

Overall there are 10 stages in various pipelines which are contributing little to the performance of the chip and are consuming power.

9.6 The Last Result Register

Register bypassing depends on different stages of the pipeline being in step and is not directly applicable in an asynchronous pipeline. However, typical programs display

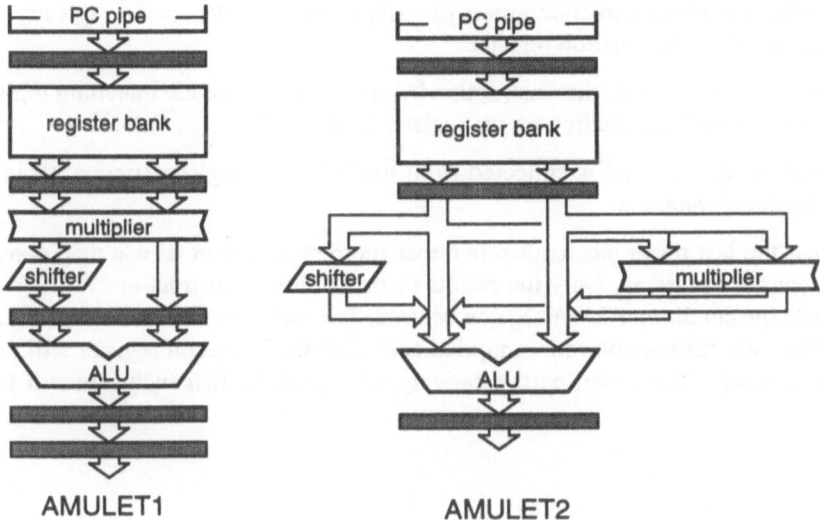

Figure 58: The proposed simplification of the AMULET2 execution pipe

frequent dependencies between consecutive instructions. Therefore an efficient way to feed results to the next instruction is needed. The 'last result register' is a proposed mechanism which is similar to register bypassing but is usable in an asynchronous pipeline.

The last result register operates as follows:

- when an instruction is decoded, the destination register is recorded in the instruction decoder

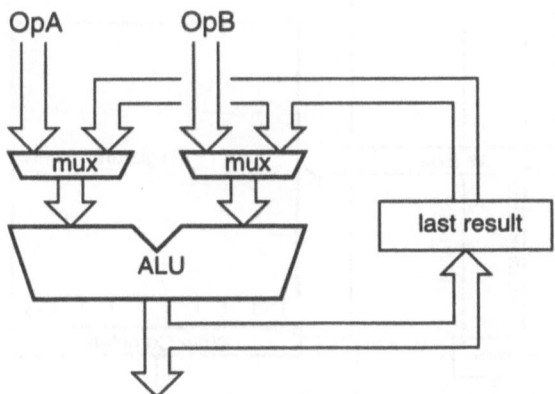

Figure 59: The last result mechanism

- when the next instruction enters decode, it compares its source registers with the previous destination register

- if a match is found, that instruction bypasses the read of the matching register, thereby avoiding stalling on the register lock

- instead the operand is collected from the last result register (figure 59) in the execution pipeline

Note that the last result mechanism is rather more restricted in its use than conventional register bypassing. Only the results of unconditional instructions (i.e. instructions with the condition ALWAYS) can be used. The inclusion of this feature will also move the optimization point for compiled code, and the last result register will make the use of dual register read ports relatively infrequent; is their inclusion still justified?

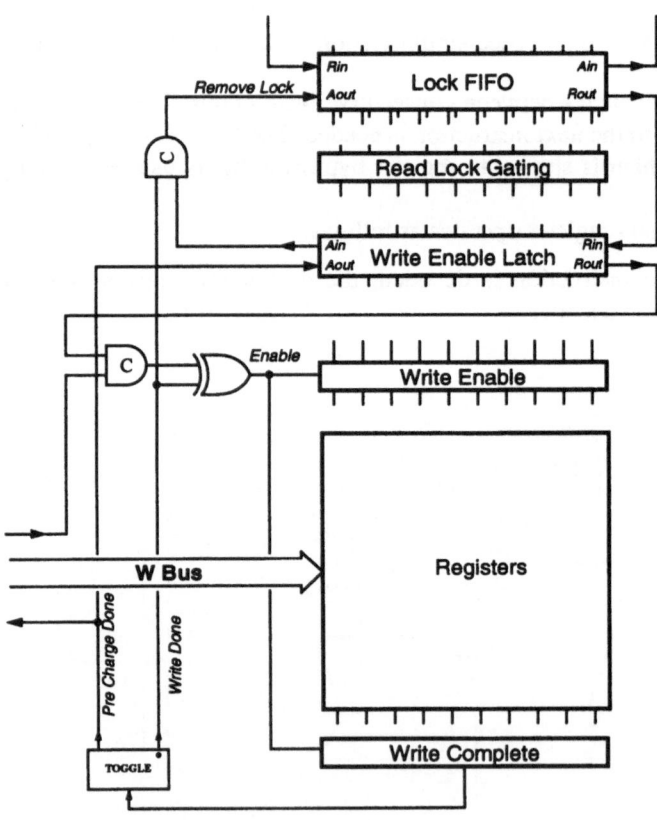

Figure 60: Improved register write logic

9.7 Improved Register Write Logic

The current register bank design releases a read which is stalled awaiting a pending write by clearing the bottom entry of the lock FIFO. As the lock FIFO also controls the register write word lines, its output must remain stable until the write has completed and the write word line disabled. The register will have been written well before this sequence terminates, so the read will have been delayed considerably longer than is necessary.

A modification to the register bank control logic (figure 60) adds a separate latch to hold the register write word line stable, allowing the bottom of the lock FIFO to be cleared as soon as the data has been written to the register. This will release the read operation significantly earlier, improving the performance on code which causes read stalls (which includes most existing ARM code).

10 Conclusions

The AMULET1 design demonstrates the feasibility of designing complex asynchronous circuits, and whilst it does not offer a direct advantage over clocked designs at this stage, there is a lot of room for improvement over the present design.

The architectural features which make synchronous processors go fast do not all transfer easily to asynchronous designs (e.g. register forwarding) and there is a need for new architectural features to be developed for asynchronous designs to replace them (e.g. the register locking FIFO and the last result register).

Micropipelines offer a good framework for the design of high-performance asynchronous circuits. They are amenable to use with conventional CAD tools, though further tool development could assist the design process.

Asynchronous techniques are enjoying a resurgence of interest amongst the VLSI design community because they offer the potential for high performance and low power whilst avoiding the increasing problem of clock skew. A major objection to asynchronous design has, until now, been the issue of the feasibility of developing asynchronous designs at levels of complexity which are representative of commercial circuits. The AMULET1 work overcomes this objection by demonstrating that such circuits are now feasible. Whilst many questions still remain, this work represents a step forward towards the commercial exploitation of asynchronous circuits.

11 Acknowledgements

The work described in this chapter involved several members of the AMULET group at Manchester University. Paul Day and Nigel Paver were the full-time research staff on the AMULET1 project. They contributed many of the ideas incorporated in the design and carried out most of the design work. Without their skill and commitment, we would not now have working silicon. Steve Temple joined the research staff towards the end of the design phase; he made a significant contribution to the final chip composition and designed and carried out the tests on the silicon. Viv Woods and Jim Garside are members of academic staff associated with the project and both made substantial contributions to the design effort.

The AMULET1 design work described in this chapter was carried out as part of ESPRIT project 5386, OMI-MAP (the Open Microprocessor systems Initiative - Microprocessor Architecture Project). Subsequent work has been supported as part of ESPRIT project 6909, OMI/DE-ARM (the Open Microprocessor systems Initiative - Deeply Embedded ARM Applications project). The author is grateful for this support from the CEC.

The author is also grateful for material support in various forms from Advanced RISC Machines Limited, Acorn Computers Limited, Compass Design Automation Limited, VLSI Technology Limited and GEC Plessey Semiconductors Limited. The encouragement and support of the OMI-MAP and OMI/DE-ARM consortia are also acknowledged.

12 References

1. Dobberpuhl, D. W. et al., "A 200-MHz 64-b Dual-Issue CMOS Microprocessor", IEEE Journal of Solid-State Circuits, Vol. 27, No. 11, Nov. 1992, pp. 1555-1565.

2. Sutherland, I.E., "Micropipelines", The 1988 Turing Award Lecture, Communications of the ACM, Vol. 32, No. 6, June, 1989, pp. 720-738.

3. Paver, N.C., "The Design and Implementation of an Asynchronous Microprocessor", PhD Thesis, University of Manchester, June 1994.

4. Furber, S.B., "VLSI RISC Architecture and Organization", Marcel Dekker Inc., New York, 1989.

5. van Someren, A., and Atack, C., "The ARM RISC Chip: A Programmer's Guide", Addison-Wesley, 1993.

6. Furber, S.B., Day, P., Garside, J.D., Paver, N.C. and Woods, J.V., "A Micropipelined ARM", Proceedings of the IFIP TC 10/WG 10.5 International Conference on Very Large Scale Integration (VLSI'93), Grenoble, France, September 1993. Ed. Yanagawa, T. and Ivey, P. A. Pub. North Holland.

7. Furber, S.B., Day, P., Garside, J.D., Paver, N.C. and Woods, J.V., "AMULET1: A Micropipelined ARM", Proceedings of the IEEE Computer Conference, March 1994.

8. Paver, N.C., Day, P., Furber, S.B., Garside, J.D. and Woods, J.V., "Register Locking in an Asynchronous Microprocessor", 1992 IEEE International Conference on Computer Design: VLSI in Computers & Processors. October 1992.

9. Garside, J.D., "A CMOS VLSI Implementation of an Asynchronous ALU", IFIP Working Conference on Asynchronous Design Methodologies, April 1993. Ed. Furber, S. B. and Edwards, M. D. Pub. North Holland.

10. Yuan, J., and Svensson, C., "High-Speed CMOS Circuit Techniques", IEEE Journal of Solid-State Circuits, Vol. 24, No. 1, February 1989, pp. 62-70.

Author Index

Published in 1990–92